A Practical Guide to
Combinatorial Chemistry

A Practical Guide to Combinatorial Chemistry

Anthony W. Czarnik, Editor
IRORI Quantum Microchemistry

Sheila H. DeWitt, Editor
Orchid Biocomputer, Inc.

American Chemical Society
Washington, DC

Library of Congress Cataloging-in-Publication Data

A practical guide to combinatorial chemistry / edited by
    Anthony W. Czarnik and Sheila Hobbs DeWitt

     p.   cm.

    Includes bibliographical references and index.

    ISBN 0–8412–3485–X

    1. Combinatorial chemistry. I. Czarnik, Anthony W.,
1957–   . II. Dewitt, Sheila Hobbs, 1960–   . III. Series:
ACS professional reference book.

RS419.P7   1997

615'.19—dc21   97–36358

                                                 CIP

The paper used in this publication meets the minimum requirements of American
National Standard for Information Sciences—Permanence of Paper for Printed Library
Materials, ANSI Z39.48-1984.

PRINTED IN THE UNITED STATES OF AMERICA

About the Editors

**Anthony W. Czarnik** was born in Appleton, Wisconsin, on November 21, 1957. He attended public schools in Appleton, Combined Locks, and Kimberly, Wisconsin, and received his B.S. (cum laude) from the University of Wisconsin, Madison, in 1977. His undergraduate major was biochemistry, but he carried out undergraduate research during that time in two organic chemistry laboratories, at the University of Wisconsin, Madison (with E. Vedejs) and Argonne National Laboratory (with M. MacCoss). Dr. Czarnik received his graduate training at the University of Illinois at Urbana-Champaign, obtaining both an M.S. in biochemistry (1980) and a Ph.D. in organic chemistry (1981) under the guidance of Nelson Leonard. From 1981 to 1983, he was a National Institutes of Health Postdoctoral Fellow at Columbia University, working with Ronald Breslow on the design of artificial enzymes. He began his academic career in 1983 at The Ohio State University on the faculty of the Department of Chemistry.

Dr. Czarnik has received both DuPont and Merck awards for new faculty, and in 1986 was presented with an American Cyanamid award in recognition of excellence in the advancement of science and the art of chemical synthesis. He was named an Eli Lilly awardee in 1988, a Fellow of the Alfred P. Sloan Foundation in 1989, and a Teacher–Scholar Fellow of the Camille and Henry Dreyfus Foundation in 1990. He is currently serving as an editor of the *Journal of Molecular Recognition*, on the editorial boards of *Organic Reactions, Current Opinion in Chemical Biology,* and *Combinatorial Chemistry,* and on the editorial advisory board of *Accounts in Chemical Research.* Dr. Czarnik also edits the World Wide Web Fluorosensor Database, located at BioMedNet.com/fluoro/. He is the author of more than 100 scientific publications.

From 1993 to 1996, Dr. Czarnik served as Director of BioOrganic Chemistry at Parke-Davis Pharmaceutical Research. In 1996, he accepted the position of

Senior Director, Chemistry, at IRORI Quantum Microchemistry in San Diego, California. His current research interests include combinatorial chemistry as a tool for drug discovery, nucleic acids as targets for small-molecule intervention, and fluorescent chemosensors for ionic and molecular recognition.

Sheila H. DeWitt was born in Medina, New York, on May 23, 1960. She grew up in Gasport, New York, where she graduated in 1978 as valedictorian at Royalton-Hartland Central School. While earning her B.A. in chemistry from Cornell University (December 1981), she began her career as a synthetic organic chemist working at FMC Agricultural Chemical Group in Middleport, New York, with Ernest Plummer and John Engel. She returned to work at FMC Agricultural Chemical Group in both Middleport and Princeton, New Jersey, for eight months before entering graduate school. She completed her Ph.D. in synthetic organic chemistry in 1986 under the direction of Daniel Sternbach at Duke University.

After her graduate work, Dr. DeWitt returned once more to FMC in the Process Research & Engineering Group (1986–1988). She then joined Parke-Davis Pharmaceutical Research in Ann Arbor, Michigan, in 1988, where she worked in medicinal chemistry (1988–1991) and bioorganic chemistry (1991–1995). Between 1995 and 1997, she served as the Vice President of Technical Development for DIVERSOMER Technologies, Inc., a start-up company which was incubated at Parke-Davis. She recently (October 1997) returned to Princeton, New Jersey, to serve as the Director of Business Development at Orchid Biocomputer, Inc.

Dr. DeWitt has been the recipient of several awards for her pioneering efforts in combinatorial chemistry and automated synthesis, including the Michigan Leading Edge Technologies Award (1993), Pioneer in Laboratory Robotics Award (1995), and Association for Laboratory Automation Outstanding Service Award (1997). She serves on the editorial boards of *Combinatorial Chemistry, Combinatorial Chemistry & High Throughput Screening, Drug Discovery Today, Laboratory Automation, Molecular Diversity,* and *Pharmaceutical News.* She is also a member of the Scientific Committee for the Association for Laboratory Automation. She has strongly contributed to the disciplines of combinatorial chemistry and automated synthesis since 1992 through publications, posters, issued patents, videos, chaired symposia, and expositions, and by organizing and serving as an instructor of workshops. Her current interests focus on enabling technologies for drug discovery, including combinatorial chemistry, automated synthesis, microfluidics, and microfabrication.

Contents

Equipment and Automation

Information Management and Biological Applications

Contributors

John J. Baldwin *page 153*
Pharmacopeia Inc., 101 College Road East, Princeton, NJ 08540

George Barany *page 51*
University of Minnesota, Department of Chemistry, 207 Pleasant Street S.E.,
Minneapolis, MN 55455

F. F. Craig *page 399*
Aurora Biosciences Corporation, 11149 North Torrey Pines Road, La Jolla,
CA 92037

Anthony W. Czarnik *page 413*
IRORI Quantum Microchemistry, 11025 North Torrey Pines Road, Suite 100,
La Jolla, CA 92037

Sheila H. DeWitt *page 413*
Orchid Biocomputer, Inc., 201 Washington Road, Princeton, NJ 08543–2197

Roland Dolle *page 153*
Pharmacopeia Inc., 101 College Road East, Princeton, NJ 08540

Michael N. Greco *page 281*
The R. W. Johnson Pharmaceutical Research Institute, Spring House, PA
19477

Michael C. Griffith *page 99*
Department of Combinatorial and Exploratory Chemistry, Houghten
Pharmaceuticals, Inc., 3550 General Atomics Court, San Diego, CA 92121

Thomas K. Hayes   *page 99*
Department of Combinatorial and Exploratory Chemistry, Houghten
Pharmaceuticals, Inc., 3550 General Atomics Court, San Diego, CA 92121

Maria Kempe   *page 51*
University of Minnesota, Department of Chemistry, 207 Pleasant Street S.E.,
Minneapolis, MN 55455
Current address: Department of Organic Chemistry 1, University of Lund,
P.O. Box 124, SE–221 00 Lund, Sweden

Christopher E. Kibbey   *page 199*
Parke-Davis Pharmaceutical Research, Division of Warner Lambert
Company, 2800 Plymouth Road, Ann Arbor, MI 48105

John S. Kiely   *page 99*
Department of Combinatorial and Exploratory Chemistry, Houghten
Pharmaceuticals, Inc., 3550 General Atomics Court, San Diego, CA 92121

Mengfen Lin   *page 123*
NMR Laboratory, Department of Central Technology, Preclinical Research,
Sandoz Research Institute, Sandoz Pharmaceuticals Corporation, East
Hanover, NJ 07936

Jonathan S. Lindsey   *page 309*
Department of Chemistry, North Carolina State University, Raleigh,
NC 27695–8204

Bruce E. Maryanoff   *page 281*
The R. W. Johnson Pharmaceutical Research Institute, Spring House,
PA 19477

Adnan M. M. Mjalli   *page 327*
Ontogen Corporation, 2325 Camino Vida Roble, Carlsbad, CA 92009
Current address: Helios Pharmaceuticals, 9800 Bluegrass Parkway,
Louisville, KY 40299

Walter H. Moos   *page 1*
Chiron Corporation, 4560 Horton Street, Emeryville, CA 94608
Current address: MitoKor, 11494 Sorrento Valley Road, San Diego, CA
92121

Steven M. Muskal   *page 357*
Affymax Research Institute, 3410 Central Expressway, Santa Clara, CA
95051

Yazhong Pei *page 99*
Department of Combinatorial and Exploratory Chemistry, Houghten
Pharmaceuticals, Inc., 3550 General Atomics Court, San Diego, CA 92121

John R. Peterson *page 177*
Chemistry Services, Panlabs, Inc., 11804 North Creek Parkway South,
Bothell, WA 98011

Ralph A. Rivero *page 281*
The R. W. Johnson Pharmaceutical Research Institute, Spring House,
PA 19477

Michael J. Shapiro *page 123*
NMR Laboratory, Department of Central Technology, Preclinical Research,
Sandoz Research Institute, Sandoz Pharmaceuticals Corporation, East
Hanover, NJ 07936

André Tartar *page 249*
Department of Chemistry, Institut Pasteur de Lille, Rue Calmette, 59000
Lille, France
Current address: Cerep, 1 Rue Calmette, 59000 Lille, France

Ted L. Underiner *page 177*
Chemistry Services, Panlabs, Inc., 11804 North Creek Parkway South,
Bothell, WA 98011
Current address: Cephalon, Inc., 145 Brandywine Parkway, West Chester, PA
19380

Peter Willett *page 17*
Krebs Institute for Biomolecular Research and Department of Information
Studies, University of Sheffield, U.K.

Xavier Williard *page 249*
Department of Chemistry, Institut Pasteur de Lille, Rue Calmette, 59000
Lille, France
Current address: Cerep, 1 Rue Calmette, 59000 Lille, France

Bing Yan *page 123*
Optical Spectroscopy, Department of Central Technology, Preclinical
Research, Sandoz Research Institute, Sandoz Pharmaceuticals Corporation,
East Hanover, NJ 07936

Preface

Combinatorial chemistry is a new subfield of chemistry with the goal of synthesizing very large numbers of chemical entities by condensing a small number of reagents together in all combinations defined by a given reaction sequence. Combinatorial chemistry is sometimes referred to as *matrix chemistry.* If a chemical synthesis route consists of three discrete steps, each employing one class of reagent to accomplish the conversion, then employing one type of each reagent class will yield $1 \times 1 \times 1 = 1$ product as the result of $1 + 1 + 1 = 3$ total reactions. Combining 10 types of each reagent class will yield $10 \times 10 \times 10 = 1,000$ products as the result of as few as $10 + 10 + 10 = 30$ total reactions; 100 types of each reagent will yield 1,000,000 products as the result of as few as 300 total reactions. While conceptually simple, considerable strategy is required to identify 1,000,000 products worth making and to carry out their synthesis in a manner that minimizes labor and maximizes the value of the resulting organized collection, called a *chemical library.*

Many of the tools useful for achieving this goal have their origins in related but distinct fields, such as peptide analysis, spectroscopy, polymer chemistry, and process optimization. The philosophy behind combinatorial chemistry surely owes much to biological methods in which a vast number of biological macromolecules are prepared and tested for activities of interest. However, combinatorial chemistry is clearly distinct from fields such as catalytic antibody research. Combinatorial chemistry requires, well, chemistry, to generate the organized collections, or libraries, of compounds that are valuable for the discovery of useful new substances. The armamentarium of organic synthetic methods is so powerful and flexible that biological approaches are unlikely ever to compare in value for the creation of diverse collections of new compounds, especially compounds that might be oral drugs. The field of combinatorial chemistry can legitimately claim to have originated in medicinal chemistry, and medicinal chemists are likely to remain among the largest groups of end users of this collection of tools.

Therein lies the motivation for this book. When we were approached by ACS Books to create a volume on combinatorial chemistry, we decided early that what was most needed was a kind of guidebook for the professional chemist—but one that would be read and used, not just bought and shelved. By aiming the presentation at an advanced undergraduate level, we felt we would avoid both offending and overwhelming practicing chemists. Furthermore, such a book might be used for teaching. We created a table of contents geared at covering the topics required to treat the subject in a scholarly fashion, then sought expert authors willing to write comprehensively, rather than just about their own work. In the end, we were very fortunate to attract scientists willing to serve the chemistry community in this way. These authors were also generous in their acceptance of the revisions of a very fine developmental editor, William Wells of BioText, Inc., who removed jargon, clarified statements not clear to the layperson, and made the writing styles more uniform. To our minds, this experiment in book production has been an unmitigated success.

The benefits of combinatorial chemistry in the drug discovery process are by now self-evident. By creating vast new sources of "unnatural" but druglike products, the great screening machines are stoked with new molecular-diversity fuel; this is known as lead generation. Once active compounds are found, libraries focused around those structures provide valuable structure–activity relationship data via what is known as lead optimization.

Combinatorial chemistry approaches to the discovery of new substances with useful properties are having an impact on many other fields. As the agricultural chemistry field evolves to higher-throughput screening methods, the smaller amounts of sample typically generated by combinatorial approaches become useful to this discipline. Moreover, just as predicting the energies of solvated host–guest complexes continues to appear to be a long-term solution, so combinatorial diversity–selection approaches may rapidly bring us closer to the discovery of synthetic receptors for a wide variety of applications, including chemosensors, artificial enzymes, and perhaps even drugs that are themselves receptors (like vancomycin). It remains impossible to predict with accuracy either the chemoselectivity or the stereoselectivity with which new reagents will act on reactants; reagents with high selectivities may be discoverable via combinatorial approaches. This also holds true for catalysts and materials. A large number of process chemists are beginning to use the enabling tools of combinatorial chemistry for reaction optimization in scale-up and manufacturing. In all of these fields, the discovery of useful new substances is important enough to warrant the use of all available methods, including that of dramatically facilitated empiricism.

We thank several individuals for their assistance in making this book a reality. Michelle Althuis was our acquisitions editor at ACS Books at the start of this project, and she first proposed it. Cheryl Shanks has done a fine job of bringing the project to fruition. Bridgett Kerr provided able administrative support at the onset of this project.

ANTHONY W. CZARNIK
SHEILA H. DEWITT
April 1997

Introduction: Combinatorial Chemistry Approaches the Next Millennium

Walter H. Moos

A collection of modern technologies referred to as "combinatorial chemistry" or "molecular diversity" (CC–MD) has spawned a rebirth of medicinal, organic, and pharmaceutical chemistry in academic, biotechnological, and pharmaceutical settings. This approach integrates several drug-discovery disciplines, including synthetic and computational chemistry, analytical methodologies, and high-throughput screening, and has the potential to reduce the development time and costs for pharmaceuticals. CC–MD is founded on solid-phase chemistry pioneered by peptide chemists in the 1960s. In the 1990s, multiple new business ventures have arisen to capitalize on the potential of CC–MD, and CC–MD technologies have become mainstream. Some have called CC–MD merely a "game of numbers", but it is clearly much more. It will change the game for drug hunters practicing their trade during the 1990s and beyond.

A Historical Perspective

Combinatorial chemistry and other synthetic approaches to molecular diversity are relatively recent additions to the toolbox of medicinal chemists. This brief introduction is intended to provide an even-handed commentary on the past and a look toward the future. A few caveats: History is often rewritten by scientists in the field as well as by historians! Although a few contributors to the work described in this chapter are highlighted by name, full references to this work are to be found only in the chapters that follow. It is impossible in this brief introduction to be all-inclusive in citing the history of the field, so I apologize at the outset for errors of omission or emphasis. Finally, this chronological retrospective emphasizes pharmaceutical applications, which may or may not reflect the emphases of the field in the future.

What is combinatorial chemistry and molecular diversity (CC–MD)? The field of molecular diversity includes all forms of chemical diversity, whether "natural", as in natural products like morphine and its congeners or in synthetic combinations of endogenous building blocks like oligopeptides or oligonucleotides, or "unnatural", as in the synthetic preparation of large libraries of novel heterocycles for drug screening in the pharmaceutical and biotechnology industries. Given the focus of this book, I will cover primarily the latter category.

There is much to be learned from the timing of various contributions to this field. To gain perspective, let us review some humbling evolutionary time frames (Table I). The earliest generation of molecular diversity on Earth must have occurred in the primordial soup, predating all life-forms. In the four billion years since then, the main source of molecular diversity has been natural products. Now, human beings are beginning to explore their ability to make and test large numbers of novel chemicals in the laboratory. The first use of natural products for medical purposes certainly occurred before the time of Hippocrates, who is often referred to as the father of modern medicine. This puts mother nature at least 2500 years in the lead in devising medically useful products by molecular diversity. Aspirin (acetylsalicylic acid), obtained from the bark of willow trees, was only recognized for its broad medicinal value about 100 years ago. Merrifield's work on solid-phase peptide synthesis (SPPS) dates to the early 1960s. It was not until the 1970s that the solid-phase synthesis of unnatural chemical structures of various classes (e.g., nonpeptide heterocycles) was explored significantly. The 1980s saw the formation of the first companies to focus solely or principally on aspects of synthetic molecular diversity. Many of these venture-backed companies have subsequently merged with larger corporations (Table II). Some of the earliest contributors to the field are listed in Table III. It is only in the present decade, however, that the field has fully taken hold. One might now predict that all major pharmaceutical, biotechnology, and university chemistry laboratories will, without exception, use combinatorial chemistry regularly by the year 2000.

TABLE I. Time Line Leading up to Modern Pharmaceutical Diversity

Event	Approximate Number of Years Since
The Earth's formation	4,500,000,000
The first human beings	5,000,000
Hippocrates' birth	2,500
Aspirin's medicinal value recognized	100
First major publications on solid-phase peptide synthesis	35
Early publications on solid-phase nonpeptide synthesis	20
First major publications in modern combinatorial chemistry	10
First journal devoted to molecular diversity	1

TABLE II. Examples of the First Companies Founded to Explore CC–MD

Early Diversity Ventures	Now Part of These Companies
Affymax	Glaxo Wellcome
Arris	Arris
Mimotopes	Chiron
Multiple Peptide Systems	Trega
Protos	Chiron
Selectide	Hoechst Marion Roussel
Sphinx (Genesis)	Lilly

TABLE III. An Incomplete Sampling of Pre-1990 CC–MD Pioneers

Early Groups Active in the Field	Country
Benkovic, Huse, Janda, Lerner, and colleagues	United States
Devlin, and colleagues	United States
Dower, Barrett, and colleagues	United States
Fodor, Pirrung, Read, and colleagues	United States
Frank, and colleagues	Germany
Furka, and colleagues	Hungary
Geysen, Maeji, Meloen, and colleagues	Australia and the Netherlands
Houghten, and colleagues	United States
Hruby, Lam, Lebl, and colleagues	United States
Ladner, and colleagues	United States
Leznoff, and colleagues.	Canada
Moos, Pavia, and colleagues	United States
Rutter, Santi, and colleagues	United States
Scott, Smith, and colleagues	United States

One strategy at the heart of combinatorial chemistry is the concept of combining readily available chemical building blocks like "beads on a chain", potentially in all possible combinations and permutations. The number of compounds that can be synthesized using this approach is demonstrated in Table IV. This concept (at least in the case where one compound is synthesized at a time) was first popularized by Merrifield, who received a Nobel prize for his work on SPPS. The SPPS concept was broadened to the synthesis of multiple compounds at one time, at first with peptide libraries by researchers like Geysen in the 1980s, and then with peptoids, robotics, and broadly based pharmaceutical chemistries in the 1990s (Figure 1). (Although combinatorial approaches are being extended to worthwhile solution-phase synthetic approaches, the seminal work in this field has been performed primarily using solid-phase strategies.) Combinatorial methods are powerful, allowing the preparation of very large numbers of compounds by combining modest numbers of building blocks simultaneously and often randomly at multiple variable positions. The strategies and concepts are closely related to those of a molecular biologist making a systematic or random set of mutations in a gene.

TABLE IV. Numbers of Compounds Possible in CC–MD

Oligomer Size	Number of Building Blocks That Can Fill Each Position	Total Number of Compounds Possible in Library
Dimer	10	$10^2 = 100$
	100	$100^2 = 10,000$
	1000	$1000^2 = 1,000,000$
Trimer	10	$10^3 = 1000$
	100	$100^3 = 1,000,000$
	1000	$1000^3 = 1,000,000,000$
Dimer	4 (e.g., nucleotides)	$4^2 = 16$
Trimer		$4^3 = 64$
Tetramer		$4^4 = 256$
Pentamer		$4^5 = 1024$
Dimer	20 (e.g., standard amino acids)	$20^2 = 400$
Trimer		$20^3 = 8000$
Tetramer		$20^4 = 160,000$
Pentamer		$20^5 = 3,200,000$

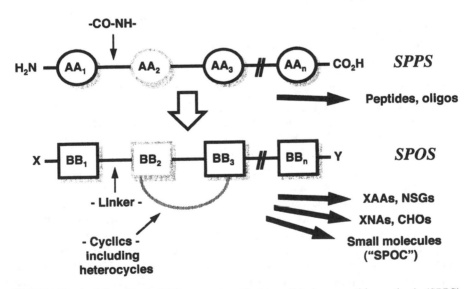

FIGURE 1. The building block (BB) concept applied to solid-phase peptide synthesis (SPPS) and solid-phase organic synthesis (SPOS). The familiar concept of stepwise peptide synthesis using amino acids (AA; top) can be extended (bottom) by using various chemical linkages to join building blocks to create oligomers of N-substituted glycines (NSGs), nucleotides (XNAs), saccharides (CHOs) or other small molecules such as heterocycles, in a process sometimes referred to as solid-phase organic chemistry (SPOC).

Characterizing mixtures of compounds is not new. For years agricultural and perhaps other companies have screened unpurified mixtures of compounds. Only active mixtures were followed up in terms of structural characterization and purification. Some of the early practitioners of SPPS may have realized that they could make large mixtures of peptides on resin beads, but they apparently thought that this would not be worthwhile.

Why has molecular diversity taken so long to become mainstream? There are at least two major factors. First, much of the early work focused on peptides or oligonucleotides, which do not generally make "good" drugs, as they have poor oral activity, short half-lives, and so on. (What makes a "good" drug is the subject of countless articles and books, and will not be dealt with further here.) Second, the idea that the chemist should construct libraries of compounds that have uncertain material composition is the antithesis of how organic chemists have been trained over the years. Organic chemists have typically been drilled to confirm and characterize the identity of each intermediate in exquisite detail, worrying about even 1% impurities. Thus, it took a decade to move from SPPS to solid-phase organic synthesis (SPOS), another decade to move to libraries of peptides and oligonucleotides, and yet another decade to achieve small-molecule chemistry that is truly useful for pharmaceutical development and can be used in a combinatorial way (Table V).

Together with several other paradigm shifts in recent decades (Table VI), combinatorial drug discovery and development technologies promise to "change the game" as drug hunters practice their trade during the 1990s and beyond. Even so, the old-fashioned way of discovering drugs will not be completely abandoned. Indeed, combinatorial chemistry is but one of many tools, taking its rightful place alongside structure-based design, biotechnologies including cell and molecular biology, and other modern methods.

Why do we need CC–MD? The need for CC–MD is underscored by the high cost, long time frames, and high rate of failure of drug research and development (Figure 2). CC–MD will allow drug discovery to proceed more successfully, faster, and more cheaply. So far, it appears to be faster and cheaper, but increasing the success rate in development may take more time.

TABLE V. Sources of Molecular Diversity in Recent Decades

Decade	Emergence of New Sources of Diversity
Pre-1960	Natural products and solution-phase organic and medicinal chemistry
1960s	Early solid-phase peptide synthesis
1970s	Early solid-phase nonpeptide synthesis
1980s	First synthesis of large libraries, primarily peptides or oligonucleotides, and recombinant protein/nucleic-acid-based technologies
1990s	True combinatorial pharmaceutical chemistry

TABLE VI. Recent Paradigm Shifts in the Pharmaceutical Industry

Decade	Representative Developments
1960s	Solid-phase peptide synthesis (SPPS) Nuclear magnetic resonance (NMR) spectroscopy
1970s	Affinity chromatography Radioligand receptor binding
1980s	Biotechnologies Structure-based design
1990s	Molecular diversity Solid-phase organic chemistry (SPOC)

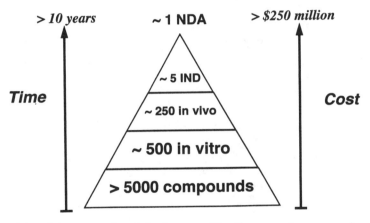

FIGURE 2. Risk and drug research and development. The discovery of every new drug starts at least 10 years and $250 million earlier, with the testing of more than 5000 compounds.

Abbreviations: IND, Investigational New Drug Application; NDA, New Drug Application.

Unfortunately, although there is great promise for CC–MD, the field is so new that there are, as yet, no fully developed success stories in clinical trials to cite. The situation was similar for molecular modeling and structure-based design in the 1980s, and other new technologies have followed similar paths. This is not surprising. Like any new tool, combinatorial chemistry is not the miracle worker that many hope it is or portray it to be. But CC–MD has already contributed significantly to many pharmaceutical and biotechnology research programs, and these successes will surely be reported in due course.

Indeed, combinatorial discovery is already creating bottlenecks in preclinical development; thus, the application of combinatorial principles to preclinical development will soon become the focus of many laboratories. Already, groups are assessing the ability of compounds to cross membranes by studying mixtures of compounds using, for example, Caco-2 monolayers and "transwell" setups, which act as models for gastrointestinal permeability and the absorption of molecules from the gut.

Through smart application of molecular diversity technologies, forward-looking academic laboratories and start-up biotechnology ventures have been able in recent years to compete head-to-head in research and discovery with much larger pharmaceutical rivals. Whether the brute force strength of major pharmaceutical companies will now begin to overpower such smaller programs remains to be seen.

With this background, let us now consider some of the interdisciplinary details of CC–MD, starting with organic chemistry.

Organic Chemistry and CC–MD

Most classical and modern-day organic transformations have now been shown to work well on the solid phase, much to the surprise of many chemists (Figure 3). General solid-phase organic synthesis (SPOS) requires appropriate access to and knowledge of the characteristics of base polymers, supports, linkers, spacers, and other features of polymer and interphase chemistries. With the basics in hand, chemists have the ability to prepare both mixture libraries, using techniques such as alternate mixing and splitting of resin, and large numbers of individual compounds using, for example, multipin methods. Efficient synthetic approaches have been developed for mixture library synthesis (Figure 4), and libraries can be prepared

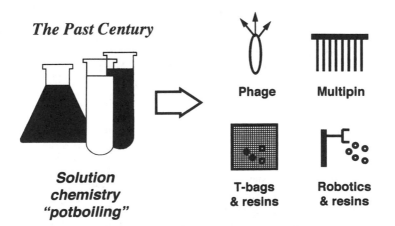

The Past Century

Phage   Multipin

Solution chemistry "potboiling"

T-bags & resins   Robotics & resins

FIGURE 3. From "pot-boiling" solution chemistry to solid-phase organic synthesis. The figure shows some of the approaches used to synthesize multiple compounds; for details see Chapter 5.

Note: Multipins from Geysen and colleagues at Chiron. T-bags = teflon bags, borrowed from the notion of tea bags, from Houghten and colleagues at Torrey Pines Institute for Molecular Studies and Multiple Peptide Systems.

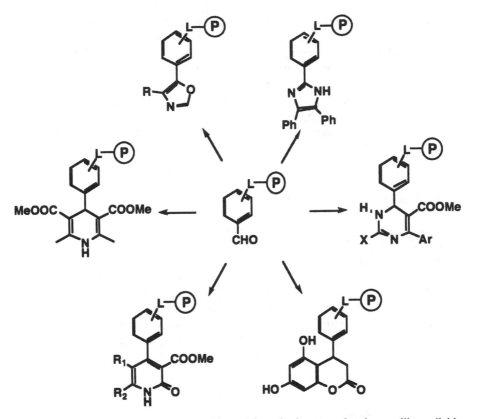

FIGURE 4. An example of an efficient combinatorial synthesis approach using readily available starting materials, as introduced by Desai and colleagues at Chiron. *P* represents the polymer, *L* the linker. Many simple building blocks can be added to the aldehyde group in the starting compound.

from existing libraries using "postmodification" approaches (Figure 5). Furthermore, several automated, robotics-based, and semiautomated systems have been developed.

Biological approaches also exist, such as the bacteriophage libraries that use the phage as a kind of "biological pin". In these libraries a genetic insert in the bacteriophage encodes a random or structured peptide or protein sequence that is then expressed on the surface of the phage. A strength of this system is its ability to amplify a minute signal when a mixture of more than 100 million peptide sequences is screened in a single experiment. A major weakness, however, is that only the 20 standard amino acids are readily available as building blocks. New delivery methods may someday overcome the weaknesses of peptides as drugs, allowing their true strengths of specificity and safety to shine through, but this will probably not happen until well after the year 2000.

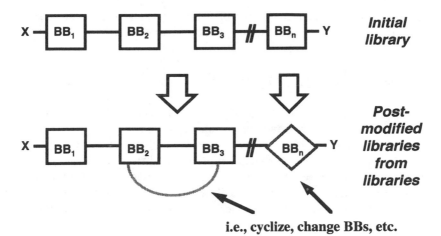

FIGURE 5. The postmodification "libraries from libraries" concept, introduced by Pei and Moos at Chiron, and Houghten and colleagues at Trega Biosciences. Modifications of the existing building blocks that depend on the context of the building block (e.g., cyclization) can greatly increase diversity with minimal additional effort.

Analytical and Process Chemistry and CC-MD

Quality assurance and quality control (QA–QC, together termed QX) have a key place in the preparation of libraries of compounds, as it is often difficult (or perhaps impossible) to determine whether each compound has been synthesized in a given library. Of course, at the end of the day, if a lead has been identified, concerns about completeness of a library are less of an issue. Nonetheless, it behooves the combinatorial chemist to establish that trial reactions using a range of reactants will run successfully under the conditions used to create the library, factoring in electron donating and withdrawing substituents, steric or charge interactions, stability of protecting groups, and so on. Fortunately, combinatorial synthetic schemes are intrinsically well suited to parallel optimization. Once the reaction characteristics are defined, the user has greater expectations that most of the planned compounds will indeed be present in the library. Otherwise, combinatorial chemistry can take on too much of the undesirable side of natural products research, where, for example, the active component in a fermentation broth may be a complete mystery.

Reaction progress can be monitored in a variety of ways, some of which do not require cleavage from the solid phase. The methods include colorimetric techniques; gel-phase nuclear magnetic resonance (GP-NMR) and infrared (IR) spectroscopies; and liquid chromatography (LC–MS) and gas chromatography (GC–MS) mass spectrometries.

A strong process chemistry effort assists combinatorial discovery by developing new synthetic methods, testing reaction conditions, preparing quantities of building blocks, and often "turning the crank" in the production of libraries or deconvolutions. After the identification of a lead, rapid scale-up to multigram quantities facilitates further study of the compound's biological activity.

Biotechnology and CC–MD

Chemistry by itself is simply not enough in our industry, and the ability to clone, sequence, express, produce, and purify human biological targets is critical to the optimal use of combinatorial diversity. Having the human homolog of a given target protein is important because nonhuman proteins can have affinities for compounds that are quite different from those of their human counterparts. This may, at least in part, explain the failure of certain clinical trials, where preclinical work in rodents, dogs, or other experimental animals identified agents that in fact had little affinity for the human protein homologs.

With human protein reagents in hand, CC–MD researchers must then develop assays and strategies to uncover active "hit pools" and lead series. The assay may measure an activity of the protein, or simply follow its binding by candidate compounds. A number of high-throughput screening (HTS) approaches have been developed to screen from one to a few thousand compounds or more per well, although if libraries are screened as mixed pools, HTS is often not necessary. A variety of approaches are available to identify hits in pools, including brute force deconvolution of hit pools into smaller and smaller pools until individual compounds are identified (Figure 6), selection of colored beads under a microscope after staining of the bound protein target, affinity chromatography, and encoding approaches (Figure 7). One very powerful method uses affinity selection plus size-exclusion chromatography (to separate bound compounds from those that have little or no affinity for the target protein), followed by mass spectroscopy (AS–MS), to identify leads that bind to the target of interest. This method is particularly useful, as it eliminates the need to separately encode compounds on beads, using instead the molecular weight of the compounds, a code that is intrinsic to all molecules. The possible redundancy of some molecular weights within a library of compounds can be a problem, but higher resolution and fragmentation MS techniques should ultimately overcome this quite effectively.

Computational Chemistry and CC–MD

How high is up? Some have called molecular diversity merely a "game of numbers" (Figure 8). The number of building blocks and compounds is certainly part of the story, but we believe that the complexity or diversity of libraries is perhaps more important (Figure 9). As a result, we have developed several ways of calculating and visualizing library diversity, using statistical, fragment-based, and graphical methods. Calculation of the structural properties of library components, followed by sta-

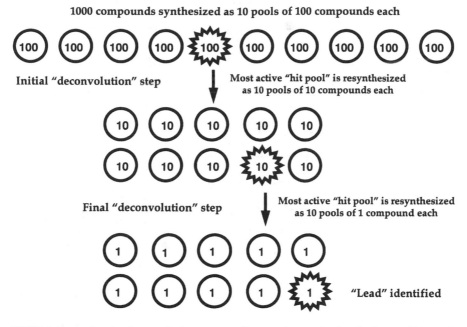

1000 compounds synthesized as 10 pools of 100 compounds each

Initial "deconvolution" step

Most active "hit pool" is resynthesized as 10 pools of 10 compounds each

Final "deconvolution" step

Most active "hit pool" is resynthesized as 10 pools of 1 compound each

"Lead" identified

FIGURE 6. Example of a deconvolution process. Repeated resynthesis of subsets of hit pools results in the identification of the active component(s) in a mixture.

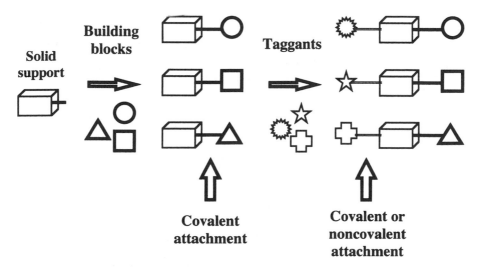

Solid support

Building blocks

Taggants

Covalent attachment

Covalent or noncovalent attachment

FIGURE 7. The encoded library concept, originated by Dollinger and colleagues at Cetus (now Chiron). Each time a building block is added, a corresponding tag is covalently or noncovalently attached to the same solid support.

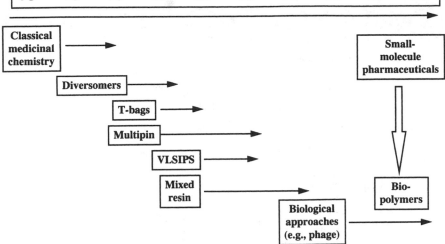

FIGURE 8. A game of numbers. The figure depicts the number of compounds that are typically generated using the various methods, with the products ranging from small-molecule pharmaceuticals to biopolymers.

Note: Diversomers originated from the work of Hobbs-DeWitt, Pavia, and colleagues at Parke-Davis. VLSIPS (very large scale immobilized polymer synthesis) is from the work of Fodor and colleagues at Affymax.

tistical selection methods, generates nonredundant libraries that more optimally sample property space (the many-dimensional "space" consisting of all possible physicochemical properties including size, shape, and charge). Our "flower plots" (Figure 10), which are a pictorial representation of the physicochemical characteristics of library members, have captured the interest of many groups for their ability to represent a many-dimensional physicochemical parameter in a world with a limited number of dimensions. We are in the process of adding true three-dimensional descriptors to these analyses, as well as coupling genetic algorithms with docking programs to enable the automated design of libraries that fit the spatial requirements of crystal structures and/or pharmacophoric hypotheses. New (bio)informatics approaches are also necessary to handle the overwhelming amount of data and numbers of compounds resulting from molecular diversity approaches.

Combinatorial Discovery vs. Combinatorial Development

There are now a number of success stories in discovery or early development phases (Figure 11), though few have been published in detail thus far. At times, we ourselves have found it relatively straightforward to uncover novel, nanomolar hits against a wide variety of biological targets. (In some cases, for reasons not yet fully understood, it is clearly not so easy!) We have also been able to use the chemical

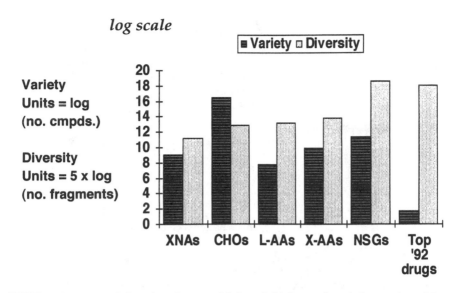

FIGURE 9. The structural diversity of successful drugs is high even though the number of drugs is low. This is in contrast to libraries generated using endogenous building blocks such as the standard amino acids. The variety scale shows the representative number of compounds that could be made given the indicated set of building blocks. The diversity scale is a measure of the range of physicochemical properties that can be sampled by the indicated library.

Note: XNAs = oligonucleotides, CHOs = oligosaccharides, L-AAs = oligopeptides generated using the standard 20 amino acids, X-AAs = oligopeptides containing nonstandard amino acids, NSGs = oligopeptoids generated using *N*-substituted glycines, top '92 drugs = 45 top-selling small-molecule drugs in 1992. Original work by Blaney, Martin, Spellmeyer, and colleagues at Chiron.

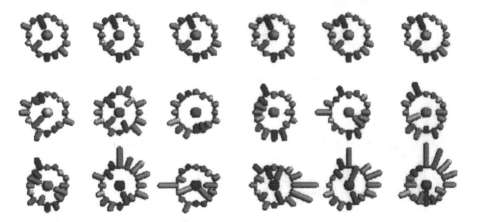

FIGURE 10. A library-design-based graphic display using "flower plots". The plots represent different physicochemical features of a molecule with a flower "petal", with the length and directionality of the petal representing the magnitude (+/−) of a given property relative to other members of a selected library. Color can be used to indicate other characteristics, such as biological activity, or simply for visual effect.

Note: Original work by Blaney, Martin, Spellmeyer, and colleagues at Chiron.

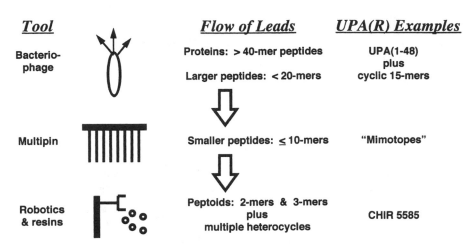

| *Tool* | | *Flow of Leads* | *UPA(R) Examples* |

FIGURE 11. Example of a success story. To arrive at a small molecule that binds the urokinase plasminogen activator receptor [UPA(R)], Rosenberg and colleagues at Chiron started with a large peptide from the urokinase plasminogen activator. Shortening and alteration of this lead eventually resulted in the identification of a small-molecule inhibitor, CHIR 5585. Each stage of lead identification and optimization can be better served by using a different set of combinatorial approaches.

methods described above to enhance both the general activity or potency of initial leads. Thus, combinatorial chemistry is finding a significant role in both the discovery and optimization of leads, and time frames can be very short in comparison to those necessary when using traditional approaches (Figure 12).

With increasing success in lead discovery, the bottleneck shifts back to pre-clinical development. The idea of studying multiple compounds simultaneously using in vivo models may, even with equimolar mixtures, seem daunting, if not theoretically impossible. We have found, however, that at least some pharmacokinetic parameters of mixtures may be studied with the assistance of ultrasensitive detection methods like accelerator mass spectrometry (XMS, or better known as AMS). In vitro models have also been used to study absorption parameters, including gastrointestinal cell monolayers (e.g., Caco-2), and in vitro methods can be used to study toxicity at the subcellular level (e.g., reporters of stress-induced gene transcription), the cellular level, and even at the level of whole organs.

Conclusions About CC–MD

As a technology base, CC–MD continues to expand, at least as measured by the number of publications in the field (Figure 13). Combinatorial chemistry has joined a formidable arsenal of research methods, including structure-based design and other tools of recent decades. No one today doubts that it is here to stay. The new

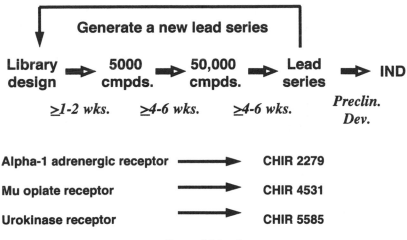

FIGURE 12. An idealized time frame for drug development using combinatorial chemistry. Shown are examples of leads that were developed in this manner by Spellmeyer, Stauber, Zuckermann, and colleagues at Chiron.

Abbreviation: IND, Investigational New Drug Application.

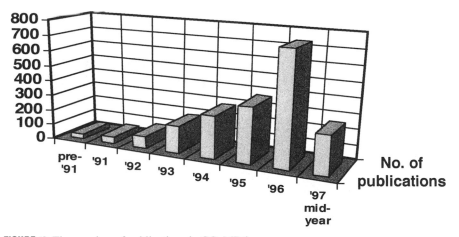

FIGURE 13. The number of publications in CC–MD by year.

Note: Graph developed using information from Lebl and other editors of the journal *Molecular Diversity*.

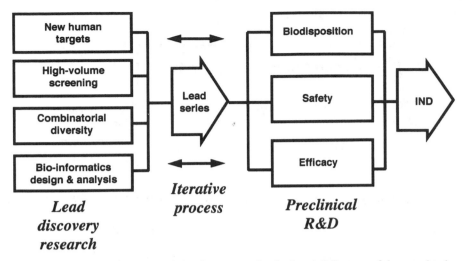

FIGURE 14. Preclinical research and development technologies. A full menu of these technologies must be pursued to fully realize the potential contribution of combinatorial chemistry to drug discovery.

methods have the potential to increase the efficiency and productivity of preclinical studies, but to make the most of them one needs a full menu of preclinical technologies (Figure 14). Only then will this new technology earn its keep in this most competitive arena.

Acknowledgments

I thank my many outstanding colleagues past and present, especially those at Chiron and Parke-Davis, for their numerous special contributions to this work.

2

Using Computational Tools To Analyze Molecular Diversity

Peter Willett

There is much interest in using computational tools to support combinatorial chemistry programs. Drawing, in large part, on approaches that were originally developed for searching and clustering chemical databases, methods have been developed for selecting compounds for addition to a library, for the design of new libraries, and for quantifying the diversity of sets of compounds, inter alia. There have already been several reports of the operational use of these methods, with many of them now available in commercial software systems for combinatorial information management.

The term "diversity" is widely used in the combinatorial chemistry literature, but has generally been discussed in a qualitative manner (along with related terms such as "different", "disparate", or "dissimilar"). There has been much interest in the development of quantitative tools for the analysis of diversity that enable the computer to assist in the search for novel bioactive molecules (*1–11*). Many of these tools are derived from those that have been developed previously for searching and clustering databases of chemical structures.

Chemical Structure Databases

Chemical structure databases (*12*) contain the machine-readable representations of either the chemical structure diagrams of molecules (termed two-dimensional (2D) databases) or their atomic coordinates (termed three-dimensional (3D) databases). The databases contain compounds that have either been reported in the literature (in the case of a public file such as the Cambridge Structural Database [CSD]), or synthesized in-house by an organization (in the case of corporate databases in the pharmaceutical and agrochemical industries). The techniques used for processing such databases can, however, be applied equally well to entities that we might call virtual databases, which contain the family of molecules that result from a combinatorial synthesis program, or a group of molecules constructed by computational approaches such as computer-aided synthesis design or de novo structure generation.

The most common form of representation for a 2D chemical structure is a connection table, which contains a list of all of the atoms within a structure (typically excepting hydrogen), together with bond information that describes the exact manner in which the individual atoms are linked together. There are many ways in which a connection table can be described in machine-readable form. One of the most common forms of connection table at present is a SMILES, a linear notation that encodes the atom and bond information in a compact character string (*13*). A connection table provides a complete and explicit description of a molecule's topology, and thus can be depicted as a graph. A graph is a mathematical construct that describes a set of objects, called nodes or vertices, and the relationships, called edges or arcs, that exist between pairs of objects. In the chemical context, the nodes represent atoms and the edges represent bonds. The equivalence between a labeled graph and a connection table means that it is possible to compare connection tables, and hence databases of 2D chemical structures, using isomorphism algorithms that were originally designed for establishing the structural relationships between pairs of graphs (*14*). Analogous graph-based techniques are also used for processing the molecules in 3D databases, where the nodes and edges of a graph denote the atoms and interatomic distances of a 3D molecule (*15, 16*)

The two most important search methods for chemical database systems are substructure searching (*12, 17*) and similarity searching (*18, 19*). Substructure searching scans the database for all molecules that contain a given substructure irrespective of the environment in which it occurs. This type of search may, for example, be used to identify all molecules that contain a pharmacophoric pattern that has been identified in a previous molecular modeling study. Substructure searching is the chemical equivalent of subgraph isomorphism, the matching of two graphs to determine whether one is contained within the other (*20*). Subgraph isomorphism is extremely demanding of computational resources. Efficient substructure searching of large databases therefore usually requires a two-level search procedure, in which an initial, very rapid screening search precedes the detailed and time-consuming subgraph-isomorphism search. The screening search identifies those molecules in a database that match the query at the screen (or fragment) level, where a screen is a substructural feature, the presence of which is necessary, but not sufficient, for a molecule to contain the query substructure. Only those few molecules that contain all of the query screens then undergo the time-consuming subgraph-isomorphism search. For example, a carbonyl group might be used as one of the screens in a substructure search for a quinone moiety.

The second type of search, similarity searching, requires the specification of an entire target structure. This is typically a molecule that has previously been shown to exhibit activity in a biological test, and the search is for those structures that have a high degree of structural resemblance to the entire target structure, rather than those that contain the partial structure as in a substructure search. A similarity search compares a set of characteristics of the target structure with the corresponding set of characteristics for each of the database structures, calculates a measure of intermolecular structural similarity based on the degree of resemblance of these two sets of characteristics, and then sorts the database structures into order of decreasing similarity with the target. The output from the search is a ranked list displaying the structures in the order of apparent similarity to the target structure. The nearest neighbors, i.e., those database structures that are most similar to the target structure, are displayed first. In some cases there may be a need for molecules that are dissimilar to the target molecule, for example in a selective compound-acquisition program (5). In such cases, the most interesting molecules may be those near the bottom of the ranked search output (21).

At the heart of any system for similarity searching is the similarity measure that is used to identify the target structure's nearest neighbors. The next section provides an overview of the measures that can be used to quantify the degree of resemblance between pairs of machine-readable chemical structure representations.

The Measurement of Structural Similarity

Any measurement of similarity involves three main components (18): the structural descriptors that are used to characterize the molecules, with similar molecules being those that have comparable sets of structural characteristics; the weighting scheme that is used to differentiate more important characteristics from less important characteristics; and the similarity coefficient that is used to quantify the degree of similarity between pairs of molecules. Many different methods for measuring chemical similarity have been reported in the literature (18, 19, 22–25); this section focuses upon those that are appropriate when very large numbers of intermolecular similarities need to be computed.

Structural Descriptors

Before molecules can be ranked or compared they must first be described in terms that a computer can process. The connection tables representing the molecules in a database can be used to generate many different types of descriptors. In practice, however, only three descriptors have been used extensively for the calculation of molecular similarity. These are fragment substructures, topological indices, and global physical properties.

Fragment Substructures

The most common similarity descriptors are the fragment substructures that are used for screening in chemical substructure searching systems, with two molecules being judged structurally similar if they have many fragment substructures in common (*18*). Screens used in 2D substructure searching are typically small features centered on an atom, bond, or ring system. Examples include the augmented atom, which consists of an atom (other than hydrogen) and all of the atoms and bonds that are immediately adjacent to it in a molecule (*26*), and the topological torsion, which is a linear sequence of four adjacent atoms, none of which are hydrogen (*27*). Screens used in 3D substructure searching typically consist of a set of two (or more) atoms, together with associated distance ranges, or bins, that represent the separation(s) of these atoms (*28, 29*); similar range-based features can be used to characterize valence and dihedral- or torsion-angle constraints (*15, 16*).

The traditional approach to screening is to define a fragment dictionary (*17*) containing a small number of fragments that are carefully selected using statistical information about how often they appear in the database to be searched, and how often their appearance coincides with that of other fragments (*see*, for example, references *30* and *31*). The presence of particular fragments (or screens) within a molecule or substructure is normally denoted by a bit-string, a binary string with each bit denoting the presence or absence of a specific screen. Predefined fragment dictionaries can only contain a limited number of screens, however, and these must be carefully selected. A more flexible approach, especially when a database is growing and new classes of molecules need to be accommodated, is exemplified by the fingerprints that are used for substructure and similarity searching in the chemical information system produced by Daylight Chemical Information Systems, Inc. (18500 Von Karman Ave. Suite 450, Irvine, CA 92715). Here, all linear sequences containing up to seven bonds in a molecule are generated exhaustively and each is then assigned a code, four or five bit positions in length, in the bit-string that is used to represent a molecule.

Topological Indices

A topological index is a single-valued integer or real number that characterizes the topology of a molecule (the way that the atoms are bonded together). The expectation is that structurally similar molecules will have similar index values, and topological indices have hence been used extensively in structure–property correlations since the publication of the first such index, that of Wiener (*32*). The numbers in this index are half the sum of the path lengths between each pair of atoms in a molecule, measured along the intervening bonds. Many different topological indices have now been reported in the literature. The most widely used are the molecular connectivity indices, χ, originally formulated by Randic (*33*), and subsequently generalized and extended by Kier and Hall (*34–36*). Although individual indices are widely used in structure–property studies, when calculating similarities it is common practice to use a number of different topological indices. Principal component analysis or some other data-reduction technique is used to obtain a reduced set of uncorrelated indices (*see*, for example, references *37* and *38*).

A related descriptor is a topological sequence, as exemplified by the path numbers described originally by Randic and Wilkins (*39*). The path number sequence for a structure is constructed by generating all self-avoiding interatomic paths, where a self-avoiding path in a chemical graph is an alternating sequence of vertices (representing atoms) and edges (representing bonds) that are adjacent (bonded), with no vertex appearing more than once. The total number of paths of each length are then assigned to corresponding elements of the sequence. The path number sequence is the list of these numbers, starting with the number of "paths" one atom in length and, unless arbitrarily truncated, finishing with the element corresponding to the longest path length in the structure. Structures with approximately the same distribution of path lengths tend to be structurally similar. Related measures of structural similarity are described by Broto et al. (*40*), Jerman-Blazic et al. (*41*) and Klopman and Raychaudhury (*42*), inter alia.

Global Physical Properties
A molecule can also be characterized by calculating a set of physical properties that describe its topological, electronic, steric, lipophilic, or geometric features. For example, standard computational-chemistry packages have been used to calculate properties (such as the heat of formation, ionization potential, number of filled orbitals, sum of total-dipole moments, difference between highest occupied molecular orbital (HOMO) and lowest unoccupied molecular orbital (LUMO) values, octanol–water partition coefficient, molar refractivity, and van der Waals volume) for similarity searching (*43*) and clustering (*44*). Kearsley et al. (*45*) have discussed the use of fragment-based similarity measures in which the atoms in the fragments are characterized by physicochemical properties, and Pepperrell et al. discuss the use of analogous atomic descriptions in distance-based 3D similarity searching (*46*).

The availability of software packages such as ADAPT (*47*) and PROFILES (*48*) facilitates the generation of property-based molecular descriptors. For example, Stanton et al. described a similarity analysis of dihydrofolate reductase (DHFR) inhibitors that used no less than 82 physical property descriptors calculated using the ADAPT package (*49*).

Weighting Schemes

When chemists look at chemical structure diagrams, it is natural for them to regard certain features as being of more importance than others. In so doing, they are, in effect, giving greater weight to those parts that they deem more important, such as functional groups. One of the unresolved questions in molecular similarity systems is whether to imitate this instinctive process by weighting the value for some descriptors, so that a match between two molecules on a highly weighted feature makes a greater contribution to the overall similarity than a match on a less important feature. A trivial example of weighting occurs with similarity measures based on fragment dictionaries: if a specific fragment present in the two molecules is in the dictionary then it has a weight of one, but if it is not present in the dictionary it has a weight of zero. More sophisticated approaches have been described in the literature (50–54), but most fragment-based similarity measures do not use weights, because of the equivocal nature of the evidence supporting their use and because of the additional computational requirements when the weights for very large numbers of compounds need to be processed.

A related topic is that of standardization, which is extensively used in multivariate analysis to ensure that all of the variables under consideration are measured on comparable scales. This is important in the context of similarity measurement since, if standardization is not carried out, the contributions of a few variables with wide ranges of values may swamp the contributions of all of the other variables. Many different approaches to standardization have been described in the literature (55, 56) but studies by Bath et al. (57) and by Turner et al. (58) suggest that the choice of standardization method does not significantly affect the outcome when using common similarity measures.

Similarity Coefficients

Once the attributes of compounds have been described, the degree of similarity between compounds can now be calculated. To do this requires the use of similarity coefficients, a definitive account of which is provided by Sneath and Sokal (59). They identify four main classes of coefficients: distance; association; correlation; and probabilistic coefficients. The first two classes are by far the most extensively used. The most common distance measure is the Euclidean distance, and many nearest-neighbor searching algorithms assume that this is what is used to measure the degree of resemblance between pairs of objects. For two molecules I and J, the Euclidean distance is given by:

$$\sqrt{\sum (x_{IK} - x_{JK})^2}$$

where x_{IK} represents the descriptor value of molecule I for the attribute K. The summation is performed over all of the attributes, K. (Similar comments apply to the two following formulae in this section.) The Euclidean distance is usually the measure of choice when processing data, such as physicochemical properties, that can be represented by real numbers.

A study by Willett and Winterman (50) suggested that association coefficients, such as the cosine and the Tanimoto coefficients, gave better results than distance coefficients for fragment-based similarity measures. Association coefficients are normalized functions of the dot product between the vector representations of the molecules I and J, with the Tanimoto and cosine coefficients defined to be:

$$\frac{\sum x_{IK} \times x_{JK}}{\sum x_{IK}^2 + \sum x_{JK}^2 - \sum x_{IK} \times x_{JK}}$$

and

$$\frac{\sum x_{IK} \times x_{JK}}{\sqrt{\sum x_{IK}^2 \times \sum x_{JK}^2}}$$

respectively. These forms of the two coefficients can be used with both binary and nonbinary variables (*18*). A simpler form of the coefficient can be used if the data is only binary, as is true for fragment bit-strings. If we assume that, in their fragment bit-strings, a target molecule and a database molecule have T and D nonzero bits, respectively, and that C of these are in common, then the formulae shown above for the Tanimoto coefficient and the cosine coefficient would reduce to:

$$\frac{C}{T + D - C}$$

and to

$$\frac{C}{\sqrt{T \times D}}$$

respectively. The complements of such coefficients can be used as measures of dissimilarity. One use for such measures is in the dissimilarity-based approaches to compound selection that are discussed later in this chapter.

Selection Methods: The Grouping of Compounds

The techniques summarized above provide efficient means of quantifying the degree of (dis)similarity between pairs of molecules. This information forms a vital component of methods for the rational selection of the sets of molecules that comprise the monomer pools in a combinatorial synthesis. The remainder of this section reviews the three main methods that are available for this purpose: cluster-based compound selection; dissimilarity-based compound selection; and partition-based compound selection.

Cluster-Based Compound Selection

Cluster analysis, or clustering, is the process of subdividing a group of objects (chemical molecules in the present context) into groups, or clusters, of objects that are similar to other objects in their cluster (i.e., showing a high degree of intracluster similarity) but dissimilar to objects in other clusters (i.e., showing a high degree of intercluster dissimilarity) (59–62). It is thus possible to obtain an overview of the range of structural types present within a data set by selecting one, or a small number, of the molecules from each of the clusters. The representative molecule (or molecules) for each cluster may be either selected at random or selected as being the closest to the cluster centroid (the hypothetical molecule with attribute values that are the arithmetic mean of those of the members of the cluster).

The requirement for intracluster similarity means that the representative molecules are expected to typify the others in the clusters from which they were drawn, while the requirement for intercluster dissimilarity means that the representative molecules are expected to differ significantly from the representative molecules from other clusters. If an appropriate clustering method has been used, molecules selected in this way should thus be excellent candidates for inclusion in the monomer pools used to generate a combinatorial library. A similar approach has been used in quantitative structure–activity relationship (QSAR) studies. Here, substituents are clustered to ensure that analogues synthesized in lead-optimization programs span the available physicochemical space (63–66). The clustering of molecules, rather than of substituents, has been used extensively to identify candidate molecules for inclusion in biological screening programs (52, 67, 68), to group the outputs of substructure searches (69) and de novo design systems (70), to aid user understanding of the results, and to organize large-scale QSAR studies (71), as well as for selective compound acquisition (5). This large body of experience in the application of clustering methods to the processing of structural information makes this approach an attractive one for compound selection.

Many different clustering methods have been described in the literature (59–62). The remainder of this section summarizes the methods that have been applied to the clustering of chemical databases; more extended discussions are provided by Willett (18) and by Downs and Willett (68).

Hierarchical Clustering Methods

Hierarchical clustering methods produce classifications in which small clusters of very similar molecules are nested within larger and larger clusters of less closely related molecules. Hierarchical agglomerative methods generate a classification in a bottom-up manner, by a series of agglomerations in which small clusters, initially containing individual molecules, are fused together to form progressively larger clusters until there is just a single cluster. Conversely, hierarchical divisive methods generate a classification in a top-down manner by progressively subdividing the single cluster representing the entire data set.

Hierarchical agglomerative methods are the most widely used type of clustering procedure, and have been the subject of intense study since their initial development and use for taxonomic applications (59). The obvious algorithm for implementing a hierarchical agglomerative method, as shown in Algorithm 1, involves calculating an initial similarity matrix that contains the similarities between all pairs

---

Algorithm 1. Stored-Matrix Algorithm for Hierarchic Agglomerative Clustering Methods

1. Calculate the similarity matrix, giving the similarities between all pairs of molecules in the data set that is to be clustered.

2. Find the most similar pair of points (where a point denotes either a single molecule or a cluster of molecules) in the matrix and merge them into a cluster to form a new single point.

3. Calculate the similarity between the new point and all remaining points.

4. Repeat steps 2 and 3 until only a single point remains, that is, until all of the molecules have been merged into one cluster.

---

of molecules in the data set that is to be clustered. Next, one repeatedly finds the most similar pair of points (where a point denotes either a single molecule or a cluster of molecules), merges them into a cluster to form a new point, and then calculates the similarity between the new point and all of the remaining points. This procedure is repeated until all of the molecules are in a single cluster.

There are many different hierarchical agglomerative methods (*62*), including the single-linkage method, the group-average method, and Ward's method, inter alia. They differ, however, merely in the ways in which the most similar pair of points is defined and in the representation of the merged pair as a single point (steps 2 and 3 in Algorithm 1), and so can all be implemented using the so-called stored-matrix algorithm shown above (*72*).

An example of a hierarchical divisive method that has been used for clustering chemical compounds is the minimum-diameter method (*44, 73*), which recursively divides the cluster with the largest diameter (i.e., the cluster with the most diverse members) such that the resulting clusters have the smallest possible diameters. The set of clusters is then reassessed to find the cluster that now has the largest diameter. This procedure is summarized in Algorithm 2.

Nonhierarchical Clustering Methods

A nonhierarchical method partitions a data set to give a set of groups having no hierarchical relationships between them. A systematic evaluation of all possible partitions is totally infeasible, but this has not prevented the development of many different methods to identify good, but possibly suboptimal, partitions. Such methods are generally much less demanding of computational resources than the hierarchical methods.

One approach that is extremely efficient is the single-pass (or Leader) method. A single molecule in the database is designated as the first cluster; this molecule can be chosen at random or on the basis of some quantitative decision procedure (*18*). Each subsequent molecule is assigned to the most similar existing clus-

---

Algorithm 2. Algorithm for the Minimum-Diameter Clustering Method

1. Calculate the dissimilarity matrix, which contains the similarities between all pairs of molecules in the data set that is to be clustered. Sort the matrix by order of dissimilarity.

2. Split the pair of molecules that is most dissimilar to form the two initial clusters. Distribute all other molecules to one of the two clusters, such that each compound joins the cluster with which it is more similar.

3. Select the cluster with the greatest diversity (the largest diameter cluster), and partition it into two clusters such that the larger cluster has the smallest possible diameter

4. Repeat step 3 for a maximum of $N - 1$ bipartitions, where N is the number of molecules.

---

ter(s), or, if the similarity does not equal or exceed a user-defined threshold similarity, is designated as the start of a new cluster. This procedure is repeated until the whole database has been processed. The similarity calculation requires that the representation of each existing cluster (usually the centroid) be updated whenever a molecule is added to it, so that its similarity to the current molecule can be calculated. The basic algorithm is detailed in Algorithm 3.

Hodes has discussed the use of a single-pass method for the cluster-based selection of compounds for biological screening in a very large scale antitumor program organized by the National Cancer Institute (*53, 74, 75*). The major problem with such methods is that the clusters are dependent on the order in which the structures are processed and upon the similarity threshold that is used to determine whether the current molecule should join an existing cluster.

Although several nonhierarchical methods for clustering molecules have been studied (*18*), the most widely used is that originally described by Jarvis and Patrick (*76*). The initial stage of this method involves creating a list of the k nearest neighbors for each of the N molecules in a database. Once these N nearest-neighbor lists have been produced, the second stage clusters molecules on the basis of their nearest-neighbor lists (rather than on the basis of common attribute values as is normally the case). Specifically, two molecules, i and j, are clustered together if each is a nearest neighbor of the other and if, additionally, they have k_{min} nearest neighbors in common, where k_{min} is a user-defined parameter in the range $1 \le k_{min} \le k$. The procedure is summarized in Algorithm 4.

The exact partition that is produced by the Jarvis–Patrick method depends on the number of nearest neighbors that are stored for each molecule (values between 10 and 20 are common) and the threshold number of common nearest neighbors. Examples of the use of this method are described by Shemetulskis et al. (*5*), Willett et al. (*67*) and Nouwen and Hansen (*71*), inter alia.

Algorithm 3. Algorithm for the Single-Pass Clustering Method

1. Designate the first molecule as the first cluster.

2. Assign the next molecule to the most similar existing cluster, or, if the similarity does not equal or exceed a user-defined threshold similarity, designate it as the start of a new cluster.

3. Repeat step 2 until all of the molecules in the database have been processed.

Algorithm 4. Algorithm for the Jarvis–Patrick Clustering Method

1. Create an N-element label array that contains a cluster label for each of the N molecules in the data set. Initialize this array by setting each element to its array position, thus assigning each molecule to its own initial cluster.

2. Compare the nearest-neighbor lists for each pair of molecules, i and j ($i < j$).

   If, for each such pair,

   > i is in the top-k nearest-neighbor list of j,
   > j is in the top-k nearest-neighbor list of i,
   > and i and j have at least k_{min} of their top k nearest neighbors in common

   then replace all occurrences of the label-array entry for j with the label-array entry for i.

3. The members of each cluster then all have the same array entry in the final label array.

Comparison of Clustering Methods

There are many clustering methods available, and quantitative criteria are needed to determine how appropriate they are for molecular diversity applications. A method can be evaluated in terms of either its effectiveness (the extent to which its use results in the selection of a diverse set of molecules) or its efficiency (the computational requirements of the method, which determine the sizes of the data sets to which it can be applied). There is generally a trade-off between the effectiveness and the efficiency of methods for database processing (*19*) and the choice of method will thus be affected by the size of the data sets that are to be processed. The discussion that follows focuses upon the comparison of clustering methods, but many of the comments are equally applicable to the dissimilarity-based and partition-based methods for compound selection that are presented later in this chapter.

Efficiency. It is helpful when comparing different computational approaches to characterize the approaches in terms of their computational complexity, where the complexity is a mathematical function that describes how the computational requirements of a particular algorithm vary with the number of molecules in the data set. This has the advantage that it removes extraneous, nonalgorithmic factors such as the language in which an algorithm has been programmed or the machine on which it has been implemented. The two main types of complexity are the time complexity and the storage complexity. If, for example, an algorithm has a time requirement proportional to the number of molecules, N, in the data set then it is said to have a time complexity of $O(N)$; similarly, the complexity is $O(N^2)$ if the requirement is proportional to the square of the number of molecules.

In making comparisons, it is important to note that in some cases there may well be several different algorithms by which any particular selection method can be implemented, and one must ensure that the most efficient algorithms are considered. In other cases, however, the method is already defined in purely algorithmic terms. This difference may be illustrated by comparing the outputs of a hierarchical agglomerative and the single-pass clustering methods. The output of the hierarchical method is a set of clusters in which the members have a particular similarity relationship to each other, with any algorithm that can yield this particular relationship being applicable. The output of the single-pass method is, however, defined solely by the algorithm that is used to identify the clusters.

Given a database of N molecules, the stored-matrix algorithm for hierarchical agglomerative methods requires $O(N^2)$ space to generate and store the full intermolecular similarity matrix and $O(N^3)$ time for the subsequent clustering stage. This is impractical for databases of nontrivial size. More sophisticated algorithms are, however, available, requiring only $O(N)$ space and $O(N^2)$ time (*68, 77*). An example is the reciprocal nearest neighbors (RNN) algorithm for clustering chemical databases using the group-average and Ward methods (*44, 78*). The minimum diameter hierarchical divisive method requires $O(N^2)$ space and $O(N^2 \log N^2)$ time, which severely limits the size of the data set to which it can be applied. The Jarvis–Patrick method requires $O(N)$ space and $O(N^2)$ time for the generation of the nearest-neighbor lists, and is thus comparable in complexity with the RNN algorithm for the hierarchical agglomerative methods. The single-pass method has an $O(N)$ space requirement and an $O(mN)$ time requirement (for a classification that yields m clusters), making it the most efficient of the methods mentioned here.

Effectiveness of Methods. The extensive literature relating to the comparison and evaluation of clustering methods describes at least three main approaches. First, theoretical analyses aim to identify methods with characteristics that match most closely a set of predefined criteria of effectiveness (*79, 80*). Second, simulation studies use artificial data sets for which the groupings are already known, and investigate the extent to which different methods are able to recover this structure (*81*). Finally, purely empirical comparisons use evaluation criteria specific to the problem (*82*). The basis for this last approach is the similar-property principle of Johnson and Maggiora (*23*), which states that structurally similar molecules should have similar properties (or activities). Thus, if a particular member of a cluster is known to exhibit some property (whether biological, chemical, or physical), the other members of that cluster are also likely to exhibit that property.

It is possible to test the similar-property principle quantitatively, using data sets for which property data are available. The property value of each molecule is assumed to be unknown, and the classification resulting from the use of some particular clustering method is scanned to identify the cluster that contains a particular molecule, A. The predicted property value for A, $P(A)$, is then defined as the arithmetic mean of the observed property values of the other compounds in that cluster. This procedure is repeated for each molecule in the data set. An overall figure of merit for the classification is obtained by calculating the product moment correlation coefficient between the sets of N observed and N predicted values. The most generally useful clustering methods will be those that give high correlation coefficients across as wide a range of data sets as possible.

This approach to the comparison of clustering methods, which was first suggested by Adamson and Bush (*83*), was used by Willett (*18*) in an extended comparison of more than 30 hierarchical and nonhierarchical clustering methods. The best results were given by Ward's hierarchical agglomerative method and by the nonhierarchical Jarvis–Patrick nearest-neighbor method. When these comparisons were carried out (in the early and mid-1980s), the available software and hardware technology was insufficient to enable the application of Ward's method to large databases, and the Jarvis–Patrick method thus became widely used for clustering databases of chemical structures. Willett's experiments involved only very small QSAR data sets in which the molecules were characterized by fragment bit-strings, and it should not be assumed that the Jarvis–Patrick method is invariably the method of choice with other types of data. For example, Downs et al. (*44*) found that this method was noticeably inferior to the Ward and group-average hierarchical methods when clustering molecules that had been characterized by calculated global molecular properties, and similar conclusions have been reached more recently by Brown and Martin (*10*) in an extended comparison of methods for selecting molecules for the construction of diverse combinatorial libraries. The work of Brown and Martin is discussed in more detail later in the chapter.

Dissimilarity-Based Compound Selection

The cluster-based approaches to compound selection described above identify a set of dissimilar molecules indirectly (since the approaches require the initial identification of clusters of similar molecules, from which dissimilar molecules can subsequently be selected), and the same is true for the partition-based approaches described in the next section. Dissimilarity-based approaches are, however, based on algorithms that identify a diverse set of molecules directly. Specifically, the approaches seek to identify the n most dissimilar molecules in a data set containing N molecules (where, typically, $n \ll N$) using some quantitative measure of dissimilarity such as the complement of the Tanimoto coefficient. Unfortunately, the identification of the maximally diverse subset is computationally infeasible, as it requires consideration of all possible n-member subsets of the database, and there are no less than

$$\frac{N!}{n!(N-n)!}$$

such subsets. Accordingly, the dissimilarity-based methods that have been described are based on efficient, but suboptimal, algorithms that are most unlikely to result in the identification of the maximally dissimilar set of molecules. A typical procedure is shown in Algorithm 5. Molecules are selected from *Database* (which initially contains N molecules) and placed in *Subset* (which finally contains n molecules). This algorithm is based on work carried out by groups at Pharmacia and Upjohn (*84–87*) and at Pfizer (*21, 88*), and has been shown to work well in practice.

An analysis of this algorithm shows that it has a worst-case time complexity of $O(n^2N)$; a fast $O(nN)$ implementation has been described by Holliday et al. (*89*). Several other algorithms for dissimilarity selection have been described. Taylor (*90*) describes a modification of the stored-matrix algorithm for hierarchical agglomerative clustering, and Wootton et al. (*91*) discuss a strategy for lead optimization in which the chosen substituents are at least a threshold distance apart as defined by the sum of the physicochemical parameters for each analogue.

As written, step 2 of Algorithm 5 is not sufficiently detailed to allow an implementation of the algorithm, as there are several ways in which the phrase "most dissimilar" can be defined (which is also true when defining the most similar pair of objects that are to be fused at each stage in the generation of a hierarchical agglomeration). Holliday and Willett (*92*) compare several different definitions of dissimilarity and show that markedly different sets of molecules can be obtained depending on the definition that is used.

How do dissimilarity-based methods compare to other selection methods? Detailed studies of dissimilarity-based selection have been reported by Lajiness and co-workers (*84–87*). One such study (*86*) compared four approaches to the selection of molecules for biological screening: random selection; dissimilarity-based selection; cluster-based selection; and a modification of cluster-based selection that eliminated from consideration all clusters containing any molecules that had previously been tested in the biological screen of interest. Simulation experiments suggested

---

Algorithm 5. Algorithm for Dissimilarity-Based Compound Selection

1. Select a molecule at random from *Database* and place it in *Subset*.

2. Identify that molecule in *Database* that is most dissimilar to the molecules already in *Subset* and add that molecule to *Subset*.

3. Repeat step 2 a total of $n - 2$ times.

that the dissimilarity and modified cluster-based selection methods were best able to summarize the range of structural classes present in a small data set. A later, more extended study used both simulated and real activity data (*87*), and concluded that the relative merits of cluster-based and dissimilarity-based selection were determined in large part by the particular characteristics of the data set that was being processed. Specifically, and hardly surprisingly, cluster-based selection gave the better results when there were large concentrations of active molecules within small regions of structural space, that is, when the data set was naturally strongly clustered, whereas dissimilarity-based selection worked better when this was not the case.

A more detailed simulation study of compound selection methods has been reported by Taylor (*90*). Here, cluster-based and dissimilarity-based selection methods similar to those shown in Algorithms 4 and 5, respectively, were used to generate a prioritized list of molecules for biological testing. A random number generator was used to assign activities to the molecules so that one molecule was designated as very active, a few were weakly active, and the majority inactive. Molecules were then checked for activity in the order suggested by the two selection methods, and the checking continued until the single, highly active molecule had been identified. Importantly, a feedback mechanism was used to modify the ordering of the molecules as the simulation proceeded, with the identification of a weakly active molecule leading to the promotion of its nearest neighbors in the prioritized list of molecules waiting to be checked for activity. Conversely, "discovery" of an inactive molecule led to the demotion of its neighbors. Taylor's results suggested that cluster-based selection was only marginally better at finding the single highly active molecule than was random selection (primarily because of the effects of the positive feedback mechanism) and that dissimilarity-based selection was worse than random selection. However, as Taylor notes, there may well be a premium attached to identifying active molecules that are very dissimilar to the rest of a database (e.g., for ease of patenting) even though the overall identification rate is lower than that achieved when using alternative approaches to compound selection.

Partition-Based Compound Selection

Cluster-based selection has been used for many years to select molecules for biological screening, and there is an extensive literature associated with its use (*5, 10, 18, 53, 67, 69, 71, 75*). Dissimilarity-based selection is a more recent development, but it is becoming increasingly well established, as exemplified by the studies reported above. Partition-based compound selection, the subject of this section, is the newest approach to compound selection, but one that is already attracting much interest.

The basic approach is detailed in Algorithm 6, and it requires the identification of a set of p characteristics that are of importance to the task in hand; in the case of library generation, these characteristics might be molecular properties that would be expected to affect binding at a receptor site. The characteristic numbered i ($1 \leq i \leq p$) has a range of possible values that is then subdivided into a set of b_i bins (or sub-ranges). The combinatorial product of all possible bins then defines the set

---

Algorithm 6. Algorithm for Partition-Based Compound Selection

1. Select a set of p properties and a set of b_i bins for each property i ($1 \leq i \leq p$).

2. Generate all possible combinations of bins, where each combination represents one of the groups in the partition.

3. Calculate the p property values for each molecule and allocate it to the appropriate group.

4. Select a number of molecules from each group.

---

of groups that make up the partition, and each molecule is assigned to the group that matches the set of binned characteristics for that molecule. Thus, for example, all molecules with high melting points and low hydrophobicity might be in one group. One (or a small number) of the molecules in each of the resulting groups is then selected for inclusion in a monomer pool.

The first report of partition-based selection was by Mason et al. (*1*). The characteristics used were six global molecular properties encoding a molecule's hydrophobicity, polarity, torsional flexibility, shape, and ability to act as a hydrogen-bond donor and acceptor. The range of possible values for each of these six properties was divided into 2–4 bins, yielding a combinatorial partition that contained 560 different groups. Each molecule was then allocated to a group on the basis of the molecule's property values, which were calculated from the SMILES notation. The resulting partition was used successfully to generate leads capable of reducing the blood concentration of low-density lipoproteins.

To generate a partition, well-defined characteristics must be assigned to each molecule. Topological indices have been shown to correlate well with a number of physicochemical properties (*34*), and the ease with which such indices can be calculated means that they are being used increasingly in similarity and diversity studies. For example, Cummins et al. (*93*) have used topological indices as the basis for a partitioning algorithm that has been developed at Glaxo Wellcome for analyzing the diversity of both public and corporate structure databases. The molecules in these databases were characterized by a total of 60 topological indices and by a computed value for the free energy of solvation, and factor analysis used to generate a reduced-dimensionality space into which the molecules were then assigned. Topological descriptors also form the basis for the BCUT approach to partitioning that has been described recently by Pearlman (*94*). This is based on work by Burden (*95*), who suggested that a "molecular ID number" could be defined for a molecule in terms of the two lowest eigenvalues of a matrix representing the connection table (excluding hydrogen atoms) of that molecule. In the BCUT system, three matrices are used. The diagonal entries of each matrix contain values describing atomic charges, atomic polarizabilities, and atomic hydrogen-bonding abilities, respectively. The pairs of eigenvalues that can be derived from these matrices form the 6D space into which molecules are projected, with the precise nature of the space determined by the computational procedures that are used to calculate the various atomic properties encoded in the matrices. A partition of the molecules would then be based on their location in 6D space.

Physicochemical parameters and topological indices, which have been used extensively in QSAR studies, provide obvious bases for the generation of a partition, but any appropriate molecular characteristics can be used. For example, Martin and co-workers (*8, 11*) have suggested a partition based on three-point pharmacophore distances. Possible pharmacophore points (specifically, hydrogen donors or acceptors, ring centroids, and positive, negative, and hydrophobic centers) are identified using a set of more than 400 substructure definitions. All of the triangle descriptors within each molecule are then generated, where a triangle descriptor contains three of the possible pharmacophore points with interpoint distances of 3–10 Å [these distances are calculated from a 3D structure produced by a structure-generation program (*96*)]. Considering all possible arrangements of all the different types of pharmacophore points, there are ~40,000 such descriptors possible, each of which represents a group of molecules that share a common potential pharmacophore and so might have a common activity at a specific macromolecular site. This process is repeated for all of the molecules in a database, yielding the frequency of occurrence of each possible triangle descriptor. A diverse set of molecules is then obtained by selecting molecules from the database that contain low-frequency triangles.

A very similar approach is taken in the pharmacophore-derived queries (PDQs) method described by Mason et al. (*9*). PDQ recognizes six types of pharmacophore points (O/N donors, O/N acceptors, aromatic centers, basic nitrogens, acidic centers, and hydrophobic groups) and six interpoint distance ranges (2–4.5 Å, 4.5–7 Å, 7–10 Å, 10–14 Å, 14–19 Å, and 10–24 Å), for which combinatorial and geometric considerations give a total of 5916 distinct groups. Each such group defines a query pharmacophore that is used in a flexible 3D database search (*15, 16*) to identify the molecules occurring within that group.

At first sight, partition-based selection would appear to be analogous to cluster-based selection, in that a set of groups is created and molecules are then selected from each of the groups. There are, however, important differences that reflect the fact that, using the terminology of pattern recognition, cluster-based and partition-based selection are examples of unsupervised and supervised classification, respectively. In partition-based approaches the groups are specified, and a molecule is allocated to one such predefined group solely on the basis of some particular characteristics of that molecule, the nature of the groupings being determined by the characteristics that are chosen for this purpose.

Once the set of groups is defined, partition-based selection has an expected time complexity of $O(N)$ for a database of N molecules (although the constant of proportionality in the complexity expression may be large in the case of a method such as PDQ where a flexible 3D substructure match is required for each molecule). It is thus potentially much faster than cluster-based selection, where the groups are only identified as the method is applied, and where the allocation of a particular molecule to a particular group is dependent upon the other molecules in the database. A further advantage of a partition is that it identifies explicitly those sections of structural space that are either underrepresented or not represented in a database, and this can provide valuable information for de novo synthesis programs. The structural and functional constraints for these programs have traditionally been obtained by characterizing the active site of a known receptor (97), but it is also possible to design molecules that satisfy particular (physico)chemical constraints (98, 99), thus allowing the generation of molecules that are very different from those already available within a database. Given these characteristics, it seems likely that partition-based selection will be important in large-scale compound selections in the future.

Applications

Having introduced the data structures and algorithms that are used for representing, searching, and selecting molecules, we now review the application of such techniques to real-world problems in molecular design. That said, it must be emphasized that the separation of principles and practice is rather artificial, given that the whole field is still at a very early stage of development.

Selecting Compounds To Increase the Diversity of Compound Databases

When the biological screening of compounds was very time consuming, a great deal of research went into methods of compound selection, with the compounds coming from within an organization's corporate database. The focus is now on selecting compounds, from whatever source, that will enhance the range of structural types available to an organization, and the selection methods described above are the main tools in this process. Sources of structural diversity include publicly available commercial databases, collaborations with specialist synthetic groups, folk medicine, and selective compound-exchange agreements. Below, we refer to any set of structures that are possible additions to an existing database as an external data set. Shemetulskis et al. (5) provide a detailed account of the selective compound acquisition procedures that have been developed at Parke-Davis to enhance the diversity of their existing corporate database. The external data sets considered in this work were two public databases (the CAST-3D database of 379,487 rigid molecules from the Chemical Abstracts Service (CAS) Registry System, and the Maybridge catalogue, which contains the structures of 41,912 molecules for which samples are commercially available), but the approach is applicable to any external data set.

A cluster-based selection approach was adopted, in which the corporate database was merged with an external data set, and the clusters within the resulting, merged database examined to identify molecules, or groups of molecules, from the external data set that appeared to be significantly different from those in the corporate database. For example, in one set of runs after the Parke-Davis database had been merged with the CAST-3D data set, 78% of the CAS structures were in clusters containing only CAS structures. Such clusters clearly represent groups of molecules that are very different from those previously available to Parke-Davis, and no less than 101,000 CAS compounds were acquired as a result of these, and related, analyses. The Daylight implementation of the Jarvis–Patrick clustering method was used for the experiments, with the molecules being characterized by their Daylight fingerprints, and with run times of up to 64 CPU days on a high-performance UNIX workstation.

Shemetulskis et al. (*5*) also characterized the molecules in their databases using three calculated properties: the logarithm of the octanol/water partition coefficient (CLOGP); the molar refractivity (CMR); and the electronic dipole moment (CDM). Here, the similarity was calculated between each Parke-Davis molecule and each molecule in one of the external data sets. The Maybridge molecules were found to be far less dissimilar than the CAS molecules. These property-based analyses were restricted to far smaller numbers of molecules than the fingerprint analyses because of limitations in the programs currently available for calculating physical properties.

Another application of the Daylight implementation of the Jarvis–Patrick method is described by Mason et al. (*9*), in a comparison of cluster-based and partition-based selection methods that used the 164,000 molecules in the corporate database of Rhone Poulenc Rorer (RPR). The Jarvis–Patrick method is known to yield many small, or even singleton, clusters, as well as a small number of very large clusters (characteristics shared by the single-linkage hierarchical agglomerative method, a close relative). The RPR group felt it was important to restrict both the size of the largest clusters and the number of singletons, and hence adopted a multi-level, or cascaded, clustering approach. The very large clusters were reclustered using 2D and 3D fragment bit-strings, and the very small clusters were reclustered after the larger clusters had been removed. The molecules nearest the centroids of both the original clusters and those obtained after the cascaded clustering then formed the basis for an 11,000-molecule subset for biological testing. Mason et al. have also considered the clustering of the database that resulted from merging the RPR corporate database with the Available Chemicals Database (ACD). Finally, they have applied their partition-based selection procedures (*1, 9*) to the RPR database, and found that the 1281 molecules selected in this way were complementary to those selected by their cluster-based procedure.

Design and Generation of Libraries

Martin et al. (*7*) provide a detailed account of the computational tools used in the design of combinatorial libraries of oligo(*N*-substituted) glycine peptoids at Chiron Corp. (*100*). On the basis of cost, available quantity, and lack of toxic or reactive functionality, 721 primary amines and 1133 carboxylic acids and acid chlorides were selected from lists of commercially available molecules. Each of these molecules was characterized in terms of lipophilicity (using CLOGP and related pro-grams), shape and branching (using principal components analysis on 81 topologi-cal descriptors to give five latent variables), chemical functionality (using multidi-mensional scaling (MDS) on the intermolecular similarity matrices resulting from pairwise comparisons of the molecules' Daylight fingerprints), and receptor-recog-nition ability (using MDS on the intermolecular similarity matrices resulting from the pairwise comparison of sets of atom-centered substructural descriptors). The dissimilarity-based selection was carried out using an interactive, iterative proce-dure that is based on "D-optimal" approaches to experimental design. The proce-dure seeks to identify small subsets of the monomer pools that are well spread out and nearly nonoverlapping in property space. It is clearly effective but, unlike clus-ter-based selection methods, it is apparently applicable only to small data sets.

Ferguson et al. (*101*) discuss a computer-based method for the design of highly dissimilar molecules that are suitable for use in lead-discovery programs, illustrating the procedure by means of a simple combinatorial amidation reaction. Given a library of available amines (e.g., those in a database such as ACD or the Aldrich Catalogue), an initial substructural filter removes molecules containing groups that would interfere with the chosen combinatorial reaction or would confer undesirable biological properties on the final reaction product. A second filter removes molecules that are too heavy (with molecular weights (MW) > 600) or too lipophilic (with CLOGP $\geq$ 10). The similarities between the remaining molecules are then calculated by means of a descriptor that is based on comparative molecular field analysis (CoMFA) steric fields (*102*), and these similarities provide the input to a cluster-based selection procedure that yields the final set of amines. A similar set of procedures is used to obtain a diverse set of acylating agents, and the products of combinatorial reaction of the two sets of molecules are computed. A further MW–CLOGP filter is applied to these products, and the final subset of molecules for synthesis is then obtained by a dissimilarity-based selection procedure. Here, the product molecules are characterized by their fragment bit-strings, and molecules are excluded if their similarity to a molecule that has already been selected exceeds 0.85, this threshold figure being based on a study by Martin et al. (*11*).

The protocols of Martin et al. (*7*) and Ferguson et al. (*101*) are excellent case studies of the traditional approach to the generation of libraries, where detailed selection procedures are employed to create very small, but structurally diverse, monomer pools that are expected to yield diverse oligomers when the monomers are reacted together. A radically different approach has been suggested by Sheridan and Kearsley (*3*), who use the searching capabilities of a genetic algorithm (GA) (*103*). A GA uses individual monomers as input. After selecting certain monomers based on various fitness functions, the GA "evolves" combinations of these monomers by reacting them together. The two different types of fitness function that are used by Sheridan and Kearsley are similarity to a molecule that is known to exhibit the bio-logical activity of interest (using a fragment bit-string measure of structural similar-ity), and predicted biological activity using a fragment-based activity weighting scheme (*104*). The algorithm is an extremely rapid way of exploring product space.

For example, only 25 generations of the GA were required to analyze a pool of 3312 primary and secondary amines to create high-fitness tripeptoids that were similar to two known CCK antagonist tetrapeptides. The generation of novel structures on the basis of similarity or activity weighting is not new (*105*), but the application to library design is both elegant and potentially very powerful. Although the basic operations (crossover and mutation) of the GA are extremely simple, the fitness function provides a powerful mechanism for utilizing whatever constraint information is available. For example, the fitness function could encompass steric, hydrophobic, and electrostatic constraints, energy criteria, and conformational flexibility, inter alia. Moreover, although Sheridan and Kearsley considered only a GA, there are several other types of combinatorial searching procedure that could also be used to generate diverse libraries (*106*).

Comparative Studies To Determine the Best Available Methods

Given the range of descriptor types, similarity measures, and clustering methods, detailed comparative testing is essential to determine the most effective procedures. A long series of experiments by Willett and his collaborators, the main results of which are summarized in reference 18, demonstrated the efficiency and general effectiveness of fragment bit-strings for the measurement of intermolecular similarities, and of the Jarvis–Patrick method for clustering chemical databases (specifically for summarizing the outputs of substructure searches and for selecting compounds for biological testing). However, as noted in the discussion on comparing methods for cluster-based compound selection, these findings have recently been reevaluated in a detailed analysis of methods for compound selection by Brown and Martin (*10*). These authors used property prediction to compare seven different structural descriptors (including 2D fragment bit-strings, both rigid and flexible 3D fragment bit-strings, and pharmacophore descriptors analogous to those described previously for use in partition-based compound selection) and four different clustering methods (Ward's and group-average hierarchical agglomerative methods, the minimum-diameter hierarchical divisive method, and the Jarvis–Patrick method). Unlike the earlier experiments, which involved small QSAR data sets containing just a few tens of compounds (*18*), Brown and Martin used four data sets containing between 1650 and 16,000 molecules for which associated bioassay data were available.

The detailed comparative experiments demonstrated that the 2D bit-strings were noticeably better than the other types of structural representation, with those bit-strings that were based on fragment dictionaries being consistently superior to those based on a fingerprinting mechanism. The best results of all were obtained with a set of 153 structural fragments from the MACCS substructure search system produced by MDL Information Systems Inc. (14600 Catalina Street, San Leandro, CA 94577), which were weighted to reflect the frequency of occurrence within each molecule. Of the four clustering methods that were tested, Ward's method gave by far the best results, with the Jarvis–Patrick method being noticeably inferior to the others. The main advantage of the latter method is its speed, which makes it feasible for use in clustering data sets containing many hundreds of thousands of structures. However, the RNN algorithm (*44, 78*) used to implement Ward's method in Brown and Martin's experiments was fast enough to permit the clustering of the bit-strings representing a 100,000-molecule Abbott internal data set in 54 CPU hours on an R8000 UNIX workstation. It thus seems reasonable to conclude that Ward's method is now the clustering procedure of choice for data sets containing up to, perhaps, a quarter of a million molecules. This is especially true given that this method tends to produce clusters that are all about the same size, thus avoiding the very disparate cluster sizes that can cause problems with the Jarvis–Patrick method. Brown and Martin note that an adequate prediction performance can only be obtained if one uses small clusters containing perhaps five molecules (*10*). This implies that as much as 20% of a database may need to be selected if one wishes to characterize its constituent structural types fully.

The use of property-based descriptors in similarity calculations provides an obvious way of characterizing the steric, hydrophobic, and electrostatic forces that determine binding. In later, unpublished work, Brown and Martin (Brown, R. D.; Martin, Y. C., Abbott Laboratories, personal communication) have used the combination of MACCS bit-strings and Ward's clustering method to predict a range of physical properties (including octanol–water partition coefficients, cyclohexane–water partition coefficients, pK_a values, surface areas and volumes, and κ shape indices) and have shown that it is possible to obtain high, and in some cases very high, correlations between observed and predicted values. This suggests that 2D fragment descriptors may suffice to characterize the principal intermolecular forces that determine binding and molecular recognition, without the need for explicit physicochemical descriptors. It has already been noted that there are many molecules for which reliable property values cannot be calculated. Brown and Martin's work suggests that simple bit-string descriptors may suffice for the inclusion of property information in similarity and diversity analyses, which would drastically simplify the characterizations needed for successful library design. This finding is, however, at variance with the work of Martin et al. (*7*), who have advocated the use of very detailed structural characterizations covering many properties, and of Shemetulskis et al. (*5*), who have noted marked discrepancies between the structure-based and property-based similarities analyzed in the Parke-Davis study.

Cribbs et al. (*107*) have recently reported a comparison of several different selection methods for the generation of combinatorial libraries of ureas. The molecules were characterized by five calculated physicochemical properties (specifically CLOGP, the total E-state sum, the total dipole, the ellipsoidal volume, and the smallest moment of inertia), and four different selection methods were applied to these descriptions: D-optimal design, clustering by Ward's method, and two partitioning procedures. The compounds selected by a particular method were projected into 2D spaces defined by pairs of the physicochemical properties, and the effectiveness of the method assessed by visual inspection of the extent to which the selected compounds covered the 2D space. The authors note the limitations of such an approach, and the need for more quantitative evaluation measures. The results were, however, sufficient to demonstrate the inferiority of the D-optimal design procedure when compared to the other three methods. Only the cluster-based approach appeared to be significantly better than random, but Cribbs et al. suggested that this was because this approach led to the selection of quite a large number of compounds from a pool of candidate molecules that had little diversity. In this test, two sets of 30 compounds were selected from pools containing 89 benzylamines and 159 isocyanates. The authors also compared the resulting sets of 900 ureas with a corresponding set that had been selected from the full library of 14,151 possible ureas (all benzylamines reacted with all isocyanates), and concluded that selection from the full library was too inefficient, both in computational and synthetic terms, for it to be practicable.

Quantifying Database Diversity

Devising a simple quantification of database diversity is important. There is interest in quantifying both the diversity of a single database, and the extent to which the acquisition of an external data set will increase this diversity. Boyd et al. (6) have recently described a program, called HookSpace, that seeks to estimate the geometric diversity of structure databases by providing a quantitative description of the arrangement of functional groups in 3D space. This program also addresses an important aspect of combinatorial synthesis: the ability to position diverse functionalities at those points that are thought to be involved in binding at a biological site. In HookSpace, each functional group has an associated vector that specifies how that group lies in 3D space when it is linked to a central structural feature. For example, the vector in a halide group links a carbon atom in the central moiety (which is referred to as the "tail" atom) to the "head" halogen atom. Each molecule in a database is matched against a dictionary of such groups to identify those that are contained within the molecule, and then the geometric relationships are calculated between every possible pair of vectors, as in distance-based and angle-based screening techniques for 3D substructure searching (15, 16). This procedure is repeated for all molecules in a database, using a common coordinate system that enables the accumulation and comparison of the results across the entire database. The result is a list of how many times a particular pair of functional groups is found at a particular distance and angle from each other.

One of the descriptions generated by Boyd et al., which they refer to as the HookSpace Index, is the ratio of the intergroup orientations that occur within a database to the orientations that could occur if all possible orientations were observed. The authors demonstrate that very different values for this index are obtained for a benzodiazepine library, for the CSD, and for the ACD, and suggest that the index can be used to characterize databases in targeted ligand-design programs. Unfortunately, it is difficult to judge the significance of the differences that were observed, as the 3D structures in the three databases used by Boyd et al. came from very different sources: a template library followed by minimization for the benzodiazepines; X-ray experimental data for the CSD; and CONCORD for the ACD.

Several diversity indices have already been described in the literature (and some of them have been briefly mentioned in this review). Martin et al. (7) have suggested a diversity index for use with fragment bit-strings. Here, the bit-strings for all of the molecules in a database are joined by a series of Boolean OR operations so as to identify all bit positions that are set at least once, thus providing information about the number of distinct types of substructure present in the database. Martin et al. have shown that the resulting union bit-strings are far less densely populated for designed combinatorial libraries than they are for even quite small sets of drug-like molecules. This implies that the libraries span only a very small subset of the possible structural space. This index is very simple but has the practical limitation that a database may be sufficiently diverse for all of the bits to be set at least once. Thus it may be impossible to increase the diversity further, irrespective of the content of any additional data set that is merged with it. Holliday et al. (89) suggest using the sum of all of the pairwise intermolecular dissimilarities for a database (so that the more diverse a set of compounds, the greater the resulting pairwise sum) and describe an efficient algorithm of $O(N)$ complexity for the calculation of this sum. Turner et al. (108) discuss the use of this index to quantify the change in diversity when an external data set is merged with an existing database, thus providing a rational basis for the selection of a particular external data set when several are available.

Partition-based approaches to compound selection provide a simple and direct estimate of the diversity of a database by counting the fraction of the possible groups for which at least some threshold number of molecules exist. Thus, Martin et al. (11) note the much greater percentages of possible three-point pharmacophores that occur in databases of typical drug molecules when compared with typical combinatorial libraries, a finding that is analogous to those reported by both Boyd et al. (6) and Martin et al. (7). Pharmacophoric coverage has also been used by Mason et al. (9) to quantify the heterogeneity of sets and subsets of compounds. A detailed comparison of partitions of the Wellcome corporate database and several databases of commercially available compounds has been reported by Cummins et al. (93), who discuss the overlap between pairs of databases in terms of the numbers of groups that are occupied by compounds from both databases.

Measures of diversity are also available for cluster-based selection procedures, such as the Parke-Davis group's use of those clusters that contain molecules only from a single source when two merged databases are clustered together. Similarly, Martin et al. (11) have suggested that, when activity data are available, the number of clusters in which all constituent molecules exhibit the same activity could be used as a measure of diversity.

Commercial Software Systems

There are several companies whose main line of business is the development of computer software to support the discovery of bioactive molecules, and these companies have rapidly brought to the marketplace a range of tools to process information pertaining to combinatorial libraries. These tools are often integrated with other sorts of software that may be required in a combinatorial research program, such as software for handling high-throughput screening data, plate management, or robot control. The accounts given here are based upon information provided by the companies themselves, and no endorsement of any particular vendor or package is intended.

The ChemDiverse module (*109*) from Chemical Design Limited (Roundway House, Cromwell Park, Chipping Norton, OX7 5SR, U.K.) seeks to maximize diversity by means of a partitioning procedure that is analogous to the three-point pharmacophore approaches described by Martin et al. (*8, 11*) and Mason et al. (*9*). Molecules are characterized in terms of the presence of hydrogen bond donors, hydrogen bond acceptors, positively charged centers, and aromatic ring centers, and the distances between pairs of such points are measured in terms of 31 bins spanning 15 Å, giving a total of no less than 494,020 distinct, geometrically feasible pharmacophores. Each molecule in a database is submitted to a fast, rule-based conformational analysis to identify some number of low-energy conformers, and the molecule is then encoded in terms of the possible pharmacophores that occur within these conformers. Molecules are selected for inclusion in a library program in one of two ways: either because they are representative of the molecules within a particular group (i.e., the molecules possessing a particular pharmacophore in common), or on the basis of a dissimilarity-based selection procedure in which the intermolecular similarity measure is the number of shared pharmacophores.

Daylight Chemical Information Systems, Inc. (18500 Von Karman Ave. Suite 450, Irvine, CA 92715) is slightly different from the other vendors listed here, in that it provides users with software tools for building information management systems, rather than supplying such systems directly. CHUCKLES is a method for representing and searching molecules at both an atomic and a monomer level (monomers are rigorously defined molecular fragments). CHUCKLES is based on the SMILES notation (*13*) and allows facile interconversion between the monomer and atom representations of molecules. This description is suitable for representing arbitrary molecule classes. A superset of CHUCKLES (CHORTLES) allows the specification of libraries by means of controlled variation at the monomer level of the representation. A detailed description of the encoding rules for CHORTLES and CHUCKLES is presented by Siani et al. (*4*), who also discuss the implementation of substructure searching on such representations.

Project Library from MDL Information Systems Inc. is a desktop, project-level computer system for supporting research programs in combinatorial chemistry. The information-management module within Project Library provides facilities for the representation and searching of building blocks, individual molecules, complete combinatorial libraries and mixtures of molecules, and for linking these representations to reaction databases and biological data files. Other modules provide computational support for tracking mixture and discrete-compound libraries, including the construction of "virtual" libraries via enumeration for the purpose of diversity assessment, and for the common techniques (spatially addressable synthesis, tagged mixture synthesis, and deconvolution) that are used to identify active molecules. The interaction of these various components is illustrated by Snyder and Crofton (110), who provide a detailed account of the use of a range of MDL tools and databases in the design of a combinatorial library for gastrin antagonist activity.

The Molecular Diversity Manager from Tripos, Inc. (1699 South Hanley Road, St. Louis, MO 63144) is a suite of software modules for the creation, selection, and management of compound libraries. LEGION, the module for creating libraries, takes as input a base molecule (which can be represented in one of several different forms) and the variations in the number, type, and position of attachments that are possible. For example, it is possible to specify that a particular position on a ring can be occupied only by members of a carefully chosen list of substituents, or that the substituents at two particular positions can, optionally, be linked to form a further ring. The output from LEGION is a definition of the library that results from combinatorial exploration of the range of variants. This definition can either be in the form of a single string, called a CSLN or Combinatorial SYBYL Line Notation [or SLN, where an SLN is a chemical coding system that is derived from the SMILES notation (13)], or in the form of the SLNs for each of the individual molecules in the library. Structural descriptors for these molecules are generated by the SELECTOR module, which can characterize molecules in terms of both 2D and 3D structural fingerprints, substituent parameters, topological indices, and physicochemical parameters, and which can process the resulting characterizations by means of principal components analysis. Structurally diverse subsets are then chosen by means of either cluster-based (using Jarvis–Patrick or hierarchical agglomerative methods) or dissimilarity-based selection. The use of the Molecular Diversity Manager tools is illustrated by the identification of over 60 nanomolar inhibitors of angiotensin converting enzyme (ACE) from a library of 1232 synthetic analogues of a known ACE inhibitor (111).

Systems such as those summarized above are under active development (112), and at least two further such systems, C^2-Diversity from Molecular Simulations (9685 Scranton Road, San Diego, CA 92121–3752) and RS^3 Discovery HTS from Oxford Molecular Limited (The Medewar Centre, Oxford Science Park, Oxford, OX4 4GA, U.K.), were announced while this chapter was being written. Increasing attempts are also being made to integrate such systems with other sorts of computational chemistry software, such as for searching reaction databases (113) and for 3D QSAR (114). The reader should thus be aware that this section is likely to date extremely rapidly.

Conclusions

Many of the procedures described in this review are based on software that was originally developed for purposes other than the creation of combinatorial libraries;

they use the same, or related, structural descriptors as those used for representing and searching individual molecules in conventional databases. Although this is a cost-effective use of existing software and expertise, the resulting algorithms and data structures may not be ideally suited to the representation and processing of the ensemble of molecules that make up a combinatorial library. There is thus a continuing need for detailed algorithmic studies to maximize the effectiveness and the efficiency of the various approaches suggested so far. An example of work in this area is a new algorithm for estimating the volume of a partitioned compound space that is occupied by a particular set of compounds (*93*).

Existing structure-handling techniques may well suffice for combinatorial libraries containing a few thousands of molecules, but they may be inadequate for the analysis of the far larger libraries that have been constructed more recently. One possible source of new techniques is highlighted by Barnard and Downs. They have noted the strong resemblances between the representations of combinatorial libraries and the generic, or Markush, structures that characterize chemical patents (*115*). In both cases, it is possible to characterize very large numbers of molecules in terms of a much smaller number of structural characteristics. Thus, in libraries one need specify only the individual monomers that are reacted together in a combinatorial manner to define the membership of the entire library. For patents one has variable substructural specifications, including variable points of attachment for substituents, lists of alternatives at a particular position, and generic descriptions (such as 'heteroaryl' or 'alkyl 2-5') that, when taken together, may well encode many millions, or even an infinite number, of molecules. The last few years have seen the development of several sophisticated systems for the representation and searching of generic chemical structures in chemical patents (*116, 117*) that avoid the need to enumerate and to process each of the individual molecules contained in a patent claim. Barnard and Downs (*115*) have suggested that similar techniques could be used for the representation and searching of combinatorial libraries. Such an approach would overcome problems such as those identified by Siani et al. (*4*), who discuss the limitations of CHORTLES for unrestricted substructure searching of combinatorial libraries.

Another area requiring further study is the comparison of the different approaches that have already been suggested for compound selection. Thus far, there have been only a few attempts to compare different approaches: specifically, the simulation studies of cluster-based and dissimilarity-based selection by Lajiness (*87*) and Taylor (*90*), and the comparison by Mason et al. (*9*) of cluster-based and partition-based selection. Detailed, quantitative comparative studies, analogous to the evaluations of clustering methods carried out by the Abbott (*10*) and Sheffield (*18*) groups, remain to be carried out. This may, at least in part, be a result of the absence of an appropriate experimental methodology. Although the similar property principle (*23*) provides an obvious basis for the quantitative evaluation of similarity studies, it is not clear that it is equally applicable to the quantitative evaluation of diversity studies. In the former case, one is seeking to establish whether a correlation exists between similarity in structure and similarity in property or activity. In the latter case, conversely, the correlation that is sought is between structural diversity (assuming that it is possible to arrive at general agreement as to what is represented by this term) and the degree of coverage of property, or activity, space. To establish such correlations will require the development and comparison of further diversity indices to characterize structural heterogeneity. Comparisons are also necessary to assess the relative merits of libraries generated by selection procedures [such as the D-optimal design algorithm used by the Chiron group (7)] and those generated by searching procedures [such as the genetic algorithm used by the Merck group (3)]. The merits of the various diversity indices, and the effect of variations in the molecular descriptors, also need to be explored. So far, most workers have used features that had been used for similarity and QSAR studies, such as fragment bit-strings, topological indices, or global physicochemical properties. However, there may well be other features that are more appropriate for studies of molecular diversity. It will be very interesting to see whether the new features that have already been suggested, such as the BCUT matrices or the three-point pharmacophore distances, or other features that have not yet been described, prove to be more appropriate for diversity analysis than the existing approaches.

Methodological developments such as those suggested above will, among other things, provide the drug designer with rational ways of identifying gaps in the coverage of an existing database, determining which of several external data sets best complements an existing database, and selecting the smallest number of molecules necessary to cover some section of structural space. There is also considerable scope for fruitful interactions between such diversity studies and other computational tools. For example, a partition-based diversity study could be used to identify structure-based and/or property-based constraints that could then be fed into a de novo structure design program. The opportunities that exist for integrating different computational tools for drug discovery have been discussed recently by Martin (118).

In conclusion, it is important to note that the whole field is still at an early stage, with the great majority of the studies discussed in this review appearing only within the last eighteen months. It thus seems reasonable to expect very rapid progress in the development of computational tools for the analysis of molecular diversity over the next few years.

Acknowledgments

I thank Val Gillet, John Holliday, and David Turner for comments on an initial draft of this chapter, and the Engineering and Physical Sciences Research Council, Glaxo Wellcome Research and Development, Pfizer Central Research, and Tripos, Inc. for funding. This chapter is a contribution from the Krebs Institute for Biomolecular Research, which has been designated as a center for biomolecular sciences by the Biotechnology and Biological Sciences Research Council.

References

1. Mason, J. S.; McLay, I. M.; Lewis, R. A. In *New Perspectives in Drug Design*; Dean, P. M.; Jolles, G.; Newton, C. G., Eds.; Academic: London, 1994; pp 225–253.
2. Siani, M. A.; Weininger, D.; Blaney, J. M. *J. Chem. Inf. Comput. Sci.* **1994**, *34*, 588–593.
3. Sheridan, R. P.; Kearsley, S. K. *J. Chem. Inf. Comput. Sci.* **1995**, *35*, 310–320.
4. Siani, M. A.; Weininger, D.; James, C. A.; Blaney, J. M. *J. Chem. Inf. Comput. Sci.* **1995**, *35*, 1026–1033.
5. Shemetulskis, N. E.; Dunbar, J. B.; Dunbar, B. W.; Moreland, D. W.; Humblet, C. J. *Comput.-Aid. Mol. Des.* **1995**, *9*, 407–416.
6. Boyd, S. M.; Beverley, M.; Norskov, L.; Hubbard, R. E. *J. Comput.-Aid. Mol. Des.* **1995**, *9*, 417–424.
7. Martin, E. J.; Blaney, J. M.; Siani, M. A.; Spellmeyer, D. C.; Wong, A. K.; Moos, W. H. *J. Med. Chem.* **1995**, *38*, 1431–1436.
8. Martin, Y. C. In *Random and Rational Drug Discovery via Rational Design and Combinatorial Chemistry*; Strategic Research Institute: New York, 1995.
9. Mason, J. S.; Lewis, R. A.; McLay, I. M.; Menard, P. R.; Pickett, S. D. In *Random and Rational Drug Discovery via Rational Design and Combinatorial Chemistry*; Strategic Research Institute: New York, 1995.
10. Brown, R. D.; Martin, Y. C. *J. Chem. Inf. Comput. Sci.* **1996**, *36*, 572–584.
11. Martin, Y. C.; Brown, R. D.; Bures, M. G. In *Combinatorial Chemistry and Molecular Diversity in Drug Design*; Gordon, E. M.; Kerwin, J. F., Eds.; in press.
12. *Chemical Structure Systems*; Ash, J. E.; Warr, W. A.; Willett, P., Eds.; Ellis Horwood: Chichester, 1991.
13. Weininger, D. *J. Chem. Inf. Comput. Sci.* **1988**, *28*, 31–36.
14. Gray, N. A. B. *Computer-Assisted Structure Elucidation*; John Wiley: New York, 1986.
15. Bures, M. G.; Martin, Y. C.; Willett, P. *Topics Stereochem.* **1994**, *21*, 467–511.
16. Good, A. C.; Mason, J. S. *Rev. Comput. Chem.* **1995**, *7*, 67–117.
17. Barnard, J. M. *J. Chem. Inf. Comput. Sci.* **1993**, *33*, 532–538.
18. Willett, P. *Similarity and Clustering in Chemical Information Systems*; Research Studies Press: Letchworth, 1987.
19. Downs, G. M.; Willett, P. *Rev. Comput. Chem.* **1995**, *7*, 1–66.
20. Sussenguth, E. H. *J. Chem. Doc.* **1965**, *5*, 36–43.
21. Bawden, D. In *Chemical Structures 2. The International Language of Chemistry*; Warr, W. A., Ed.; Springer-Verlag: Heidelberg, 1993; pp 383–388.
22. Johnson, M. A. *J. Math. Chem.* **1989**, *3*, 117–145.
23. *Concepts and Applications of Molecular Similarity*; Johnson, M. A.; Maggiora, G. M., Eds.; John Wiley: New York, 1990.
24. Rouvray, D. H. *J. Chem. Inf. Comput. Sci.* **1992**, *32*, 580–586.
25. *Molecular Similarity in Drug Design*; Dean, P. M., Ed.; Blackie: Edinburgh, 1995.
26. Dittmar, P. G.; Farmer, N. A.; Fisanick, W.; Haines, R. C.; Mockus, J. *J. Chem. Inf. Comput. Sci.* **1983**, *23*, 93–102.
27. Nilakantan, R.; Bauman, N.; Dixon, J. S.; Venkataraghavan, R. *J. Chem. Inf. Comput. Sci.* **1987**, *27*, 82–85.
28. Jakes, S. E.; Willett, P. *J. Mol. Graphics* **1986**, *4*, 12–20.
29. Good, A. C.; Ewing, T. J. A.; Gschwend, D. A.; Kuntz, I. D. *J. Comput.-Aid. Mol. Des.* **1995**, *9*, 1–12.
30. Hodes, L. *J. Chem. Inf. Comput. Sci.* **1976**, *16*, 88–93.
31. Cringean, J. K.; Pepperrell, C. A.; Poirrette, A. R.; Willett, P. *Tetrahedron Comput. Methodol.* **1990**, *3*, 37–46.
32. Wiener, H. *J. Am. Chem. Soc.* **1947**, *69*, 17–20.

33. Randic, M. *J. Am. Chem. Soc.* **1975**, *97*, 6609–6615.
34. Kier, L. B.; Hall, L. H. *Molecular Connectivity in Structure–Activity Analysis*; John Wiley: New York, 1986.
35. Hall, L. H.; Kier, L. B. *Rev. Comput. Chem.* **1991**, *2*, 367–422.
36. Hall, L. H.; Mohney, B.; Kier, L. B. *J. Chem. Inf. Comput. Sci.* **1991**, *31*, 76–82.
37. Basak, S. C.; Magnuson, V. R.; Niemi, G. J.; Regal, R. R. *Discrete Appl. Math.* **1988**, *19*, 17–44.
38. Basak, S. C.; Grunwald, G. D. *SAR QSAR Environ. Res.* **1995**, *3*, 265–277.
39. Randic, M.; Wilkins, C. L. *J. Chem. Inf. Comput. Sci.* **1979**, *19*, 31–37.
40. Broto, P.; Moreau, G.; Vandycke, C. *Eur. J. Med. Chem.* **1984**, *19*, 66–70.
41. Jerman-Blazic, B.; Fabric-Petrac, I.; Randic, M. *Chemomet. Intell. Lab. Syst.* **1989**, *6*, 49–63.
42. Klopman, G.; Raychaudhury, C. *J. Chem. Inf. Comput. Sci.* **1990**, *30*, 12–19.
43. Fisanick, W.; Cross, K. P.; Rusinko, A. *Tetrahedron Comput. Methodol.* **1990**, *3*, 635–652.
44. Downs, G. M.; Willett, P.; Fisanick, W. *J. Chem. Inf. Comput. Sci.* **1994**, *34*, 1094–1102.
45. Kearsley, S. K.; Sallamack, S.; Fluder, E. M.; Andose, J. D.; Mosley, R. T.; Sheridan, R. P. *J. Chem. Inf. Comput. Sci.* **1996**, *36*, 118–127.
46. Pepperrell, C. A.; Willett, P.; Taylor, R. *Tetrahedron Comput. Methodol.* **1990**, *3*, 575–593.
47 Stuper, A. J.; Brugger, W. E.; Jurs, P. C. *Computer-Assisted Studies of Chemical Structure and Biological Function*; John Wiley: New York, 1979.
48. Glen, R. C.; Rose, V. S. *J. Mol. Graphics* **1987**, *5*, 79–86.
49. Stanton, D. T.; Murray, W. J.; Jurs, P. C. *Quant. Struct.-Activ. Relat.* **1993**, *12*, 239–245.
50. Willett, P.; Winterman, V. *Quant. Struct.-Activ. Relat.* **1986**, *5*, 18–25.
51. Moock, T. E.; Grier, D. L.; Hounshell, W. D.; Grethe, G.; Cronin, K.; Nourse, J. G. *Tetrahedron Comput. Methodol.* **1988**, *1*, 117–128.
52. Downs, G. M.; Poirrette, A. R.; Walsh, P.; Willett, P. In *Chemical Structures 2. The International Language of Chemistry*; Warr, W. A., Ed.; Springer Verlag: Heidelberg, 1993; pp 409–421.
53. Hodes, L. *J. Chem. Inf. Comput. Sci.* **1989**, *29*, 66–71.
54. Hagadone, T. R. *J. Chem. Inf. Comput. Sci.* **1992**, *32*, 515–521.
55. Milligan, G. W.; Cooper, M. C. *J. Classif.* **1988**, *5*, 181–204.
56. Todeschini, R. *Chemomet. Intell. Lab. Syst.* **1989**, *6*, 213–220.
57. Bath, P. A.; Morris, C. A.; Willett, P. *J. Chemomet.* **1993**, *7*, 543–550.
58. Turner, D. B.; Willett, P.; Ferguson, A. M.; Heritage, T. W. *SAR QSAR Environ. Res.* **1995**, *3*, 101–130.
59. Sneath, P. H. A.; Sokal, R. R. *Numerical Taxonomy*; WH Freeman: San Francisco, 1973.
60. Hartigan, J. A. *Clustering Algorithms*; John Wiley: New York, 1975.
61. Dubes, R.; Jain, A. K. *Patt. Recog.* **1976**, *11*, 235–254.
62. Everitt, B. S. *Cluster Analysis*; Edward Arnold: London, 1993.
63. Hansch, C., Unger, S. H.; Forsythe, A. B. *J. Med. Chem.* **1973**, *16*, 1217–1222.
64. Dunn, W. J.; Greenberg, M. J.; Callejas, S. S. *J. Med. Chem.* **1976**, *19*, 1299–1301.
65. van de Waterbeemd, H.; Tayar, N. E.; Carrupt, P.-A.; Testa, B. *J. Comput.-Aid. Mol. Des.* **1989**, *3*, 111–132.
66. Pleiss, M. A.; Unger, S. H. In *Comprehensive Medicinal Chemistry 4*; Ramsden, C. A., Ed.; Pergamon: New York, 1990; pp 561–587.
67. Willett, P.; Winterman, V.; Bawden, D. *J. Chem. Inf. Comput. Sci.* **1986**, *26*, 109–118.
68. Downs, G. M.; Willett, P. In *Advanced Computer-Assisted Techniques in Drug Discovery*; van de Waterbeemd, H., Ed.; VCH: New York, 1994; pp 111–130.
69. Barnard, J. M.; Downs, G. M. *J. Chem. Inf. Comput. Sci.* **1992**, *32*, 644–649.
70. Clark, D. E.; Murray, C. W. *J. Chem. Inf. Comput. Sci.* **1995**, *35*, 914–923.

71. Nouwen, J.; Hansen, B. *SAR QSAR Environ. Res.* **1995**, *4*, 1–10.
72. Lance, G. N.; Williams, W. T. *Comput. J.* **1967**, *9*, 373–380.
73. Guenoche, A.; Hansen, P.; Jaumard, B. *J. Classif.* **1991**, *8*, 5–30.
74. Whaley, R.; Hodes, L. *J. Chem. Inf. Comput. Sci.* **1991**, *31*, 345–347.
75. Hodes, L.; Feldman, A. *J. Chem. Inf. Comput. Sci.* **1991**, *31*, 347–350.
76. Jarvis, R. A.; Patrick, E. A. *IEEE Trans. Comput.* **1973**, *C-22*, 1025–1034.
77. Murtagh, F. *Comput. J.* **1983**, *26*, 354–359.
78. Murtagh, F. *Comput. Stat. Quart.* **1984**, *1*, 101–114.
79. Fisher, L.; van Ness, J. W. *Biometrika* **1971**, *58*, 91–104.
80. Jardine, N.; Sibson, R. *Mathematical Taxonomy*; John Wiley: New York, 1971.
81. Milligan, G. W. *Multi. Behav. Res.* **1981**, *16*, 379–407.
82. Everitt, B. S. *Biometrics* **1979**, *35*, 169–181.
83. Adamson, G. W.; Bush, J. A. *Inf. Stor. Retr.* **1973**, *9*, 561–568.
84. Johnson, M. A.; Lajiness, M. S.; Maggiora, G. In *QSAR: Quantitative Structure–Activity Relationships in Drug Design*; Fauchere, J. L., Ed.; Alan R. Liss: New York, 1989; pp 167–171.
85. Lajiness, M. S.; Johnson, M. A.; Maggiora, G. In *QSAR: Quantitative Structure–Activity Relationships in Drug Design*; Fauchere, J. L., Ed.; Alan R. Liss: New York, 1989; pp 173–176.
86. Lajiness, M. In *Computational Chemical Graph Theory*; Rouvray, D. H., Ed.; Nova Science: New York, 1990; pp 299–316.
87. Lajiness, M. In *QSAR: Rational Approaches to the Design of Bioactive Compounds*; Silipo, C.; Vittoria, A., Eds.; Elsevier Science: Amsterdam, 1991; pp 201–204.
88. Bawden, D. In *Concepts and Applications of Molecular Similarity*; Johnson, M. A.; Maggiora, G. M., Eds.; John Wiley: New York, 1990; pp 65–76.
89. Holliday, J. D.; Ranade, S. S.; Willett, P. *Quant. Struct.–Activ. Relat.* **1996**, *14*, 501–506.
90. Taylor, R. *J. Chem. Inf. Comput. Sci.* **1995**, *35*, 59–67.
91. Wootton, R.; Cranfield, R.; Sheppey G. C.; Goodford, P. J. *J. Med. Chem.* **1975**, *18*, 607–613.
92. Holliday, J. D.; Willett, P. *J. Biomolec. Screening* **1996**, *1*, 145–151.
93. Cummins, D. J.; Andrews, C. W.; Bentley, J. A.; Cory, M. *J. Chem. Inf. Comput. Sci.* **1996**, *36*, 750–763.
94. Pearlman, R. S. *Novel software tools for addressing chemical diversity*, accessible via WWW at URL *http://www.awod.com/netsci/Issues/Jun96/feature1.html.*
95. Burden, F. R. *J. Chem. Inf. Comput. Sci.* **1989**, *29*, 225–227.
96. Sadowski, J.; Gasteiger, J. *Chem. Rev.* **1993**, *93*, 2567–2581.
97. Lewis, R. A.; Leach, A. R. *J. Comput.-Aid. Mol. Design* **1994**, *8*, 467–475.
98. Kier, L. B.; Hall, L. H. *Quant. Struct.–Activ. Relat.* **1993**, *12*, 383–388.
99. Glen, R. C.; Payne, A. W. R. *J. Comput.-Aid. Mol. Design* **1995**, *9*, 181–202.
100. Zuckermann, R. N.; Kerr, J. M.; Kent, S. B. H.; Moos, W. H. *J. Am. Chem. Soc.* **1992**, *114*, 10646–10647.
101. Ferguson, A. M.; Patterson, D. E.; Garr, C. D.; Underiner, T. L. *J. Biomolec. Screening* **1996**, *1*, 65–73.
102. Cramer, R. D.; Patterson, D. E.; Munce, J. D. *J. Am. Chem. Soc.* **1988**, *110*, 5959–5967.
103. Goldberg, D. E. *Genetic Algorithms in Search, Optimization, and Machine Learning*; Addison-Wesley: New York, 1991.
104. Carhart, R. E.; Smith, D. H.; Venkataraghavan, R. *J. Chem. Inf. Comput. Sci.* **1985**, *25*, 64–73.
105. Nilakantan, R.; Bauman, N.; Venkataraghavan, R. *J. Chem. Inf. Comput. Sci.* **1991**, *31*, 527–530.
106. *Modern Heuristic Techniques for Combinatorial Problems*; Reeves, C. R., Ed.; McGraw-Hill: London, 1995.

107. Cribbs, C.; Menius, A.; Cummins, D.; Scoffin, R.; Young, S. Presented at the 211th National Meeting of the American Chemical Society.
108. Turner, D. B.; Tyrrell, S. M.; Willett, P. *J. Chem. Inf. Comput. Sci.* **1997**, *37*, 18–22.
109. Davies, E. K.; Briant, C. *Combinatorial chemistry library design using pharmacophore diversity*, accessible via WWW at URL *http://www.edisto.awod.com/netsci/Issues/July95/feature6.html.*
110. Snyder, R. W.; Crofton, G. *Planning a Combinatorial Synthesis. Optimizing the Benzodiazepine Scaffold for Gastrin Antagonist Activity*; MDL Information Systems Inc.: San Leandro, 1995.
111. *Molecular Diversity Manager Generates Lead Followup Synthesis Candidates*; Tripos Inc.: St Louis, 1995.
112. Rees, P. *Sci. Comput. World* **1996**, February, p 25.
113. *Modern Approaches to Chemical Reaction Searching*; Willett, P., Ed.; Gower: Aldershot, 1986.
114. *3D QSAR in Drug Design*; Kubinyi, H., Ed.; ESCOM: Leiden, 1993.
115. Barnard, J. M.; Downs, G. M. *Chem. Des. Automat. News* **1995**, *10*, 36.
116. *Computer Handling of Generic Chemical Structures*; Barnard, J. M., Ed.; Gower: Aldershot, 1984.
117. Fisanick, W. *J. Chem. Inf. Comput. Sci.* **1990**, *30*, 145–154.
118. Martin, Y. C. *Opportunities for computational chemists afforded by the new strategies in drug discovery: an opinion*, accessible via WWW at URL *http://www.awod.com/netsci/Issues/Jun96/feature2.html.*

Solid-Phase Strategies

The Context of Solid-Phase Synthesis

George Barany and Maria Kempe

Multistep organic synthesis can be substantially facilitated by conducting reactions on substrates that are covalently attached to insoluble supports. The principles of such "solid-phase synthesis" were enunciated first by Merrifield in the 1960s, in the context of assembling amino acid building blocks to prepare peptides. Solid-phase approaches have reached full fruition for peptides and oligonucleotides, and there has been significant progress in generalizations to other chemical families. This chapter identifies key issues of solid-phase synthesis, including the choice of solid support, anchoring step, coupling chemistries, appropriate protection strategies, and suitable analytical methods for monitoring the course and outcome of individual and cumulative solid-phase reactions. We offer both a historical perspective and an overview of the current state-of-the-art. An appreciation and understanding of the concepts and various methodologies that were originally developed for solid-phase synthesis of peptides and oligonucleotides are expected to be invaluable to practitioners of small organic molecule solid-phase combinatorial chemistry.

Solid-phase synthetic methods were originally developed to answer the need for rapid, efficient, reliable, and general ways to chemically synthesize peptides in sufficient amounts and purities for their myriad applications. Prior to Merrifield's seminal first publication in 1963 (*1, 2*), the chemical preparation of peptides had to be carried out in solution by procedures that were time-consuming and labor-intensive, requiring the efforts of teams of scientists with specialized training in the field (*3–5*). After each chemical step is carried out in solution, an individual purification procedure must be worked out; this is a nonroutine task made all the more challenging by the unpredictable solubility characteristics of intermediates. Merrifield's key insight was that stepwise chemical assembly of amino acid building blocks to form peptide chains could be carried out while these were attached covalently to an insoluble support, and that this arrangement would lead to much faster synthesis with greatly reduced labor and effort compared to classical solution methods. Driving the repetitive steps to completion by the use of excess soluble reagents, and carrying out purification by simple filtrations and washings of the growing peptide-resins, are critical to the overall success of this plan. Within a half-dozen years, the initial instruments to automatically carry out the repetitive steps of solid-phase peptide synthesis (SPPS) had been designed and built, and one such instrument was applied to assemble the 124-residue enzyme ribonuclease A (6).

State-of-the-art peptide synthesizers are computer-controlled and operate in either batchwise or continuous-flow modes at various scales. Many of the original criticisms (some justified and others hyperbolic) of the solid-phase approaches have faded as a result of rigorous optimizations of the underlying chemistries, and the development of suitable polymeric supports. Key contributions to SPPS came from Merrifield and his school (many in their independent laboratories), as well as Atherton, Li, Meienhofer, and Sheppard, among others (7–16). The multiplexing of SPPS (in concert with rapid biological screening) has been of escalating interest since the mid-1980s (17–19), and has been driven by the pioneering contributions of Geysen (20), Furka (21), Houghten (22, 23), Lam and Hruby (24), and a research group from Affymax (25).

Solution methods are still important in the peptide field, particularly for large-scale production in the pharmaceutical industry. (Almost nothing is published on this point because of proprietary considerations, but for a recent, relatively rare example of disclosure at a symposium, see reference 26.) Two of the central ideas of solid-phase synthesis (SPS), that is, that repetitive reactions can be driven to completion by excesses *and* that purifications can be "facilitated" on the basis of the physical properties of a carrier, have been elegantly carried over to solution modes (5, 27, 28). Furthermore, extensive work has provided a rationale and protocols for enzyme-mediated synthesis under exceptionally mild stereospecific and regioselective reaction conditions (29–34). Nevertheless, the solid-phase mode has become the preferred one by far, and the only one accessible to the nonspecialist. The maturity of the field was recognized by the Nobel Prize committee with its 1984 award in chemistry to Merrifield (10).

Another field of science that has been revolutionized by automated solid-phase methodology is molecular biology, specifically with regard to the synthesis of oligodeoxyribonucleotides (DNA) that can be used—among other applications—as primers applied to DNA sequencing (35–38), site-directed mutagenesis (39, 40), and the polymerase chain reaction (PCR) (41); as probes in sensor devices (42, 43); and in the preparation of artificial genes for insertion into plasmid vectors that can be used to transform host cells (39). The pioneers in DNA solid-phase synthesis were Letsinger, Sheppard, Gait, Itakura, Caruthers, and Ogilvie, with the breakthrough papers on phosphoramidite chemistry published in the early 1980s (37, 44), and reliable workhorse instruments becoming widely available soon thereafter (37,

45–47). More recently, the solid-phase chemistry and associated instrumentation have been extended to oligoribonucleotides (RNA) (*44, 48–50*) and analogues of peptide nucleic acids (PNA) (*51, 52*), among other oligomers that are relevant to modern biological research.

The recent explosion of interest in solid-phase organic synthesis (SPOS) for combinatorial applications (*53–55*), reflected in part by the existence of this book, raises the question of why solid-phase synthesis was not performed years ago. In fact, there were isolated but significant (in retrospect) research efforts carrying out organic synthesis on polymeric supports as far back as the early 1970s by Crowley and Rapoport, Patchornik, Leznoff, and others (*56–64*). The more general applications of solid-phase concepts to organic synthesis require a greater range of reaction types than those used in peptide and oligonucleotide synthesis, and issues of solvent compatibility and phase homogeneity must be considered. Added are the challenges of carrying out reactions (manually or automated) at a vastly extended temperature range and often under anhydrous conditions or with an inert atmosphere. Questions of chemo-, regio-, and stereoselectivity in general organic synthesis are sometimes more complicated than for syntheses of repetitive oligomers. On the other hand, yields per step need not be quite as high as the > 99% per step required for solid-phase oligomer synthesis, because the organic targets of SPOS are usually made in rather few (e.g., 3–10) steps. The past two decades of research in SPPS have seen a substantial expansion in the possible options for solid-phase supports and companion linker chemistries; these ongoing developments facilitate the further development of SPOS. Relevant concepts and their optimal experimental manifestations are covered in this review.

General Solid-Phase Synthetic Strategies and Concepts

A scheme for the general stepwise solid-phase synthesis of linear oligomers from monomeric building blocks is shown in Figure 1. Monomers have (at least) two functional groups, one of which requires blocking with a "temporary" protecting group, whereas the other is left unprotected so as to allow coupling reactions to take place. For those monomers that contain a third (or more) functional group(s), "permanent" protecting groups are often required to avoid unwanted branching or other side reactions. A bifunctional spacer, called a *handle* (or *linker*), is coupled covalently to an insoluble *support* (resin carrier), which either inherently contains a suitable functional group X, or has been functionalized to contain such a group X. The handle must be designed so as to incorporate features of a smoothly cleavable protecting group on the end which is not used for attachment to the support. Next, the first (*terminal*) monomer is bound covalently to the free end of the handle (*anchoring*). Synthesis proceeds by removal of the temporary protecting group, followed by stereospecific addition of the next protected monomer. The oligomer chain is elongated by a series of such *deprotection–coupling* steps, interspersed by filtration–washing steps to remove excess building blocks and reagents. Yields from the repet-

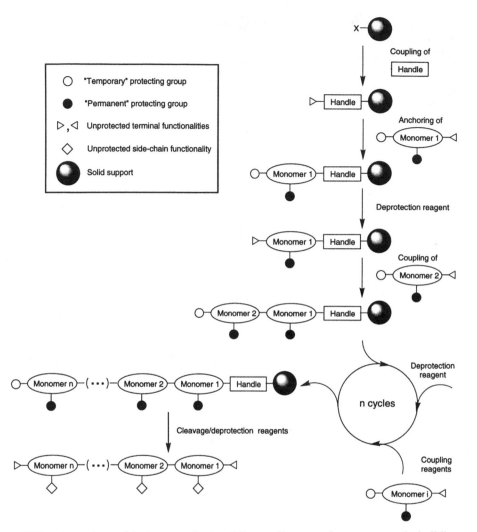

FIGURE 1. Stepwise solid-phase synthesis of linear oligomers from monomeric building blocks. Symbols are explained in the inset. Several of the steps can be carried out separately, or in a different order, as outlined in the text.

itive steps must be extraordinarily high, because low-level by-products can accumulate from incomplete reactions or a variety of side reactions, and these products will have only subtle chemical and physical differences from the desired product. Hence, analytical methods that allow reliable *monitoring* of the synthesis while it is still in progress are recommended if possible (*see* Chapter 5). Once chain assembly has been completed, the "permanent" protecting groups are removed and the oligomer is released from the support (*cleavage*); these operations are often, but not necessarily, carried out as a single step.

The general stepwise scheme (Figure 1) can be modified in a number of ways, depending on the goals of the synthesis (7). A widely used and important alternative anchoring strategy involves initial attachment of the terminal monomer to the handle, followed by coupling of the resultant preformed handle to the support (discussed in more detail later in this review). Further versatility is achieved when anchoring is not through the terminus, but through a side-chain functional group or even the oligomer backbone (7, 65, 66). Such strategies circumvent certain side reactions, provide access to a wider range of termini, and are valuable when targets for synthesis are cyclic.

A number of practical choices may critically affect the success of stepwise solid-phase synthetic schemes (Figure 1), and some of the more complicated variations. These include selection of the solid support, development of an effective anchoring regime, identification of reagents and procedures for efficient, stereospecific coupling, and, perhaps most significantly, definitions of dependable temporary and permanent protecting groups. The protection scheme dictates the chemical mechanism and precise conditions used for repeated removal of the temporary group, which in turn determine the severity of conditions required for final removal of the permanent groups. Graduated lability schemes enjoy considerable popularity, but the reagents necessary for the final cleavage often can promote partial destruction of sensitive structures found in peptides or other oligomers (particularly many of the side-chain moieties corresponding to post-translational modifications). Consequently, there has been a strong impetus to devise strategies that involve milder reaction conditions. The concept of *orthogonal* protection (the use of completely independent classes of protecting groups, which can be removed in any order and in the presence of all other classes) represents an important approach to reach this goal (67). This is because in the orthogonal case, selectivity can be achieved on the basis of differences in chemical mechanisms rather than in reaction rates, allowing protecting group structures and corresponding deprotection conditions to be modulated so as to achieve rapid rates. Orthogonal chemistries can be extended into three or even four dimensions (68–70), and are particularly advantageous when the final synthetic targets are of greater structural complexity than linear oligomers, as in cyclic, branched, and conjugate structures (71). Finally, synthetic schemes providing partially protected segments (*see* the next paragraph), or taking advantage of side-chain or backbone anchoring concepts (*see* the preceding paragraph), invariably benefit from a flexible array of protecting groups. Specific examples of graduated lability as well as orthogonal schemes are illustrated later in this chapter. It should be stressed that, regardless of the chemistry used, the products of solid-phase synthesis invariably require purification and characterization after cleavage.

The focus of the discussion so far has been on stepwise synthesis of linear oligomers, using sequential incorporation of suitably protected small building blocks. A significant achievement of researchers in these fields has been the development of effective protocols to achieve good reaction rates, excellent yields, and high stereospecificities during the assemblies of both protected amino acid building blocks to make peptides, and protected nucleoside phosphoramidates to make DNA and RNA. However, the stepwise approaches provide little tolerance for "failure" sequences, for reasons already discussed. Consequently, for some oligomers, particularly large ones, it is worth considering the relative merits of convergent strategies. In the peptide field (*7, 72, 73*), such *segment condensation* approaches use purified, partially protected peptides as the building blocks, where the peptides are typically 5 to 15 residues in length. Peptide segments can be made by either solution or solid-phase methods, and are combined further either in solution or on polymeric supports, as dictated by the overall plan for synthesis. These segments generally couple more slowly than amino acid building blocks, and couplings of peptides are often accompanied by some level (usually manageable) of racemization. A major justification for segment condensation strategies lies in the straightforward removal of the by-products of incomplete coupling. Far less work along these lines has been done in nonpeptide fields, but convergent approaches should not be overlooked by those contemplating work in these latter areas.

The solid-phase mode of synthesis offers some advantages and possibilities beyond the most obvious ones already stated, that is, simplicity, use of excesses to drive reactions to completion, facilitated purification by filtration–washing, and amenability to automation. By carrying out synthesis in a single vessel, chemical reactions can take place under a controlled environment, and the "purification" by filtration and washing occurs without manipulative losses. Certain procedures that are difficult to accomplish in solution because of physical properties of reagents, generation of reactive intermediates that must be scavenged immediately, generation of soluble side products that might contaminate the desired product, and stoichiometry considerations, are actually more suited to solid-phase synthesis. Substrates that are bound covalently to polymeric supports are less likely to undergo intermolecular reactions because of the relative isolation of reactive sites (*7, 56*), a kinetic phenomenon that Mazur has termed pseudo-dilution (*74*). Thus, those intramolecular cyclizations that are thermodynamically favored can often be carried out efficiently in the solid-phase mode. The balance between intra- and intermolecular processes is subtle, and, with time, activated functional groups from molecules bound to separate sites on the support do find each other, resulting in the formation of dimers and oligomers. Instructive examples from the peptide field (*71*) include side-chain lactamization (*75–77*), head-to-tail cyclization (*78–81*), and disulfide-bridge formation (*82, 83*). Applications of the pseudo-dilution principle in SPOS have been reported as well (*84–86*).

For peptide targets, protection is needed for the repeatedly exposed N^α-amino group (temporary protection) and for certain amino acid residue side-chains (permanent protection). The challenge in these syntheses is to create amide linkages without racemization. The principles of SPS apply in the syntheses of other oligomeric targets such as depsipeptides (*87*), polyamides (*88, 89*), oligoureas (*90*), oligocarbamates (*91, 92*), peptoids (*93, 94*), vinylogous sulfonyl peptides (*95*), oligosulfones (*92*), oligosulfoxides (*92*), carbohydrates (*96–99*), DNA (*44*), RNA (*44, 49*) and PNA (*51, 52*) (Figure 2). Practical solid-phase schemes have been worked out, and

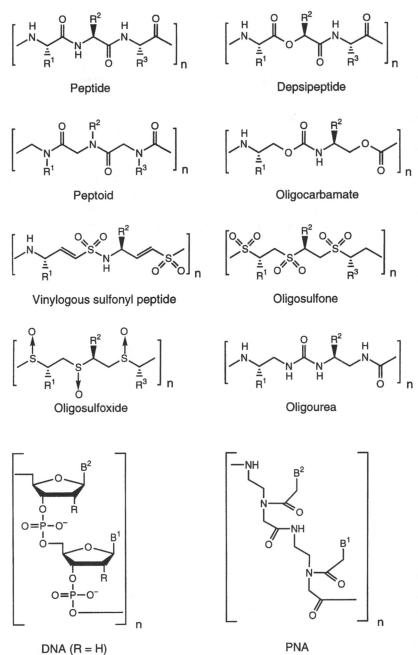

FIGURE 2. Some oligomers that can be synthesized by stepwise solid-phase procedures. Structural diagrams have been aligned to illustrate homologies.

in some cases optimized, for all of these (details are in the cited literature). Hybrid oligomers have also been described (*100–107*).

Specific Implementations of Solid-Phase Synthesis

Supports for Solid-Phase Synthesis

Part of the inspiration for SPPS came from the successful use of insoluble resin matrices for chromatographic resolution and purification. The molecules separated by these methods ranged widely in mass and included many proteins. A number of chromatographic supports and synthetic insoluble polymers (e.g., polystyrenes with low or high cross-linking, celluloses (including cotton), dextrans, silicas with high surface areas, controlled pore glass, membranes, and various grafts) have been evaluated for solid-phase synthesis, first of peptides, and then later of oligonucleotides and general organic compounds (Table I). Quite a few materials—with widely varying covalent structures, architectures, and polarities—have been found to serve admirably; many more have been tried and abandoned. Criteria for what constitutes an optimal set of features for a suitable SPS support are controversial, and much of the conventional wisdom in the field is based on empirical observations. It is imperative that the supports be absolutely insoluble (i.e., there must not be any "leaking"), and possess excellent physical and chemical stabilities. Beads are preferable but not essential. Many applications require reasonably high capacities, often in the 0.15–1.5 mmol g^{-1} range, although for syntheses of DNA and RNA oligonucleotides, substantially lower capacities of < 0.1 mmol g^{-1} are tolerable and perhaps even preferable. The evolution of support technology occurred independently of, but in parallel with, better definitions of milder synthesis chemistries.

While the term "solid-phase" might imply a static resin support, it turns out that this is not at all the case with those materials that serve best for peptide synthesis (*108*). Reactions commonly occur on mobile, well-solvated, and reagent-accessible polymer strands throughout the interiors of the supports; relatively few of the sites are in surface regions. Differentiation between interior and surface sites on microporous beaded supports has been achieved by a selective proteolysis procedure termed "shaving," which revealed the dynamic nature of surfaces and also confirmed that the large majority of sites were enzyme inaccessible and on the inside (*109*). Rigid, macroporous materials, of either polymeric or inorganic origin, are found to have low capacities despite their high surface areas. For these latter materials, SPS reactions occur on the internal surfaces.

Resins used in SPPS, both without and with pendant peptide chains, change size appreciably in the presence of solvent. This phenomenon is termed "swelling". With the best supports, reaction rates approach, but generally do not reach, those attainable in solution. Supports usually have the minimal level of cross-linking consistent with stability, so that they can form the well-solvated gels within which solid-phase chemistry takes place. As has been pointed out with differing points of

TABLE I. Solid Supports for Syntheses of Peptides, DNA, and Small Organic Molecules

Solid Support	Description and Comment	References
Synthetic Polymers—Polystyrenes		
Polystyrene[a]	Poly(styrene-*co*-divinylbenzene), microporous, 0.5–2% cross-linking (1% optimal)	7 // 193, 194
Polystyrene	Poly(styrene-*co*-divinylbenzene), macroporous, 8–50% cross-linking	7 // 195
Polystyrene–Kel-F	Pellicular impermeable core surrounded by a mobile layer of linear polystyrene chains	196–198
PEPS films	Polystyrene grafted onto polyethylene films	199
Synthetic Polymers—Polyamides		
Polyamide (Pepsyn)	Poly(*N,N*-dimethylacrylamide-*co*-*N,N'*-ethylenebisacrylamide-*co*-*N*-(Boc-β-alanyl)-*N'*-acryloylhexamethylenediamine)	13, 110, 200 // 201, 202
Pepsyn K[a]	Poly(*N,N*-dimethylacrylamide-*co*-*N,N'*-ethylene-bisacrylamide-*co*-*N*-acryloylsarcosine methyl ester) within macropores of kieselguhr matrix	13, 166, 203
Polyhipe[a]	Poly(*N,N*-dimethylacrylamide-*co*-*N*-acryloyl-sarcosine methyl ester) within macropores of 50% cross-linked poly(styrene-*co*-divinylbenzene)	204
Sparrow amide resin	Poly(*N,N*-dimethylacrylamide-*co*-*N,N'*-bis-acryloyl-1,3-diaminopropane-*co*-allylamine)	205
Expansin	Poly(*N*-acryloylpyrrolidine-*co*-*N,N'*-ethylenebis-acrylamide-*co*-*N*-acryloyl-β-alanine methyl ester), ethylenediamine derivatized	206
Synthetic Polymers—Poly(ethylene glycol)-containing		
PEG–PS[a]	Polyethylene glycol grafted covalently onto 1% cross-linked microporous poly(styrene-*co*-divinylbenzene)	207–209 // 210
POE–PS (TentaGel)[a]	Polyethylene glycol polymerized onto poly(styrene-*co*-divinylbenzene)	211, 212 // 213–215
PEGA	Poly(*N,N*-dimethylacrylamide-*co*-bisacrylamido-polyethylene glycol-*co*-monoacrylamido-polyethylene glycol)	216, 217
TEGDA–PS	Poly(styrene-*co*-tetraethyleneglycol diacrylate)	218
PEO–PEPS	3,6,9-Trioxadecanoic acid coupled to PEPS	219
CLEAR[a]	Poly(trimethylolpropane ethoxylate (14/3 EO/OH) triacrylate-*co*-allylamine)	220
Synthetic Polymers—Miscellaneous		
HPA–PP membrane	Polypropylene coated with polyhydroxypropylacrylate	221
Polyethylene pins	Rods with polyethylene crowns, grafted with acrylic acid or 2-hydroxyethyl methacrylate	222, 223
EVAL membrane	Poly(ethylene-*co*-vinyl alcohol)	224
ASPECT	Chemically modified polyolefin particles	225

Continued on next page

TABLE I—*Continued*

Solid Support	Description and Comment	References
Natural polymers		
Sephadex	—	226 // 227
Cellulose (cotton fabric)	—	228
Cellulose (paper, filter)	—	229–231 // 232, 233
Cellulose (Perloza beads)	—	234
Chitin	—	235
Bovine serum albumin (protein)	—	236
Inorganic materials		
Silica, glass, and controlled pore glass (CPG)	—	7, 237 // 144, 145, 169

NOTE: Supports are listed by type according to the following categories: synthetic polymers (subdivided further into the type of polymer chain supporting solid-phase synthesis), natural polymers, and inorganic materials. Within each category, the order corresponds to when that support was first introduced for solid-phase applications. References to the left of the // describe use of the support for SPPS; to the right of the // for solid-phase DNA synthesis. Current research is focused on which supports are appropriate for SPOS, which in turn depends on the reactions to be carried out.

[a]Shown in Figure 3.

emphasis by Sheppard (*15, 110*) and Kent (*12, 111*), supports also change their size, solvation, and other physical properties once a substantial peptide content has been incorporated. Particularly for applications in combinatorial chemistry that involve on-bead screening, it is helpful to have support materials that swell in both the organic solvents preferred for synthesis and the water-based milieus used for biological testing. The generally inorganic rigid supports that are used preferentially in DNA and RNA synthesis, and occasionally in SPPS or SPOS, do not swell.

The most common and versatile support materials for SPPS are based on cross-linked synthetic copolymers. The classical example, with a distinguished history of applications, is Merrifield's original polystyrene (PS) microporous beads cross-linked with 1% divinylbenzene (Figure 3a). Atherton and Sheppard developed a polyamide support based on the concept that the support and the peptide backbone should have similar polarities (Figure 3b). Results have usually been good with these materials, both in their gel and rigid, encapsulated forms. Similar reasoning led to Polyhipe supports (Figure 3c). Supports containing both poly(ethylene glycol) (PEG) and an approximately equal amount (by weight) of low-cross-linked polystyrene (PEG–PS in which PEG is covalently grafted onto PS; TentaGel in which the ethylene oxide is polymerized onto PS) have found favor because they swell in a range of solvents, provide a favorable milieu for "difficult" couplings, and have excellent physical and mechanical properties (Figure 3d and 3e). The related PEGA and CLEAR (Figure 3f) supports contain multiple oxyethylene units, accounting for their hydrophilic characters and superb swelling characteristics. The cross-linked ethoxylate acrylate resin (CLEAR) family of supports is highly cross-linked (90% or more of their weight is derived from a branched cross-linker) and have reasonable

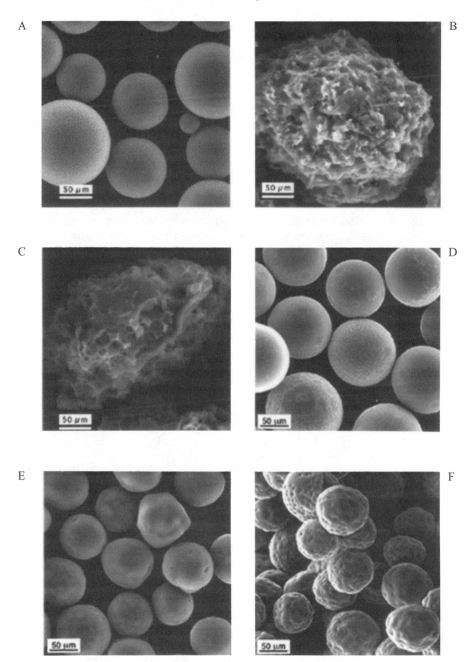

FIGURE 3. Scanning electron micrographs (accelerating voltage: 10 kV) showing the shape and texture of (A) microporous polystyrene, (B) Pepsyn K, (C) Polyhipe, (D) PEG–PS, (E) Tenta-Gel, and (F) CLEAR.

loadings, so their properties contrast with those of most highly cross-linked materials, which are rigid and macroporous, with low overall capacities. Finally, supports in which linear polystyrene chains have been grafted covalently onto dense Kel-F particles or onto polyethylene sheets are valuable for some applications.

Anchoring Chemistries

The resin supports for solid-phase synthesis described in the previous section may, or may not, have functional groups (e.g., amino, carboxyl, hydroxyl, or chloromethyl groups) as part of their matrix structure. Appropriate functional groups can be made intrinsic to the matrix by including appropriate prefunctionalized monomers in the polymerization process that generates the support. When the parent support lacks such groups, a separate functionalization reaction, or sequence of reactions, must be carried out (7). Examples, some of which are shown in Figure 4, include Friedel–Crafts substitution of the aromatic rings of polystyrene using chloromethyl methyl ether, which leads to chloromethyl ($ClCH_2$) functions (112);

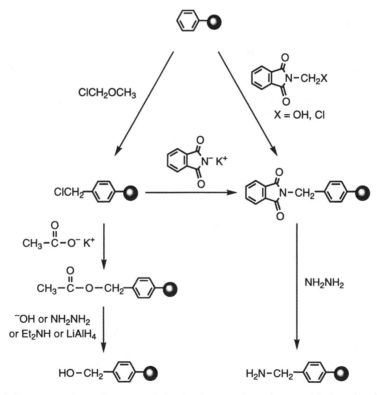

FIGURE 4. Representative polystyrene derivatization reactions that provide functional groups that can serve as starting points for solid-phase synthesis. (Primary references and reviews for these transformations are cited in the text.)

reaction of chloromethyl resin with potassium phthalimide (*113, 114*), or direct amidomethylation of polystyrene by the Tscherniac–Einhorn reaction with chloromethylphthalimide or hydroxymethylphthalimide (*115, 116*) (this introduces phthalimidomethyl moieties, which are later hydrazinolyzed to yield aminomethyl (H_2NCH_2) functions); and transformation of ester moieties by reactions such as saponification, aminolysis with bifunctional reagents, or reduction to provide useful functional sites.

In some cases, the functionalized support (prepared as described in the previous paragraph) resembles a reagent for protecting the C^α-carboxyl (or some other) end group of the planned product of the solid-phase synthesis (*7*). Derivatization of such a functionalized resin with an activated building block results in anchoring. For example, esterification of an N^α-protected amino acid to a chloromethyl–polystyrene support (termed "Merrifield resin") gives a polymeric benzyl ester, or coupling of an N^α-protected amino acid to a 4-methylbenzhydrylamide (MBHA) resin (*see* the structure in Table II) gives a benzhydrylamide linkage. Greater control and generality are offered by handles (also termed linkers), which allow the anchoring process to be carried out in two discrete chemical steps. One end of a handle resembles a protecting group for the functional group to be anchored, whereas the other end (often an activated carboxyl group) allows facile coupling to any polymeric support that has been previously functionalized (often with amino groups) (*7*). The handle approach makes possible precise tuning of both the stability of the anchoring linkage, as well as the chemical mechanism and reaction conditions for final cleavage of the anchor. Often, the handle is connected to the support first, before attachment of the initial amino acid. The preferred but more involved preformed variation (*7, 117, 118*) reverses the order of these steps, and the building-block-handle intermediate can be purified in solution. This approach has the advantage of removing by-products (including epimers) that arise from the first step. Subsequently, attachment to the support can often be achieved quantitatively, thereby allowing complete control over the loading level, and circumventing problems associated with extraneous polymer-bound functional groups (*see* Figures 5 and 6 for elegant examples of this approach giving rise to *p*-(carbamoylmethyl)benzyl (PAM) ester and *p*-alkoxybenzyl (PAB) ester linkages, respectively). An important variation is the universal handle concept, exemplified by the tris(alkoxy)benzylamide (PAL) handle (*119*) and its relatives, in which a single reagent is attached to a support, any terminal building block can be coupled, and a desired end group is obtained upon final cleavage.

Table II shows the names of some of the most widely used, commercially available derivatized supports and handles, together with general structures of the subsequently obtained anchored building block-linker-resins that are the starting points for SPS. Table II is extracted from a much more extensive, but by no means comprehensive, table provided as an appendix to this chapter (the longer version includes quite a few entries that are not as available but illustrate important principles or show considerable promise for future work). Both Table II and the appendix table give examples of anchors created by direct reaction with a terminal building block, as well as those resulting from handle approaches as outlined in the preced-

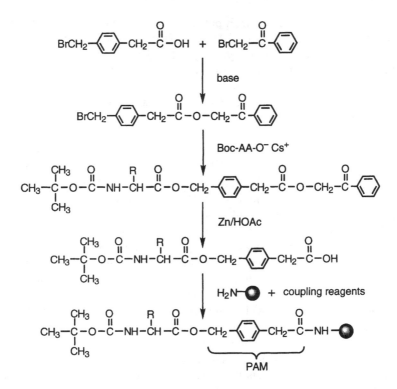

FIGURE 5. "Preformed handle" strategy for preparation of a Boc–aminoacyl–PAM–resin. The compound on the second line, 4-(bromomethyl)phenylacetic acid phenacyl ester, is condensed with a Boc–amino acid as its cesium salt to provide an intermediate, which is reduced directly to remove the phenacyl ester. The resultant Boc–aminoacyl–4-(oxymethyl)phenylacetates are first purified and then attached quantitatively onto amino-functionalized supports. (References are cited in Table II, with PAM.)

ing paragraph. Furthermore, the tables define reagents and mechanisms by which the anchoring linkages are cleaved, and specify the end groups of the SPS products that are then released from the support. Depending on the experimental design, the product can be a carboxylic acid, thioacid, amide, hydrazide, alcohol, or arene, either completely free or retaining some or all of the remaining protecting groups. Mechanisms for cleavage of anchoring linkages include acid, base, light, fluoride ion, palladium(0) in the presence of nucleophilic acceptors, and nucleophilic thiols. For anchors that are susceptible to acidolysis, distinctions are drawn between those cleaved by strong acid (e.g., hydrogen fluoride [HF], trifluoromethanesulfonic acid [TFMSA]), moderate-strength acid (e.g., trifluoroacetic acid [TFA]), and weak or dilute acid. Some anchors can be cleaved in more than one way and/or can give rise to several possible products. "Safety-catch" linkers are a special case; in these linkers an otherwise stable anchor is activated in a separate orthogonal step prior to cleavage (*120–124*). Other valuable concepts include "multidetachable" supports,

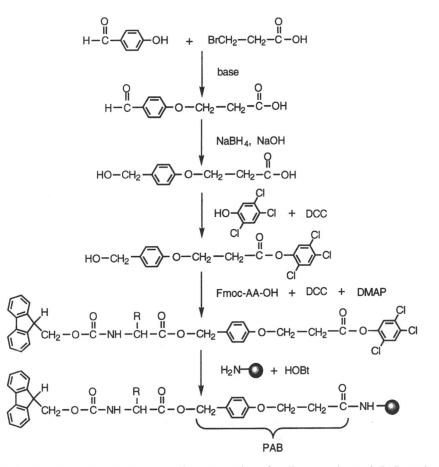

FIGURE 6. "Preformed handle" strategy for preparation of an Fmoc–aminoacyl–PAB–resin (*117*). Reduction of 3-(4'-formylphenoxy)propionic acid gives a 4'-hydroxymethyl function, while the pendant carboxyl is esterified with 2,4,5-trichlorophenol. Racemization-free DCC–DMAP-mediated coupling of Fmoc–amino acids to the 2,4,5-trichlorophenyl 3'-(4"-hydroxymethylphenoxy)propionate intermediate gives preformed handles, which are then purified and attached quantitatively onto amino-functionalized supports. Note how the 2,4,5-trichlorophenyl ester serves the dual roles of protecting the handle carboxyl while the 4"-hydroxymethyl group is acylated, and then activating that same carboxyl to facilitate attachment to the support. In other methodologies (e.g., the one outlined in Figure 5), carboxyl protection–deprotection and activation are carried out as discrete steps with differing chemistries.

Abbreviations: DCC, *N,N'*-dicyclohexylcarbodiimide; DMAP, 4-dimethylaminopyridine.

TABLE II. Selected Anchoring Linkages for Solid-Phase Syntheses of Peptides, DNA, RNA, and Small Organic Molecules

Anchoring Linkage, Cleavage Reaction, and Product	References
p-Alkylbenzyl ester (from chloromethyl resin ≡ Merrifield resin)[a]	7
p-(Carbamoylmethyl)benzyl ester (PAM) resin (from 4-(bromomethyl)phenylacetyl handles)[b]	114, 116, 238
4-Methylbenzhydrylamide (MBHA) resin	239
Rink amide resin (4-(2′,4′-dimethoxyphenylaminomethyl)-phenoxymethyl resin)	240
Tris(alkoxy)benzylamide (PAL) resin (from 5-(4-aminomethyl-3,5-dimethoxyphenoxy)valeric acid handle)	119, 241, 242
Wang resin (4-alkoxybenzyl alcohol resin)[c]	243
SASRIN resin (2-methoxy-4-alkoxybenzyl alcohol resin)	244

p-Alkylbenzyl ester (from chloromethyl resin ≡ Merrifield resin)[a]

$$R-\overset{O}{\underset{\|}{C}}-O-CH_2-\text{(aryl)}\text{(resin)} \xrightarrow{\text{strong acid}} R-\overset{O}{\underset{\|}{C}}-OH$$

p-(Carbamoylmethyl)benzyl ester (PAM) resin (from 4-(bromomethyl)phenylacetyl handles)[b]

$$R-\overset{O}{\underset{\|}{C}}-O-CH_2-\text{(aryl)}-CH_2-\overset{O}{\underset{\|}{C}}-NH-\text{(resin)} \xrightarrow{\text{strong acid}} R-\overset{O}{\underset{\|}{C}}-OH$$

4-Methylbenzhydrylamide (MBHA) resin

$$R-\overset{O}{\underset{\|}{C}}-NH-CH-\text{(aryl)}\text{(resin)} \xrightarrow{\text{strong acid}} R-\overset{O}{\underset{\|}{C}}-NH_2$$
$$CH_3$$

Rink amide resin (4-(2′,4′-dimethoxyphenylaminomethyl)-phenoxymethyl resin)

$$R-\overset{O}{\underset{\|}{C}}-NH-CH-\text{(aryl)}-O-CH_2-\text{(aryl)}\text{(resin)} \xrightarrow{\text{moderate acid}} R-\overset{O}{\underset{\|}{C}}-NH_2$$

Tris(alkoxy)benzylamide (PAL) resin (from 5-(4-aminomethyl-3,5-dimethoxyphenoxy)valeric acid handle)

$$R-\overset{O}{\underset{\|}{C}}-NH-CH_2-\text{(aryl)}-O-(CH_2)_4-\overset{O}{\underset{\|}{C}}-NH-\text{(resin)} \xrightarrow{\text{moderate acid}} R-\overset{O}{\underset{\|}{C}}-NH_2$$

Wang resin (4-alkoxybenzyl alcohol resin)[c]

$$R-\overset{O}{\underset{\|}{C}}-O-CH_2-\text{(aryl)}-O-CH_2-\text{(aryl)}\text{(resin)} \xrightarrow{\text{moderate acid}} R-\overset{O}{\underset{\|}{C}}-OH$$

SASRIN resin (2-methoxy-4-alkoxybenzyl alcohol resin)

$$R-\overset{O}{\underset{\|}{C}}-O-CH_2-\text{(aryl)}-O-CH_2-\text{(aryl)}\text{(resin)} \xrightarrow{\text{weak acid}} R-\overset{O}{\underset{\|}{C}}-OH$$

Continued on next page

TABLE II—*Continued*

Anchoring Linkage, Cleavage Reaction, and Product	*References*
2-Chlorotrityl resin	245

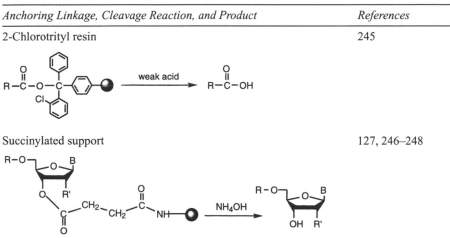

Succinylated support	127, 246–248

NOTE: This table has been abstracted and somewhat reorganized from the more extensive table in the appendix. See the chapter text as well as corresponding entries and footnotes in the appendix table for further information.

[a]These esters are also cleaved by base and a range of nucleophiles, as well as by catalytic hydrogenolysis (regular and transfer modes). Nucleophilic cleavages can be set up to give various functionalities, including free carboxyls, primary and secondary carboxamides, hydrazides, and esters.

[b]*See* Figure 5 for further description and discussion.

[c]Handle variations are shown in Figure 6.

where two handles are joined in series (*125, 126*), and inert "spacers" (at least 25 atoms in length), which are thought to provide additional distance from the support backbone and hence facilitate synthesis of some target molecules (*127, 128*). Most (but not all) of the supports or linkers listed in the appendix table (and all of the ones in the much shorter Table II) were developed originally for SPPS, but quite a few of them are valuable for more general applications in SPOS.

Coupling Chemistries

The major chemical methods for peptide-bond formation (*3, 5, 7–11, 14, 16, 129, 130*), which are also applicable to forming the amide bonds in PNA and other targets of interest (Figure 2), can be divided into two general categories. First, the nucleophilic N^α-amino group of a peptide can be reacted with an amino acid (Figure 7) or peptide pre-activated at its C^α-carboxyl group, and second, an in situ condensing agent (Figure 8) can be added to a mixture of the amino and carboxy components. A large variety of anhydrides (symmetrical, mixed, and internal), acyl azides, fluorides, "active" esters and thioesters can be used in the chemical coupling procedures. The classical example of an in situ coupling reagent is *N,N'*-dicyclohexylcar-

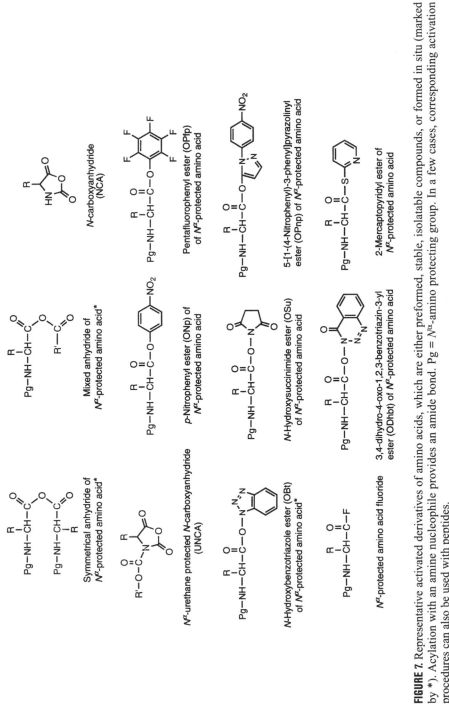

FIGURE 7. Representative activated derivatives of amino acids, which are either preformed, stable, isolatable compounds, or formed in situ (marked by *). Acylation with an amine nucleophile provides an amide bond. Pg = N^α-amino protecting group. In a few cases, corresponding activation procedures can also be used with peptides.

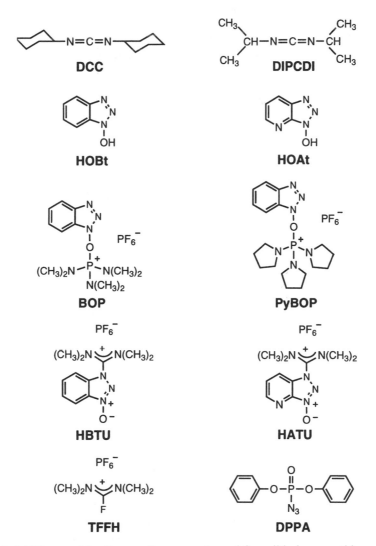

FIGURE 8. Additives and in situ coupling reagents used for solid-phase peptide synthesis. HBTU and HATU are shown as the guanidinium *N*-oxide isomers that were revealed by X-ray structural analysis (134), and not as the *N,N,N′,N′*-tetramethyluronium salts commonly presented in the literature. TBTU is the same as HBTU, except the counterion is tetrafluoroborate.

bodiimide (DCC), introduced by Sheehan and Hess in 1955 (*131*) and still widely used. The related *N,N'*-diisopropylcarbodiimide (DIPCDI) is more convenient to use under some circumstances, because of the improved solubility characteristics of the urea coproduct from the couplings. A significant contribution by König and Geiger relates to the use of 1-hydroxybenzotriazole (HOBt) as an additive that accelerates carbodiimide-mediated couplings and suppresses a variety of side reactions (*132*). Carpino has described the HOBt analogue 7-aza-1-hydroxybenzotriazole (HOAt), which is even more effective, most likely because of a 7-position-specific neighboring group effect of the extra nitrogen (*133, 134*). Protocols involving Castro's benzotriazol-1-yl-oxy-tris(dimethylamino)phosphonium hexafluorophosphate (BOP) reagent (*135*), Dourtoglou's (and Knorr's) *N*-[(1*H*-benzotriazol-1-yl)-(dimethylamino)methylene]-*N*-methylmethanaminium hexafluorophosphate *N*-oxide (HBTU) reagent (*136–138*), Carpino's *N*-[(1*H*-benzotriazol-1-yl)(dimethyl-amino)methylene]-*N*-methylmethanaminium hexafluorophosphate (HATU) (*133, 139*), and their congeners have deservedly achieved popularity (*see* the legend of Figure 8 for an explanation of the structures and nomenclature of HBTU and HATU, which were originally believed to be uronium salts). HBTU and HATU require a tertiary amine, for example *N*-methylmorpholine (NMM) or *N,N*-diisopropylethylamine (DIEA), for optimal efficiency, and in some cases addition of HOBt or HOAt is found to help. Tetramethylfluoroformamidinium hexafluorophosphate (TFFH), which generates acid fluoride intermediates in situ (*140*), can be used as a convenient coupling reagent. The action of diphenylphosphoryl azide (DPPA), a reagent widely used to mediate the formation of macrocyclic lactams, presumably involves generation of acyl azide intermediates.

Racemization is not normally a problem during incorporation of protected amino acid derivatives in current state-of-the-art stepwise SPPS (*7*). In part, this is a result of the remarkable properties of N^α-urethane-type protecting groups (*see* the next section). However, a wide range of racemization levels, in part sequence-dependent, are observed when the aforementioned methods and reagents are applied to segment condensation in solution or on a polymeric support. Slow couplings are particularly prone to racemization, as racemization most commonly proceeds via an oxazolone intermediate, a unimolecular (concentration-independent) process, whereas coupling is normally bimolecular (second-order kinetics). Consequently, careful attention to the possibility of racemization is always necessary, first by use of accurate analytical criteria to determine quantitatively whether or not the level of racemization is under a tolerable threshold, and subsequently by changes in reaction conditions or reagents.

The most reliable coupling procedure for stepwise synthesis of DNA and RNA oligonucleotides uses appropriately protected building blocks that have been activated as phosphoramidites (Figure 9). By virtue of the trivalent phosphorus (activated further in situ with 1*H*-tetrazole), couplings are much faster than in the classical pentavalent phosphate triester methods. After each coupling cycle, the intermediate phosphate triester is oxidized quantitatively with aqueous iodine, in the presence of a weak base to neutralize hydrogen iodide (HI), to provide the protected

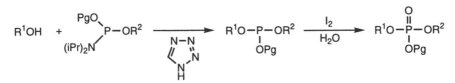

FIGURE 9. One coupling cycle of phosphoramidite chemistry. Reactions take place in dry acetonitrile, in the presence of $1H$-tetrazole (shown) or related activators. Oxidation with aqueous iodine after each cycle gives the protected phosphate triester (alternatively, treatment with appropriate sulfurizing reagents gives phosphorothioates), and removal of the protecting group Pg occurs upon conclusion of chain assembly. The Pg is usually $N{\equiv}CCH_2CH_2$ (β-cyanoethyl (CNE), which is removed by base), although numerous other possibilities are known (*44, 144, 145*). For optimal results, a "capping" step, which blocks unreacted hydroxyl groups, is carried out between each coupling and oxidation step.

pentavalent phosphate triester (*37, 44*) [sulfurization with appropriate reagents gives phosphorothioates (*44, 141–143*)]. Further details of state-of-the-art oligonucleotide synthesis have been reviewed (*44, 144, 145*). These reviews include details of the alternative applications of H-phosphonate chemistry, which allows the synthesis of phosphate-modified oligonucleotide analogs without the need for protection of the phosphate backbone.

For the synthesis of other regular oligomeric targets (Figure 2), high yield and stereospecific (when relevant) coupling reactions for the repetitive elongation steps have been described and exploited. These include N- or S-alkylations of α-haloacetyl derivatives (in turn formed by acylation of the amine precursor), as well as reactions of amines with preformed O-carbonates or with isocyanates formed in situ.

Protection Schemes

For stepwise SPPS, chain assembly is carried out in the C → N direction, and selection of the "temporary" N^α-amino protecting group defines much of the subsequent strategy. The classical "standard Merrifield" system uses "Boc chemistry", in a scheme based on graduated lability to acid (Figure 10). The acidolyzable N^α-*tert*-butyloxycarbonyl (Boc) group is stable to alkali and nucleophiles and is removed by inorganic and organic acids; TFA (20–50% v/v) in dichloromethane is usually applied. One intermediate, a *tert*-butyl carbonium ion, undergoes either an E1 process to give isobutylene or is trapped in an S_N1 process; meanwhile the other intermediate, a carbamate, spontaneously decarboxylates to furnish the protonated amine. A separate diffusion-controlled neutralization step with a tertiary amine (e.g., 5% N,N-diisopropylethylamine (DIEA) in CH_2Cl_2) is required to liberate the nucleophilic amine for reaction with the next incoming acylating agent. The most common "permanent" side-chain protecting groups in Boc chemistry are ether, ester, and urethane derivatives based on substituted benzyl alcohols (Bzl). In some instances, p-toluenesulfonyl (Tos) or cyclohexyl (cHex) derivatives are used as well.

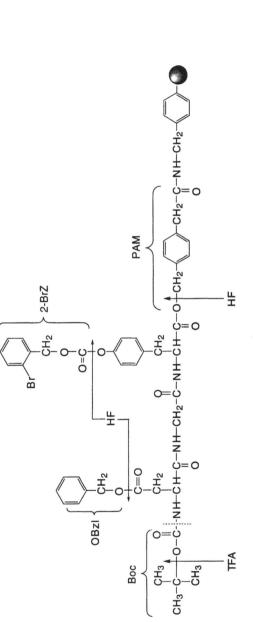

FIGURE 10. "Boc chemistry" for SPPS, illustrating graduated acid lability. The "temporary" N^α-Boc group is removed at each cycle by the moderately strong acid TFA. Upon completion of chain assembly, permanent Bzl-based side-chain protecting groups (fine-tuned as needed by electron-withdrawing or electron-donating groups) and the PAM anchoring linkage are cleaved simultaneously. Cleavage is accomplished with HF or another strong acid, in the presence of anisole and other scavengers, and yields the deprotected free peptide acid.

These are cleaved at the same time as the peptide is cleaved from a suitable resin. (A PAM linkage is best for peptide acids, whereas MBHA is best for peptide amides (Table II).) Strong acids, such as anhydrous HF or TFMSA, are used for such cleavages, preferably in the presence of effective carbonium ion scavengers such as anisole. Side reactions accompanying such cleavages, and details for the management of certain sensitive amino acid residues, are beyond the scope of this chapter and have been reviewed extensively (*5, 7, 14*).

The most popular orthogonal scheme for stepwise SPPS is "Fmoc chemistry" (Figure 11). The "temporary" N^α-9-fluorenylmethyloxycarbonyl (Fmoc) group is removed by secondary amine catalyzed β-elimination, which gives the free amine directly. The reaction is promoted by piperidine (20–50% v/v) in *N,N*-dimethylformamide (DMF) or *N*-methylpyrrolidone (NMP). A highly reactive fulvene intermediate is formed, but fortunately it reacts with excess deprotecting agent rather than damaging newly deblocked chains. A stronger base, 1,8-diazabicyclo[5.4.0]undec-7-ene (DBU, used at 2% v/v) in DMF, has also been found to be effective for Fmoc removal; some piperidine (2%) is still needed to scavenge the fulvene. Side-chain protecting groups on trifunctional amino acids in Fmoc chemistry are based on ester, ether, and urethane derivatives of *tert*-butanol; acidolyzable trityl (Trt), 9*H*-xanthen-9-yl (Xan), 2,4,6-trimethoxybenzyl (Tmob), or substituted tosyl derivatives (e.g., 2,2,5,7,8-pentamethylchroman-6-sulfonyl (Pmc) or 2,2,4,6,7-pentamethyl-dihydrobenzofuran-5-sulfonyl (Pbf) for the guanidino function of Arg) are also used. Final release of the peptide from the solid support, and removal of the side-chain protecting groups used in the Fmoc/*t*Bu strategy, is accomplished by TFA treatment in the presence of one or more suitable scavengers (e.g., 1,2-ethanedithiol (EDT), dimethyl sulfide, thioanisole, phenol, anisole, triethylsilane or tri(iso-propyl)silane), plus relatively small volumes of water. A veritable alphabet soup of acid/scavenger cocktails has been recommended, including reagents A (*119*), B (*146*), K (*147*), M (*118*), and R (*119*). As with Boc chemistry (*see* the preceding discussion), specialized considerations in Fmoc chemistry have been reviewed elsewhere (*11, 14, 139, 148–150*).

A widely used protection scheme for the synthesis of DNA, carried out in the 3′ → 5′ direction, relies on an acidolyzable 4,4′-dimethoxytrityl (DMTr) "temporary" group on the 5′-hydroxyl, and base-labile acyl (e.g., Ac, Bz) and β-cyanoethyl (CNE) "permanent" groups on the purine and pyrimidine exocyclic amines and phosphate backbone, respectively (Figure 12). Dichloroacetic acid or trichloroacetic acid is used to remove the DMTr group. The succinate anchoring linkage is cleaved quickly, and the permanent groups subsequently, during the final treatment with concentrated aqueous ammonium hydroxide at 55 °C (shorter treatment times and lower temperatures can now be used in conjunction with the latest generation of protecting groups such as phenoxyacetyl [PAC] and dimethylformamidine [dmf]). RNA synthesis proceeds similarly, with the breakthrough for this family of oligonucleotides coming in the development of reliable 2′-hydroxyl protection that can be removed without accompanying isomerization (3′ → 2′ phosphate migrations via cyclic phosphodiester intermediates provide the unwanted 2′ → 5′ phosphodiester

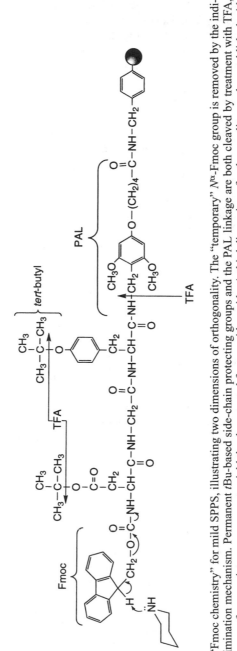

FIGURE 11. "Fmoc chemistry" for mild SPPS, illustrating two dimensions of orthogonality. The "temporary" N^α-Fmoc group is removed by the indicated β-elimination mechanism. Permanent tBu-based side-chain protecting groups and the PAL linkage are both cleaved by treatment with TFA, in the presence of appropriate scavengers, to yield the deprotected free peptide amide. A third dimension of orthogonality can be established with photolabile or allyl-type anchoring linkages (*see* appendix table for structures and references).

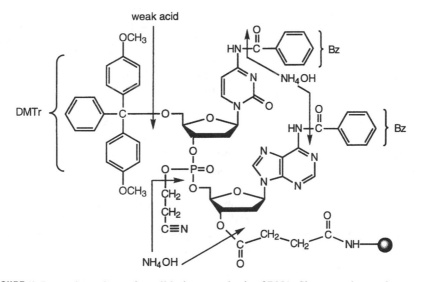

FIGURE 12. Protection scheme for solid-phase synthesis of DNA. Shown are benzoyl protecting groups on adenosine and cytidine; guanosine may be protected with *iso*-butyryl, and thymidine does not require protection. Similar chemistry is used for RNA, but more base-labile protecting groups are used on the purine and pyrimidine bases, and appropriate protection is used for the 2'-hydroxyl (e.g., the *tert*-butyldimethylsilyl (tBdms) group).

linkage) or strand scission. The subtleties of oligonucleotide synthesis, with respect to optimal reagents, alternative protection strategies, direction of chain elongation, and consideration of side reactions, have been described extensively in primary articles and reviews (*37, 44–50, 144, 145*).

Analytical Methods

A crucial issue for stepwise solid-phase synthesis of long oligomers is the repetitive yield per deprotection–coupling cycle. (This is somewhat less crucial for SPOS, where the number of steps is relatively small.) A number of methods for monitoring these steps are available, including those that enable "real-time" feedback based on the kinetics of appearance or disappearance of appropriate soluble chromophores measured in a flow-through system. In SPPS, qualitative and quantitative tests for the presence of unreacted amines after an acylation step include reactions or interactions with ninhydrin (the "Kaiser test") (*151*), isatin (*152*), chloranil (*153*), fluorescamine (*154, 155*), 2,4,6-trinitrobenzenesulfonic acid (*156*), bromphenol blue (*157, 158*), picric acid (*159, 160*), or quinoline yellow (*161*). Ideally, a negative outcome from such a test should be obtained before chain assembly is allowed to proceed further. For certain active ester methods, the leaving group has "self-indicating" properties, meaning that a colored complex is observed for as long as unreacted amino groups remain on the support (*162, 163*). In Fmoc SPPS, addition of Fmoc-amino-acids and deprotection can be monitored spectrophotometrically (*164–166*), or by conductivity measurements (*167, 168*). In solid-phase oligonucleotide synthesis, the best way to monitor progress is to quantify the DMTr released after each deprotection step (*169*).

Failure of a deprotection–coupling cycle in the solid-phase synthesis of either peptides or oligonucleotides results in deletion sequences, which may be quite difficult to remove from the desired products because of very similar physical properties. A precise way to check for deletions in long peptides is "preview" sequencing, introduced by Niall and Tregear (*170, 171*). Some workers choose to "cap" unreacted chains that arise during SPPS by treatments after each coupling cycle with acetic anhydride or equivalent reagents, thereby substituting a family of terminated peptides for a family of deletion peptides. In principle, this does not simplify the purification problem, but it offers a new opportunity for purification: those chains that have remained growing all the way to the end of the synthesis can be modified at their free amino termini by addition of a reversible blocking group. When such a terminating reagent includes a moiety amenable to affinity purification, and can be later removed selectively, excellent purification can be achieved (*172–179*). In oligonucleotide synthesis, it is routine to cap with acetic anhydride–4-dimethylaminopyridine (DMAP) (*180*) or acetic anhydride–*N*-methylimidazole (*181, 182*) after each cycle of base addition. In these syntheses, the separation of terminated chains is easier. Moreover, if the 4,4′-dimethoxytrityl (DMTr) protecting group on the 5′-hydroxyl group is retained at the stage when the anchor and side-chain protecting groups are removed, the resultant "trityl-ON" blocked full-size sequence is retained substantially longer upon reversed-phase chromatography, further simplifying the required purification.

Final products or intermediates synthesized on solid phases can be analyzed either after cleavage from the support, or while they are still bound to the support. Particularly for SPOS, noninvasive methods for determining complete or partial structures of resin-bound materials are desirable (*see* Chapter 5). This can be achieved by gel-phase ^{13}C or ^{31}P nuclear magnetic resonance (NMR) (*183–185*), high-resolution magic-angle spinning (HR-MAS) NMR (*186, 187*), MAS heteronuclear multiple quantum coherence (HMQC) and total correlation spectroscopy (TOCSY) NMR (*188*), MAS ^{13}C–^{1}H correlation experiments (*189*), and Fourier transform infrared (FTIR) spectroscopy on single beads (*190–192*).

Products assembled by solid-phase synthesis, once released from the support into solution, require initial characterization, careful purification, and scrupulous recharacterization. This is because low-level by-products can accumulate during repetitive steps of chain assembly, or from side reactions during the cleavage and deprotection step. Available purification tools include those based on liquid chromatography (including reversed-phase, normal phase, ion-exchange, gel filtration, and partition chromatography). The same separation principles can be applied at the analytical scale, and capillary zone and gel electrophoresis can be added as discriminating tests of product purity. Homogeneity should be demonstrated by at least two independent chromatographic or electrophoretic criteria. Characterization—both of "crude" and "purified" materials—by a variety of mass spectrometric techniques (fast atom bombardment mass spectrometry [FABMS], electrospray mass spectrometry [ESMS], and matrix-assisted laser desorption/mass ionization time-of-flight mass spectrometry [MALDI-TOF MS]) is highly recommended. NMR spectroscopy, FTIR spectroscopy, amino acid analysis, and sequence analysis are also of considerable value when appropriate. Several of the aforementioned techniques are exquisitely sensitive and can be used to evaluate material corresponding to a single or a small number of synthesis bead(s).

Summary and Conclusions

This chapter has aimed to provide a focused account of the most important, well-established general principles and specific chemical methods for the solid-phase synthesis of peptides, oligonucleotides, and other macromolecules of biological interest. The tremendous success and impact of these fields is a testament to the ingenuity and skills of a distinguished line of chemists beginning with Merrifield. Recent generalizations to solid-phase combinatorial chemistry targeted at small organic molecules build on the repertoire of solid supports, linker technologies, coupling methods, protection schemes, and sensitive analytical methods that originated and have been proven over the years in the peptide and DNA arenas. It is hoped that those active in SPOS chemistry will benefit from the perspectives and experimental precedents outlined here.

Acknowledgments

This chapter is dedicated to R. B. Merrifield of The Rockefeller University on the occasion of his 75th birthday. We thank members of our research group, particularly Lin Chen and Chongxi Yu, for sharing their extensive knowledge of the solid-phase synthesis literature and other invaluable assistance, and we are grateful to Fernando Albericio, Robert P. Hammer, Derek Hudson, and Karin Musier–Forsyth for helpful comments throughout the process of writing this review. The scanning electron micrographs (Figure 3) were taken at the Microscopy Facility, Center for Interfacial Engineering, University of Minnesota, with the assistance of Stuart McKernan. Maria Kempe held a Hans Werthén Fellowship from The Royal Swedish Academy of Engineering Sciences. Our research was supported by National Institutes of Health grants GM 42722, 43552, and 51628.

APPENDIX. Anchoring Linkages for Solid-Phase Syntheses of Peptides, DNA, RNA, and Small Organic Molecules

Anchoring Linkage, Cleavage Reaction, and Product	References

p-Alkylbenzyl ester (from chloromethyl resin ≡ Merrifield resin)[a]   7, 249–251

$$R-\overset{\overset{\displaystyle O}{\|}}{C}-O-CH_2-\bigcirc\!\!-\!\!\bullet \xrightarrow{\text{strong acid}} R-\overset{\overset{\displaystyle O}{\|}}{C}-OH$$

Carbamate resin[b]   252–257

$$R-NH-\overset{\overset{\displaystyle O}{\|}}{C}-O-CH_2-\bigcirc\!\!-\!\!\bullet \xrightarrow{\text{strong acid}} R-NH_2$$

p-(Carbamoylmethyl)benzyl ester (PAM) resin   114, 116, 238
   (from 4-(bromomethyl)phenylacetyl handles)[c]

$$R-\overset{\overset{\displaystyle O}{\|}}{C}-O-CH_2-\bigcirc\!\!-CH_2-\overset{\overset{\displaystyle O}{\|}}{C}-NH-\bullet \xrightarrow{\text{strong acid}} R-\overset{\overset{\displaystyle O}{\|}}{C}-OH$$

Benzhydrylamide (BHA) resin,[d] Y = H   239, 258–260
   4-Methylbenzhydrylamide (MBHA) resin, Y = CH$_3$

$$R-\overset{\overset{\displaystyle O}{\|}}{C}-NH-CH-\bigcirc\!\!-\!\!\bullet \xrightarrow{\text{strong acid}} R-\overset{\overset{\displaystyle O}{\|}}{C}-NH_2$$

Thioester resin [from 4-(α-mercaptobenzyl)phenylacetic acid handle ($n = 0$)   261–263
   or 4-(α-mercaptobenzyl)phenoxyacetic acid handle ($n = 1$)]

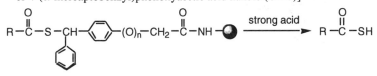

Rink amide resin (4-(2′,4′-dimethoxyphenylaminomethyl)-   240, 264–267
   phenoxymethylresin)[e]

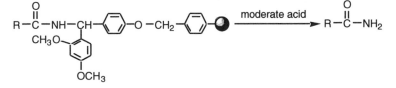

APPENDIX—*Continued.*

Anchoring Linkage, Cleavage Reaction, and Product	References

Tris(alkoxy)benzylamide resins*f* (PAL resin, R′ = H;
   R(PAL) resin, R′ = alkyl; BAL resin, R′ = peptidyl, etc.)

66, 119, 241,
242, 268, 269

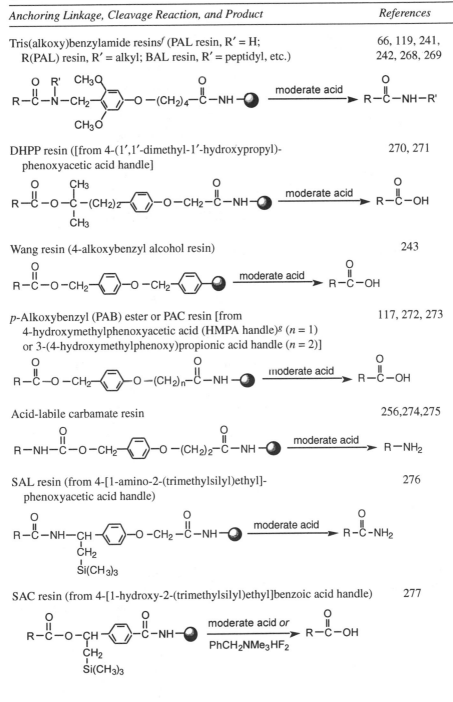

DHPP resin ([from 4-(1′,1′-dimethyl-1′-hydroxypropyl)-
   phenoxyacetic acid handle]

270, 271

Wang resin (4-alkoxybenzyl alcohol resin)

243

p-Alkoxybenzyl (PAB) ester or PAC resin [from
   4-hydroxymethylphenoxyacetic acid (HMPA handle)*g* (*n* = 1)
   or 3-(4-hydroxymethylphenoxy)propionic acid handle (*n* = 2)]

117, 272, 273

Acid-labile carbamate resin

256, 274, 275

SAL resin (from 4-[1-amino-2-(trimethylsilyl)ethyl]-
   phenoxyacetic acid handle)

276

SAC resin (from 4-[1-hydroxy-2-(trimethylsilyl)ethyl]benzoic acid handle)

277

APPENDIX—*Continued.*

Anchoring Linkage, Cleavage Reaction, and Product	References
Dihydropyran-functionalized resin (from 6-hydroxymethyl-3,4-dihydro-2*H*-pyran handle)	278
CHA resin (*n* = 1 or 4) (from 5-{[(*R,S*)-5-amino-10,11-dihydro-dibenzo[*a,d*]cyclohepten-2-yl]oxy}acetic or valeric acid handles)	279, 280
XAL resin (from 5-(9-aminoxanthen-3-oxy)valeric acid handle][h]	150, 281–284
SASRIN resin (2-methoxy-4-alkoxybenzyl alcohol resin)[i]	244, 285–287
2-Chlorotrityl resin[j]	245, 288–292
Ellman "traceless" resin	293
Veber "traceless" resin	294

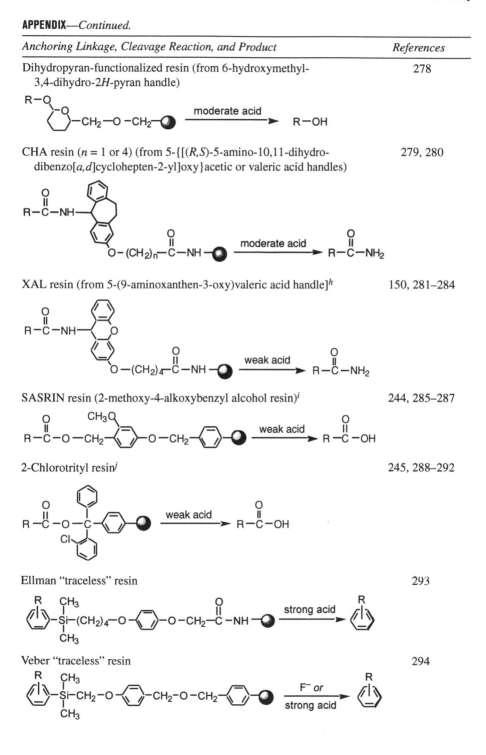

Anchoring Linkage, Cleavage Reaction, and Product	References
Showalter "traceless" resin	295

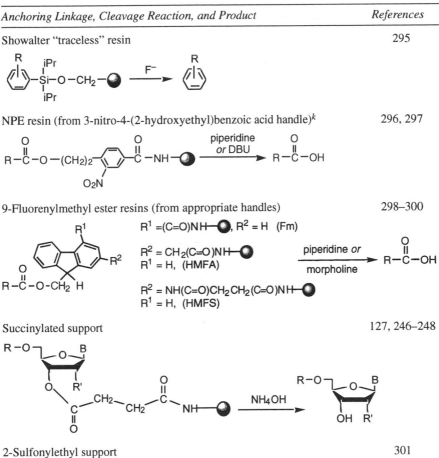

NPE resin (from 3-nitro-4-(2-hydroxyethyl)benzoic acid handle)[k]   296, 297

9-Fluorenylmethyl ester resins (from appropriate handles)   298–300

Succinylated support   127, 246–248

2-Sulfonylethyl support   301

Semicarbazone resin (from appropriate semicarbazone carboxylic acid handle)   302

APPENDIX—*Continued.*

Anchoring Linkage, Cleavage Reaction, and Product	*References*
Weinreb-type amide resin	303, 304

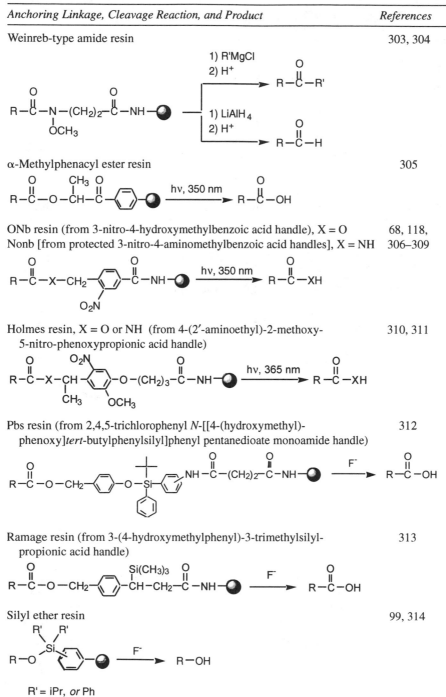

α-Methylphenacyl ester resin — 305

ONb resin (from 3-nitro-4-hydroxymethylbenzoic acid handle), X = O — 68, 118,
Nonb [from protected 3-nitro-4-aminomethylbenzoic acid handles], X = NH — 306–309

Holmes resin, X = O or NH (from 4-(2′-aminoethyl)-2-methoxy-
   5-nitro-phenoxypropionic acid handle) — 310, 311

Pbs resin (from 2,4,5-trichlorophenyl *N*-[[4-(hydroxymethyl)-
   phenoxy]*tert*-butylphenylsilyl]phenyl pentanedioate monoamide handle) — 312

Ramage resin (from 3-(4-hydroxymethylphenyl)-3-trimethylsilyl-
   propionic acid handle) — 313

Silyl ether resin — 99, 314

R′ = iPr, *or* Ph

APPENDIX—*Continued.*

Anchoring Linkage, Cleavage Reaction, and Product	References

HPDI resin (from 2-hydroxypropyl-dithio-2'-isobutyric acid handle)   315

Disulfide-containing support (DTT = dithiothcitol)   316–318

HYCRAM resin (from 4-bromocrotonic acid handle)   319–321

Allyl ester (from *cis*-4-hydroxy-but-2-enyloxy acetic acid handle)   322–324

Universal DNA linker   325
   (from 1-*O*-(4,4'-dimethoxytrityl)-2-*O*-succinoyl-3-*N*-
   allyloxycarbonylpropane linker)

Phenylhydrazide resin   326

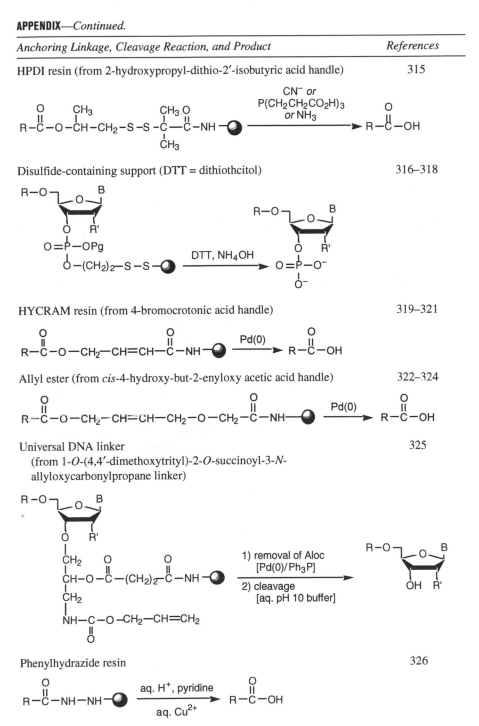

APPENDIX—*Continued.*

Anchoring Linkage, Cleavage Reaction, and Product	*References*
Kaiser 4-nitrobenzophenone oxime resin[l]	327–330

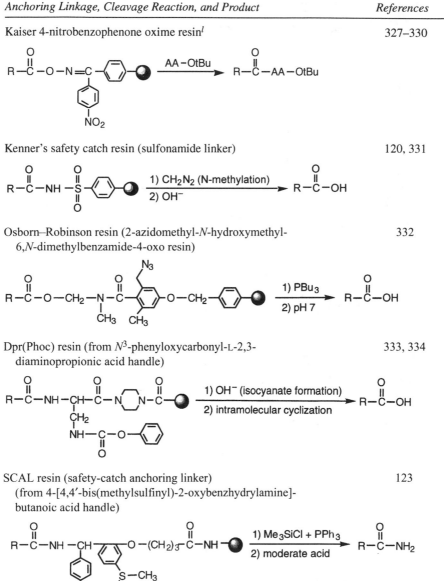

Kenner's safety catch resin (sulfonamide linker)   120, 331

Osborn–Robinson resin (2-azidomethyl-*N*-hydroxymethyl-   332
6,*N*-dimethylbenzamide-4-oxo resin)

Dpr(Phoc) resin (from N^3-phenyloxycarbonyl-L-2,3-   333, 334
diaminopropionic acid handle)

SCAL resin (safety-catch anchoring linker)   123
(from 4-[4,4′-bis(methylsulfinyl)-2-oxybenzhydrylamine]-
butanoic acid handle)

APPENDIX—*Continued.*

Anchoring Linkage, Cleavage Reaction, and Product	References
DSB resin (from 4-(2,5-dimethyl-4-methylsulfinylphenyl)-4-hydroxybutanoic acid handle)	335

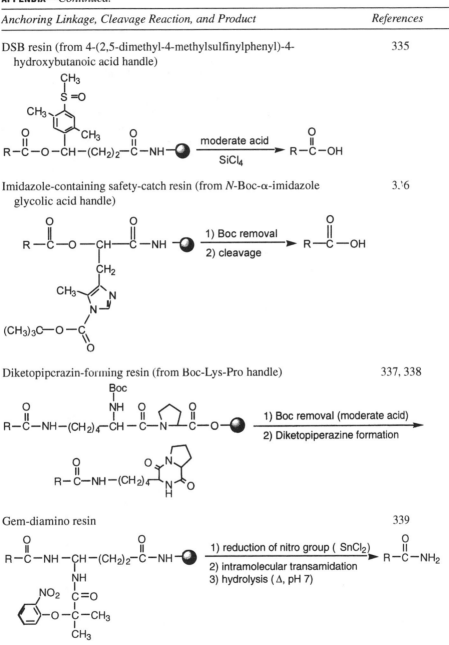

DSB resin (from 4-(2,5-dimethyl-4-methylsulfinylphenyl)-4-hydroxybutanoic acid handle)   335

Imidazole-containing safety-catch resin (from *N*-Boc-α-imidazole glycolic acid handle)   3.'6

1) Boc removal
2) cleavage

Diketopiperazin-forming resin (from Boc-Lys-Pro handle)   337, 338

1) Boc removal (moderate acid)
2) Diketopiperazine formation

Gem-diamino resin   339

1) reduction of nitro group (SnCl₂)
2) intramolecular transamidation
3) hydrolysis (Δ, pH 7)

NOTE: The organization of this table and general principles are given in the main text, which also contains a substantially abstracted and somewhat reorganized version, Table II. Each entry gives the name and structure of the appropriate derivative at the stage *after* attachment to the solid support, as well as

APPENDIX—*Continued.*

---

after creation of the anchoring linkage connected to the functional group at the left of the formula. The solid support is indicated by the shaded circle at the right of the formula, and if polystyrene, may also include the right-most benzene ring. Over the reaction arrow are the relevant cleavage conditions, and the structure of the product following cleavage is given next. When anchors can be cleaved in more than one way and/or give alternative products, or when closely related anchors can give other functional end groups than the one drawn, such information is conveyed in further footnotes. The literature cited in the References column to the right of the table provides detailed methods, or leading references from reviews, for carrying out attachment and/or cleavage chemistries relating to the structures as drawn, or to closely related structures that may be either direct resin derivatives or handle variants. No attempt has been made to provide an exhaustive listing; see refs. 7, 11, 14, 16, and 72 for other listings.

[a]These esters are also cleaved by base and a range of nucleophiles, as well as by catalytic hydrogenolysis (regular and transfer modes). Nucleophilic cleavages can be set up to give free carboxyls, primary and secondary carboxamides, hydrazides, esters, etc., as has been reviewed in detail elsewhere.

[b]Under some circumstances, cleavage occurs with a base (*257*).

[c]See Figure 5 for further description and discussion.

[d]A handle variant, Boc-aminobenzhydryloxyacetic acid (BHA-linker), can be coupled onto amino-functionalized supports and used in the same ways.

[e]Numerous resin derivatives and handles on the benzhydrylamine theme are available, associated with the names of Knorr and Breipohl, or with acronyms such as Dod, DAL, and Linker-AM (details are in the cited literature, except for Knorr, which is commercially available but was never published).

[f]PAL and related anchoring linkages can be established by coupling of a carboxylic acid to an amino-handle support, or by reductive amination of an amine to the appropriate aldehyde precursor (details in the cited literature).

[g]Compare to Figure 6.

[h]Shown is the 3-XAL$_4$ handle variant (substituent is on the 3-position of the xanthenyl system; 4 carbon spacer). The 2-XAL$_1$, 2-XAL$_4$, and 3-XAL$_1$ handles have also been described, as has a resin-bound variation in which the 3-substituent oxygen is attached directly to Merrifield resin. Also, the 2-XAL system has been used for side-chain anchoring of thiols as their polymer-supported *S*-xanthenyl derivatives.

[i]Handle variants, 4-(4-hydroxymethyl-3-methoxyphenoxy)butyric acid (HMPA) and 5-(4-hydroxymethyl-3,5-dimethoxyphenoxy)valeric acid (HAL) (which has an extra methoxy substituent) have been described (details in cited literature).

[j]Barlos's chlorotrityl resin, as shown, is optimized for preparation of protected peptide acids. Numerous trityl variants, in terms of substituents to optimize stability/lability, and involving either direct resin derivatives or handles, are covered in the references. Furthermore, the approach can be extended to anchor alcohols as trityl ethers (acidolyzable) and thiols as trityl thioethers (providing disulfides directly upon oxidative cleavage), among others.

[k]Drawn to illustrate base-labile anchoring of carboxyl groups. The same principle has been applied in oligonucleotide chemistry, with base-labile anchoring of the 3'-hydroxyl group as a carbonate.

[l]A wide range of nucleophiles have been applied successfully to cleave the oxime esters.

References

1. Merrifield, R. B. *J. Am. Chem. Soc.* **1963**, *85*, 2149–2154.
2. Merrifield, R. B. *Life During a Golden Age of Peptide Chemistry: The Concept and Development of Solid-Phase Peptide Synthesis*; American Chemical Society: Washington, DC, 1993.

3. Wünsch, E. Synthese von Peptiden. In Houben-Weyl's *Methoden der Organischen Chemie,* 4th ed., Vol. 15, parts 1 and 2. Müller, E., Ed.; Georg Thieme Verlag: Stuttgart, 1974.

4. Wieland, T.; Bodanszky, M. *The World of Peptides: A Brief History of Peptide Chemistry*; Springer-Verlag: Berlin, Germany, 1991.

5. Bodanszky, M. *Principles of Peptide Synthesis,* 2nd ed.; Springer-Verlag: Berlin, 1993.

6. Gutte, B.; Merrifield, R. B. *J. Biol. Chem.* **1971,** *246,* 1922–1941.

7. Barany, G.; Merrifield, R. B. In *The Peptides: Analysis, Synthesis, Biology,* Vol. 2; Gross, E.; Meienhofer, J., Eds.; Academic: New York, 1979; pp 1–284, and earlier reviews cited therein.

8. Meienhofer, J. *Biopolymers* **1981,** *20,* 1761–1784. (review article)

9. Stewart, J. M.; Young, J. D. *Solid Phase Peptide Synthesis,* 2nd ed.; Pierce Chemical Co.: Rockford, IL, 1984.

10. Merrifield, R. B. *Science (Washington, DC)* **1986,** *232,* 341–347. (review article)

11. Barany, G; Kneib-Cordonier, N.; Mullen, D. G. *Int. J. Pept. Protein Res.* **1987,** *30,* 705–739. (review article)

12. Kent, S. B. H. *Annu. Rev. Biochem.* **1988,** *57,* 957–989. (review article)

13. Atherton, E.; Sheppard, R. C. *Solid Phase Peptide Synthesis: A Practical Approach*; IRL: Oxford, 1989.

14. Fields, G. B.; Tian, Z.; Barany, G. In *Synthetic Peptides. A User's Guide*; Grant, G. A., Ed.; W. H. Freeman and Co.: New York, 1992; pp 77–183. (review article)

15. Sheppard, R. C. In *Peptides 1994, Proceedings of the Twenty-Third European Peptide Symposium*; Maia, H.L.S., Ed.; Escom Science: Leiden, 1995; pp 3–17. (review article)

16. Merrifield, B. In *Peptides: Synthesis, Structures and Applications*; Gutte, B., Ed.; Academic: San Diego, 1995; pp 93–169. (review article)

17. Jung, G.; Beck-Sickinger, A. G. *Angew. Chem., Int. Ed. Engl.* **1992,** *31,* 367–383. (review article)

18. Gallop, M. A.; Barrett, R. W.; Dower, W. J.; Fodor, S. P. A.; Gordon, E. M. *J. Med. Chem.* **1994,** *37,* 1233–1251. (review article)

19. Lebl, M.; Krchñák, V.; Sepetov, N. F.; Seligmann, B.; Strop, P.; Felder, S.; Lam, K. S. *Biopolymers: Peptide Science* **1995,** *37,* 177–197. (review article)

20. Geysen, H. M.; Meloen, R. H.; Barteling, S. J. *Proc. Natl. Acad. Sci. U.S.A.* **1984,** *81,* 3998–4002.

21. Furka, A.; Sebestyen, F.; Asgedom, M; Dibo, G. *Int. J. Pept. Protein Res.* **1991,** *37,* 487–493.

22. Houghten, R. A. *Proc. Natl. Acad. Sci. U.S.A.* **1985,** *82,* 5131–5135.

23. Houghten, R. A.; Pinilla, C.; Blondelle, S. E.; Appel, J. R.; Dooley, C. T.; Cuervo, J. H. *Nature (London)* **1991,** *354,* 84–86.

24. Lam, K. S.; Salmon, S. E.; Hersh, E. M.; Hruby, V. J.; Kazmierski, W. M.; Knapp, R. J. *Nature (London)* **1991,** *354,* 82–84.

25. Fodor, S. P. A.; Read, J. L.; Pirrung, M. C.; Stryer, L.; Lu, A. T.; Solas, D. *Science (Washington, DC)* **1991,** *251,* 767–773.

26. Johansson, C.; Blomberg, L.; Hlebowicz, E.; Nicklasson, H.; Nilsson, B.; Andersson, L. In *Peptides 1994, Proceedings of the Twenty-Third European Peptide Symposium*; Maia, H.L.S., Ed.; Escom Science: Leiden, 1995; pp 34–35.

27. Mutter, M.; Bayer, E. In *The Peptides: Analysis, Synthesis, Biology,* Vol. 2; Gross, E.; Meienhofer, J., Eds.; Academic: New York, 1980; pp 286–332. (review article)

28. Kisfaludy, L. In *The Peptides: Analysis, Synthesis, Biology,* Vol. 2; Gross, E.; Meienhofer, J., Eds.; Academic: New York, 1980; pp 418–440. (review article)

29. Bergmann, M.; Fraenkel-Conrat, H. *J. Biol. Chem.* **1938,** *124,* 1–6.

30. Fruton, J. S. *Adv. Prot. Chem.* **1949,** *5,* 1–82. (review article)

31. Chaiken, I. M.; Komoriya, A.; Ohno, M.; Widmer, F. *Appl. Biochem. Biotechnol.* **1982,** *7,* 385–399. (review article)

32. Jakubke, H.-D. In *The Peptides: Analysis, Synthesis, Biology*, Vol. 9; Udenfriend, S.; Meienhofer, J., Eds.; Academic: New York, 1987; pp 103–165. (review article)

33. Kullmann, W. *Enzymatic Peptide Synthesis*; CRC Press: Boca Raton, FL, 1987.

34. Wong, C.-H.; Wang, K.-T. *Experientia* **1991**, *47*, 1123–1129. (review article)

35. Sanger, F.; Nicklen, S.; Coulson, A. R. *Proc. Natl. Acad. Sci. U.S.A.* **1977**, *74*, 5463–5467.

36. Wood, W. I.; Gitschier, J.; Lasky, L. A.; Lawn, R. M. *Proc. Natl. Acad. Sci. U.S.A.* **1985**, *82*, 1585–1588, and references therein.

37. Caruthers, M. H. *Science (Washington, DC)* **1985**, *230*, 281–285, and references therein.

38. Church, G. M.; Kieffer-Higgins, S. *Science (Washington, DC)* **1988**, *240*, 185–188.

39. Sambrook, J.; Fritsch, E. F.; Maniatis, T. *Molecular Cloning—A Laboratory Manual*, 2nd ed.; Cold Spring Harbor Laboratory: New York, 1989.

40. Smith, M. *Angew. Chem., Int. Ed. Engl.* **1994**, *33*, 1214–1221.

41. Mullis, K. B. *Angew. Chem., Int. Ed. Engl.* **1994**, *33*, 1209–1213.

42. Van Ness, J.; Kalbfleisch, S.; Petrie, C. R.; Reed, M. W.; Tabone, J. C.; Vermeulen, N. M. J. *Nucleic Acids Res.* **1991**, *19*, 3345–3350.

43. Xu, X.-H.; Bard, A. J. *Am. Chem. Soc.* **1995**, *117*, 2627–2631.

44. Beaucage, S. L.; Iyer, R. P. *Tetrahedron* **1992**, *48*, 2223–2311. (review article)

45. Alvarado-Urbina, G.; Sathe, G. M.; Liu, W.-C.; Gillen, M. F.; Duck, P. D.; Bender, R.; Ogilvie, K. K. *Science (Washington, DC)* **1981**, *214*, 270–274.

46. Cook, R. M.; Hudson, D.; Mayrand, E.; Ott, J. In *Chemical and Enzymatic Synthesis of Gene Fragments: A Laboratory Manual*; Gassen. H. G.; Lang, A., Eds.; Verlag Chemie: Weinheim, 1982, pp 111–123.

47. Hunkapiller, M.; Kent, S.; Caruthers, M.; Dreyer, W.; Firca, J.; Giffin, C.; Horvath, S.; Hunkapiller, T.; Tempst, P.; Hood, L. *Nature (London)* **1984**, *310*, 105–111. (review article)

48. Usman, N.; Ogilvie, K. K.; Jiang, M.-Y.; Cedergren, R. J. *J. Am. Chem. Soc.* **1987**, *109*, 7845–7854.

49. Wincott, F.; DiRenzo, A.; Shaffer, C.; Grimm, S.; Tracz, D.; Workman, C.; Sweedler, D.; Gonzalez, C.; Scaringe, S.; Usman, N. *Nucleic Acids Res.* **1995**, *23*, 2677–2684, and references therein.

50. Sproat, B.; Colonna, F.; Mullah, B.; Tsou, D.; Andrus, A.; Hampel, A.; Vinayak, R. *Nucleosides & Nucleotides* **1995**, *14*, 255–273, and references therein.

51. Egholm, M.; Buchardt, O.; Nielsen, P. E.; Berg, R. H. *J. Am. Chem. Soc.* **1992**, *114*, 1895–1897.

52. Thomson, S. A.; Josey, J. A., Cadilla, R.; Gaul, M. D.; Hassman, C. F.; Luzzio., M. J.; Pipe, A. J.; Reed, K. L.; Ricca, D. J..; Wiethe, R. W.; Noble, S. A. *Tetrahedron* **1995**, *52*, 6179–6194, and references therein.

53. Gordon, E. M.; Barrett, R. W.; Dower, W. J.; Fodor, S. P. A.; Gallop, M. A. *J. Med. Chem.* **1994**, *37*, 1385–1401. (review article)

54. Thompson, L. A.; Ellman, J. A. *Chem. Rev.* **1996**, *96*, 555–600. (review article)

55. Früchtel, J. S.; Jung, G. *Angew. Chem., Int. Ed. Engl.* **1996**, *35*, 17–42. (review article)

56. Crowley, J. I.; Rapoport, H. *J. Am. Chem. Soc.* **1970**, *92*, 6363–6365.

57. Patchornik, A.; Kraus, M. *J. Am. Chem. Soc.* **1970**, *92*, 7587–7589.

58. Leznoff, C. C.; Wong, J. Y. *Can. J. Chem.* **1972**, *50*, 2892–2893.

59. Kawana, M.; Emoto, S. *Tetrahedron Lett.* **1972**, *48*, 4855–4858.

60. Camps, F.; Castells, J.; Pi, J. *Anales Quim.* **1974**, *70*, 848–849.

61. Crowley, J. I.; Rapoport, H. *Acc. Chem. Res.* **1976**, *9*, 135–144. (review article)

62. Erickson, B. W.; Merrifield, R. B. In *The Proteins*, 3rd ed., Vol. 2; Neurath, H.; Hill, R. L., Eds.; Academic: New York, 1976; pp 255–527. (review article)

63. Leznoff, C. C. *Acc. Chem. Res.* **1978**, *11*, 327–333. (review article)

64. Fréchet, J. M. J. *Tetrahedron* **1981**, *37*, 663–683. (review article)
65. Albericio, F.; Van Abel, R.; Barany, G. *Int. J. Pept. Protein Res.* **1990**, *35*, 284–286, and references therein.
66. Jensen, K. J.; Songster, M. F.; Vágner, J.; Alsina, J.; Albericio, F.; Barany, G. In *Peptides: Chemistry, Structure & Biology, Proceedings of the Fourteenth American Peptide Symposium*; Kaumaya, P. T. P.; Hodges, R. S., Eds.; Mayflower Scientific Ltd.: Kingswinford, England, 1996; pp 30–32.
67. Barany, G.; Merrifield, R. B. *J. Am. Chem. Soc.* **1977**, *99*, 7363–7365.
68. Barany, G.; Albericio, F. *J. Am. Chem. Soc.* **1985**, *107*, 4936–4942.
69. Bloomberg, G. B.; Askin, D.; Gargaro, A. R.; Tanner, M. J. A. *Tetrahedron Lett.* **1993**, *34*, 4709–4712.
70. Fields, C. G.; Grab, B.; Lauer, J. L.; Miles, A. J.; Yu, Y.-C.; Fields, G. B. *Lett. Pept. Sci.* **1996**, *3*, 3–16.
71. Kates, S. A.; Solé, N. A.; Albericio, F.; Barany, G. In *Peptides: Design, Synthesis, and Biological Activity*; Basava, C.; Anantharamaiah, G. M., Eds.; Birkhäuser: Boston, MA, 1994; pp 39–58. (review article)
72. Lloyd-Williams, P.; Albericio, F.; Giralt, E. *Tetrahedron* **1993**, *49*, 11065–11133. (review article)
73. Benz, H. *Synthesis* **1994**, 337–358. (review article)
74. Mazur, S.; Jayalekshmy, P. *J. Am. Chem. Soc.* **1979**, *101*, 677–683.
75. Schiller, P. W.; Nguyen, T. M.-D.; Miller, J. *Int. J. Pept. Protein Res.* **1985**, *25*, 171–177.
76. Felix, A. M.; Wang, C.-T.; Heimer, E. P.; Fournier, A. *Int. J. Pept. Protein Res.* **1988**, *31*, 231–238.
77. Plaué, S. *Int. J. Pept. Protein Res.* **1990**, *35*, 510–517.
78. McMurray, J. S. *Tetrahedron Lett.* **1991**, *32*, 7679–7682.
79. Trzeciak, A.; Bannwarth, W. *Tetrahedron Lett.* **1992**, *33*, 4557–4560.
80. Kates, S. A.; Solé, N. A.; Johnson, C. R.; Hudson, D.; Barany, G.; Albericio, F. *Tetrahedron Lett.* **1993**, *34*, 1549–1552.
81. Lee, J.; Griffin, J. II.; Nicas, T. I. *J. Org. Chem.* **1996**, *61*, 3983–3986.
82. Albericio, F.; Hammer, R. P.; García-Echeverría, C.; Molins, M. A.; Chang, J. L.; Munson, M. C.; Pons, M.; Giralt, E.; Barany, G. *Int. J. Pept. Protein Res.* **1991**, *37*, 402–413, and references therein.
83. Andreu, D.; Albericio, F.; Solé, N. A.; Munson, M. C.; Ferrer, M.; Barany, G. In *Methods in Molecular Biology*, Vol. 35: *Peptide Synthesis Protocols*; Pennington, M. W.; Dunn, B. M., Eds.; Humana: Totowa, NJ, 1994; pp 91–169. (review article)
84. Virgilio, A. A.; Ellman, J. A. *J. Am. Chem. Soc.* **1994**, *116*, 11580–11581.
85. Beebe, X.; Shore, N. E.; Kurth, M. J. *J. Org. Chem.* **1995**, *60*, 4196–4203.
86. Hiroshige, M.; Hauske, J. R.; Zhou, P. *J. Am. Chem. Soc.* **1995**, *117*, 11590–11591.
87. Gisin, B. F.; Merrifield, R. B.; Tosteson, D. C. *J. Am. Chem. Soc.* **1969**, *91*, 2691–2695.
88. Rothe, M.; Dunkel, W. *J. Polym. Sci., Part B* **1967**, *5*, 589–593.
89. Kusch, P. *Angew. Chem.* **1966**, *78*, 611.
90. Burgess, K.; Linthicum, D. S.; Shin, H. *Angew. Chem., Int. Ed. Engl.* **1995**, *34*, 907–909.
91. Cho, C. Y.; Moran, E. J.; Cherry, S. R.; Stephans, J. C.; Fodor, S. P. A.; Adams, C. L.; Sundaram, A.; Jacobs, J. W.; Schultz, P. G. *Science (Washington, DC)* **1993**, *261*, 1303–1305.
92. Moran, E. J.; Wilson, T. E.; Cho, C. Y.; Cherry, S. R.; Schultz, P. G. *Biopolymers: Peptide Science* **1995**, *37*, 213–219.
93. Simon, R. J.; Kania, R. S.; Zuckermann, R. N.; Huebner, V. D.; Jewell, D. A.; Banville, S.; Ng, S.; Wang, L.; Rosenberg, S.; Marlowe, C. K.; Spellmeyer, D. C.; Tan, R.; Frankel, A. D.; Santi, D. V.; Cohen, F. E.; Bartlett, P. A. *Proc. Natl. Acad. Sci. U.S.A.* **1992**, *89*, 9367–9371.

94. Zuckermann, R. N.; Kerr, J. M.; Kent, S. B. H.; Moos, W. H. *J. Am. Chem. Soc.* **1992**, *114*, 10646–10647.
95. Gennari, C.; Nestler, H. P.; Salom, B.; Still, W. C. *Angew. Chem., Int. Ed. Engl.* **1995**, *34*, 1763–1768.
96. Fréchet, J. M.; Schuerch, C. *Carbohydr. Res.* **1972**, *22*, 399–412.
97. Guthrie, R. D.; Jenkins, A. D.; Stehlícek, J. *J. Chem. Soc. (C)* **1971**, 2690–2696.
98. Danishefsky, S. J.; McClure, K. F.; Randolph, J. T.; Ruggeri, R. B. *Science (Washington, DC)* **1993**, *260*, 1307–1309.
99. Randolph, J. T.; McClure, K. F.; Danishefsky, S. J. *J. Am. Chem. Soc.* **1995**, *117*, 5712–5719.
100. Nielsen, J.; Brenner, S.; Janda, K. D. *J. Am. Chem. Soc.* **1993**, *115*, 9812–9813.
101. Needels, M. C.; Jones, D. G.; Tate, E. H.; Heinkel, G. L.; Kochersperger, L. M.; Dower, W. J.; Barrett, R. W.; Gallop, M. A. *Proc. Natl. Acad. Sci. U.S.A.* **1993**, *90*, 10700–10704.
102. Schuster, M.; Wang, P.; Paulson, J. C.; Wong, C.-H. *J. Am. Chem. Soc.* **1994**, *116*, 1135–1136.
103. de la Torre, B.; Aviño, A.; Tarrason, G.; Piulats, J.; Albericio, F.; Eritja, R. *Tetrahedron Lett.* **1994**, *35*, 2733–2736.
104. Bergmann, F.; Bannwarth, W. *Tetrahedron Lett.* **1995**, *36*, 1839–1842.
105. Bergmann, F.; Bannwarth, W.; Tam, S. *Tetrahedron Lett.* **1995**, *36*, 6823–6826.
106. Robles, J.; Pedroso, E.; Grandas, A. *J. Org. Chem.* **1995**, *60*, 4856–4861.
107. Cook, R. M.; Hudson, D. In *Innovations and Perspectives in Solid Phase Synthesis & Combinatorial Libraries, 1996*; Epton, R., Ed.; Mayflower Worldwide Ltd.: Birmingham, England, 1996; pp 19–26.
108. Merrifield, B. *Br. Polym. J.* **1984**, *16*, 173–178. (review article)
109. Vágner, J.; Barany, G.; Lam, K. S.; Krchñák, V.; Sepetov, N. F.; Ostrem, J. A.; Strop, P.; Lebl, M. *Proc. Natl. Acad. Sci. U.S.A.* **1996**, *93*, 8194–8199.
110. Atherton, E.; Clive, D. L. J.; Sheppard, R. C. *J. Am. Chem. Soc.* **1975**, *97*, 6584–6585.
111. Sarin, V. K.; Kent, S. B. H.; Merrifield, R. B. *J. Am. Chem. Soc.* **1980**, *102*, 5463–5470.
112. Merrifield, R. B. *Biochemistry* **1964**, *3*, 1385–1390.
113. Sparrow, J. T. *J. Org. Chem.* **1976**, *41*, 1350–1353.
114. Mitchell, A. R.; Erickson, B. W.; Ryabtsev, M. N.; Hodges, R. S.; Merrifield, R. B. *J. Am. Chem. Soc.* **1976**, *98*, 7357–7362.
115. Mitchell, A. R.; Kent, S. B. H.; Erickson, B. W.; Merrifield, R. B. *Tetrahedron Lett.* **1976**, 3795–3798.
116. Mitchell, A. R.; Kent, S. B. H.; Engelhard, M.; Merrifield, R. B. *J. Org. Chem.* **1978**, *43*, 2845–2852, and references therein.
117. Albericio, F.; Barany, G. *Int. J. Pept. Protein Res.* **1985**, *26*, 92–97, and references therein.
118. Kneib-Cordonier, N.; Albericio, F.; Barany, G. *Int. J. Pept. Protein Res.* **1990**, *35*, 527–538, and references therein.
119. Albericio, F.; Kneib-Cordonier, N.; Biancalana, S.; Gera, L.; Masada, R. I.; Hudson, D.; Barany, G. *J. Org. Chem.* **1990**, *55*, 3730–3743, and references therein.
120. Kenner, G. W.; McDermott, J. R.; Sheppard, R. C. *J. Chem. Soc., Chem. Commun.* **1971**, 636–637.
121. Marshall, D. L.; Liener, I. E. *J. Org. Chem.* **1970**, *35*, 867–868.
122. Flanigan, E.; Marshall, G. R. *Tetrahedron Lett.* **1970**, *27*, 2403–2406.
123. Pátek, M.; Lebl, M. *Tetrahedron Lett.* **1991**, *32*, 3891–3894.
124. Pátek, M. *Int. J. Pept. Protein Res.* **1993**, *42*, 97–117. (review article)
125. Tam, J. P.; Tjoeng, F. S.; Merrifield, R. B. *J. Am. Chem. Soc.* **1980**, *102*, 6117–6127, and references therein.
126. Lebl, M.; Pátek, M.; Kocis, P.; Krchnák, V.; Hruby, V. J.; Salmon, S. E.; Lam, K. S. *Int. J. Pept. Protein Res.* **1993**, *41*, 201–203.

127. Adams, S. P.; Kavka, K. S.; Wykes, E. J.; Holder, S. B.; Gallupi, G. R. *J. Am. Chem. Soc.* **1983**, *105*, 661–663.
128. Katzhendler, J.; Cohen, S.; Rahamim, E.; Weisz, M.; Ringel, I.; Deutsch, J. *Tetrahedron* **1989**, *45*, 2777–2792.
129. Carpino, L. A.; Beyermann, M.; Wenschuh, H.; Bienert, M. *Acc. Chem. Res.* **1996**, *29*, 268–274. (review article)
130. Fuller, W. D.; Goodman, M.; Naider, F. R.; Zhu, Y-F. *Biopolymers: Peptide Science* **1996**, *40*, 183–205. (review article)
131. Sheehan, J. C.; Hess, G. P. *J. Am. Chem. Soc.* **1955**, *77*, 1067–1068.
132. König, W.; Geiger, R. *Chem. Ber.* **1970**, *103*, 788–798.
133. Carpino, L. A. *J. Am. Chem. Soc.* **1993**, *115*, 4397–4398.
134. Abdelmoty, I.; Albericio, F.; Carpino, L. A.; Foxman, B. M.; Kates, S. A. *Lett. Pept. Sci.* **1994**, *1*, 57–67.
135. Castro, B.; Dormoy, J. R.; Evin, G.; Selve, C. *Tetrahedron Lett.* **1975**, 1219–1222.
136. Dourtoglou, V.; Ziegler, J.-C.; Gross, B. *Tetrahedron Lett.* **1978**, 1269–1272.
137. Dourtoglou, V.; Gross, B.; Lambropoulou, V.; Zioudrou, C. *Synthesis* **1984**, 572–574.
138. Knorr, R.; Trzeciak, A.; Bannwarth, W.; Gillessen, D. *Tetrahedron Lett.* **1989**, *30*, 1927–1930.
139. Carpino, L. A.; El-Faham, A.; Minor, C. A.; Albericio, F. *J. Chem. Soc., Chem. Commun.* **1994**, 201–203
140. Carpino, L. A.; El-Faham, A. *J. Am. Chem. Soc.* **1995**, *117*, 5401–5402.
141. Iyer, R. P.; Phillips, L. R.; Egan, W.; Regan, J. B.; Beaucage, S. L. *J. Org. Chem.* **1990**, *55*, 4693–4699.
142. Xu, Q.; Musier-Forsyth, K.; Hammer, R. P.; Barany, G. *Nucleic Acids Res.* **1996**, *24*, 1602–1607, and references therein.
143. Xu, Q.; Barany, G.; Hammer, R. P.; Musier-Forsyth, K. *Nucleic Acids Res.* **1996**, *24*, 3643–3644, and references therein.
144. Blackburn, G. M.; Gait, M. J. In *Nucleic Acids in Chemistry and Biology,* 2nd ed.; Blackburn, G. M.; Gait, M. J., Eds.; Oxford University: Oxford, 1996; pp 83–145. (review article)
145. Beaucage, S. L.; Caruthers, M. H. In *Bioorganic Chemistry: Nucleic Acids*; Hecht, S. M., Ed.; Oxford University: New York, 1996; pp 36–74. (review article)
146. Van Abel, R. J.; Tang, Y-Q.; Rao, V. S. V.; Dobb, C. H.; Tran, D.; Barany, G.; Selsted, M. E. *Int. J. Pept. Protein Res.* **1995**, *45*, 401–409.
147. King, D.; Fields, C.; Fields, G. *Int. J. Pept. Protein Res.* **1990**, *36*, 255–266.
148. Fields, G. B.; Noble, R. L. *Int. J. Pept. Protein Res.* **1990**, *35*, 161–214. (review article)
149. Barany, G.; Solé, N.; Van Abel, R. J.; Albericio, F.; Selsted, M. E. In *Innovations and Perspectives in Solid Phase Synthesis. Peptides, Polypeptides and Oligonucleotides,1992*; Epton, R., Ed.; Intercept Ltd.: Andover, UK, 1992; pp 29–38.
150. Han, Y.; Bontems, S. L.; Hegyes, P.; Munson, M. C.; Minor, C. A.; Kates, S. A.; Albericio, F.; Barany, G. *J. Org. Chem.* **1996**, *61*, 6326–6339, and references to earlier work by this research group cited therein.
151. Kaiser, E.; Colescott, R. L.; Bossinger, C. D.; Cook, P. I. *Anal. Biochem.* **1970**, *34*, 595–598.
152. Kaiser, E.; Bossinger, C. D.; Colescott, R. L.; Olsen, D. B. *Anal. Chim. Acta* **1980**, *118*, 149–151.
153. Christensen, T. *Acta Chem. Scand.* **1979**, *33 B*, 763–766.
154. Felix, A. M.; Jimenez, M. H. *Anal. Biochem.* **1973**, *52*, 377–381.
155. Grandas, A.; Pedroso, E.; Giralt, E.; Granier, C.; Van Rietschoten, J. *Tetrahedron* **1986**, *42*, 6703–6711.
156. Hancock, W. S.; Battersby, J. E. *Anal. Biochem.* **1976**, *71*, 260–264.
157. Krchnák, V.; Vágner, J.; Safár, P.; Lebl, M. *Coll. Czech. Chem. Commun.* **1988**, *53*, 2542–2548.

158. Flegel, M.; Sheppard, R. C. *J. Chem. Soc., Chem Commun.* **1990**, 536–538.
159. Gisin, B. F. *Anal. Chim. Acta* **1972**, *58*, 248–249.
160. Hodges, R. S.; Merrifield, R. B. *Anal. Biochem.* **1975**, *65*, 241–272.
161. Salisbury, S. A.; Tremeer, E. J.; Davies, J. W.; Owen, D. E. I. A. *J. Chem. Soc., Chem. Commun.* **1990**, 538–540.
162. Cameron, L.; Holder, J. L.; Meldal, M.; Sheppard, R. C. *J. Chem. Soc. Perkin Trans. I* **1988**, 2895–2901, and references therein.
163. Hudson, D. In *Peptides: Chemistry, Structure and Biology, Proceedings of the Eleventh American Peptide Symposium*; Rivier, J. E.; Marshall, G. R., Eds.; Escom Science: Leiden, 1990; pp 914–915.
164. Meienhofer, J.; Waki, M.; Heimer, E. P.; Lambros, T. J.; Makofske, R. C.; Chang, C.-D. *Int. J. Pept. Protein Res.* **1979**, *13*, 35–42.
165. Chang, C.-D.; Felix, A. M.; Jimenez, M. H.; Meienhofer, J. *Int. J. Pept. Protein Res.* **1980**, *15*, 485–494.
166. Dryland, A.; Sheppard, R. C. *J. Chem. Soc. Perkin Trans. I* **1986**, 125–137.
167. Nielsen, C. S.; Hansen, P. H.; Lihme, A.; Heegaard, P. M. H. *J. Biochem. Biophys. Meth.* **1989**, *20*, 69–75.
168. McFerran, N. V.; Walker, B.; McGurk, C. D.; Scott, F. C. *Int. J. Pept. Protein Res.* **1991**, *37*, 382–387.
169. Caruthers, M. H.; Barone, A. D.; Beaucage, S. L.; Dodds, D. R.; Fisher, E. F.; McBride, L. J.; Matteucci, M.; Stabinsky, Z.; Tang, J.-Y. *Methods Enzymol.* **1987**, *154*, 287–313.
170. Tregear, G. W.; van Rietschoten, J.; Sauer, R.; Niall, H. D.; Keutmann, H. T.; Potts, J. T., Jr. *Biochemistry* **1977**, *16*, 2817–2823.
171. Matsueda, G. R.; Haber, E.; Margolies, M. N. *Biochemistry* **1981**, *20*, 2571–2580.
172. Krieger, D. E.; Erickson, B. W.; Merrifield, R. B. *Proc. Natl. Acad. Sci. U.S.A.* **1976**, *73*, 3160–3164.
173. Merrifield, R. B.; Bach, A. E. *J. Org. Chem.* **1978**, *43*, 4808–4816.
174. Funakoshi, S.; Fukuda, H.; Fujii, N. *Proc. Natl. Acad. Sci. U.S.A.* **1991**, *88*, 6981–6985.
175. Ramage, R.; Raphy, G. *Tetrahedron Lett.* **1992**, *33*, 385–388.
176. Bianchi, E.; Sollazzo, M.; Tramontano, A.; Pessi, A. *Int. J. Pept. Protein Res.* **1993**, *42*, 93–96.
177. Sucholeiki, I.; Lansbury, P. T. *J. Org. Chem.* **1993**, *58*, 1318–1324.
178. García-Echeverría, C. *J. Chem. Soc., Chem. Comm.* **1995**, 779–780.
179. Ball, H. L.; Mascagni, P. *Int. J. Pept. Protein Res.* **1996**, *48*, 31–47, and earlier publications by these authors.
180. Chow, F.; Kempe, T.; Palm, G. *Nucleic Acids Res.* **1981**, *9*, 2807–2817.
181. Eadie, J. S.; Davidson, D. S. *Nucleic Acids Res.* **1987**, *15*, 8333–8349.
182. Farrance, I. K.; Eadie, J. S.; Ivarie, R. *Nucleic Acids Res.* **1989**, *17*, 1231–1245.
183. Giralt, E.; Rizo, J.; Pedroso, E. *Tetrahedron* **1984**, *40*, 4141–4152, and other publications from this research group.
184. Look, G. C.; Holmes, C. P.; Chinn, J. P.; Gallop, M. A. *J. Org. Chem.* **1994**, *59*, 7588–7590, and references therein.
185. Johnson, C.; Zhang, B. R. *Tetrahedron Lett.* **1995**, *36*, 9253–9256.
186. Fitch, W. L.; Detre, G.; Holmes, C. P.; Shoolery, J. N.; Keifer, P. *J. Org. Chem.* **1994**, *59*, 7955–7956.
187. Sarkar, S. K.; Garigipati, R. S.; Adams, J. L.; Keifer, P. A. *J. Am. Chem. Soc.* **1996**, *118*, 2305–2306.
188. Anderson, R. C.; Stokes, J. P.; Shapiro, M. J. *Tetrahedron Lett.* **1995**, *36*, 5311–5314.
189. Anderson, R. C.; Jarema, M. A.; Shapiro, M. J.; Stokes, J. P.; Ziliox, M. *J. Org. Chem.* **1995**, *60*, 2650–2651.
190. Yan, B.; Kumaravel, G.; Anjaria, H.; Wu, A.; Petter, R. C.; Jewell, C. F.; Wareing, J. R. *J. Org. Chem.* **1995**, *60*, 5736–5738.

191. Yan, B.; Kumaravel, G. *Tetrahedron* **1996**, *52*, 843–848.
192. Yan, B.; Sun, Q.; Wareing, J. R.; Jewell, C. F. *J. Org. Chem.* **1996**, *61*, 8765–8770.
193. Ito, H.; Ike, Y.; Ikuta, S.; Itakura, K. *Nucleic Acids Res.* **1982**, *10*, 1755–1769.
194. Bardella, F.; Giralt, E.; Pedroso, E. *Tetrahedron Lett.* **1990**, *31*, 6231–6234.
195. McCollum, C.; Andrus, A. *Tetrahedron Lett.* **1991**, *33*, 4069–4072.
196. Tregear, G. W. In *Chemistry and Biology of Peptides*; Meienhofer, J., Ed.; Ann Arbor Sci.: Ann Arbor, MI, 1972; pp 175–178.
197. Kent, S. B. H.; Merrifield, R. B. *Isr. J. Chem.* **1978**, *17*, 243–247.
198. Albericio, F.; Ruiz-Gayo, M.; Pedroso, E.; Giralt, E. *React. Polym.* **1989**, *10*, 259–268.
199. Berg, R. H.; Almdal, K.; Batsberg Pedersen, W.; Holm, A.; Tam, J. P.; Merrifield, R. B. *J. Am. Chem. Soc.* **1989**, *111*, 8024–8026.
200. Arshady, R.; Atherton, E.; Clive, D. L. J.; Sheppard, R. C. *J. Chem. Soc., Perkin Trans. I* **1981**, 529–537.
201. Atherton, E.; Gait, M. J.; Sheppard, R. C.; Williams, B. J. *Bioorg. Chem.* **1979**, *8*, 351–370.
202. Duckworth, M. L.; Gait, M. J.; Goelet, P.; Hong, G. F.; Singh, M.; Titmas, R. C. *Nucleic Acids Res.* **1981**, *9*, 1691–1706.
203. Atherton, E.; Brown, E.; Sheppard, R. C.; Rosevear, A. *J. Chem. Soc., Chem. Commun.* **1981**, 1151–1152.
204. Small, P. W.; Sherrington, D. C. *J. Chem. Soc., Chem. Commun.* **1989**, 1589–1591.
205. Kanda, P.; Kennedy, R. C.; Sparrow, J. T. *Int. J. Pept. Protein Res.* **1991**, *38*, 385–391, and references therein.
206. Mendre, C.; Sarrade, V.; Calas, B. *Int. J. Pept. Protein Res.* **1992**, *39*, 278–284.
207. Zalipsky, S.; Albericio, F.; Barany, G. In *Peptides: Structure and Function, Proceedings of the Ninth American Peptide Symposium*; Deber, C. M.; Hruby, V. J.; Kopple, K. D., Eds.; Pierce: Rockford, IL, 1985; pp 257–260.
208. Barany, G.; Albericio, F.; Solé, N. A.; Griffin, G. W.; Kates, S. A.; Hudson, D. In *Peptides 1992: Proceedings of the Twenty-Second European Peptide Symposium*; Schneider, C. H.; Eberle, A. N., Eds.; ESCOM Science: Leiden, 1993; pp 267–268.
209. Zalipsky, S.; Chang, J. L.; Albericio, F.; Barany, G. *React. Polym.* **1994**, *22*, 243–258, and references therein.
210. Barany, G.; Albericio, F.; Biancalana, S.; Bontems, S. L.; Chang, J. L.; Eritja, R.; Ferrer, M.; Fields, C. G.; Fields, G. B.; Lyttle, M. H.; Solé, N. A.; Tian, Z.; Van Abel, R. J.; Wright, P. B.; Zalipsky, S.; Hudson, D. In *Peptides—Chemistry and Biology: Proceedings of the Twelfth American Peptide Symposium*; Smith, J. A.; Rivier, J. E., Eds.; ESCOM Science: Leiden, 1992; pp 603–604.
211. Bayer, E.; Hemmasi, B.; Albert, K.; Rapp, W.; Dengler, M. In *Peptides: Structure and Function, Proceedings of the Eighth American Peptide Symposium*; Hruby, V. J., Rich, D. H., Eds.; Pierce: Rockford, IL, 1983; pp 87–90.
212. Bayer, E.; Rapp, W. In *Poly(ethylene Glycol) Chemistry: Biotechnical and Biomedical Applications*; Harris, J. M., Ed.; Plenum: New York, NY, 1992; pp 325–345. (review article)
213. Gao, H.; Gaffney, B. L.; Jones, R. A. *Tetrahedron Lett.* **1991**, *32*, 5477–5480.
214. Wright, P.; Lloyd, D.; Rapp, W.; Andrus, A. *Tetrahedron Lett.* **1993**, *34*, 3373–3376.
215. Bayer, E.; Bleicher, K.; Maier, M. *Z. Naturforsch., B* **1995**, *50*, 1096–1100.
216. Meldal, M. *Tetrahedron Lett.* **1992**, *33*, 3077–3080.
217. Auzanneau, F.-I.; Meldal, M.; Bock, K. *J. Pept. Sci.* **1995**, *1*, 31–44.
218. Renil, M.; Nagaraj, R.; Pillai, V. N. R. *Tetrahedron* **1994**, *50*, 6681–6688.
219. Winther, L.; Almdal, K.; Batsberg Pedersen, W.; Kops, J.; Berg, R. H. In *Peptides: Chemistry, Structure and Biology, Proceedings of the Thirteenth American Peptide Symposium*; Hodges, R. S.; Smith, J. A., Eds.; Escom Science: Leiden, 1994; pp 872–873.
220. Kempe, M.; Barany, G. *J. Am. Chem. Soc.* **1996**, *118*, 7083–7093.

221. Daniels, S. B.; Bernatowitcz, M. S.; Coull, J. M.; Köster, H. *Tetrahedron Lett.* **1989**, *30*, 4345–4348.
222. Valerio, R. M.; Bray, A. M.; Campbell, R. A.; Dipasquale, A.; Margellis, C.; Rodda, S. J.; Geysen, H. M.; Maeji, N. J. *Int. J. Pept. Protein Res.* **1993**, *42*, 1–9.
223. Maeji, N. J.; Valerio, R. M.; Bray, A. M.; Campbell, R. A.; Geysen, H. M. *React. Polym.* **1994**, *22*, 203–212.
224. Wen, J. J.; Klein, E.; Spatola, A. F. In *Peptides: Chemistry, Structure and Biology, Proceedings of the Thirteenth American Peptide Symposium*; Hodges, R. S.; Smith, J. A., Eds.; Escom Science: Leiden, 1994; pp 153–155.
225. Hudson, D.; Cook, R. M. In *Peptides: Chemistry, Structure & Biology, Proceedings of the Fourteenth American Peptide Symposium*; Kaumaya, P. T. P.; Hodges, R. S., Eds.; Mayflower Scientific Ltd.: Kingswinford, England, 1996; pp 39–41.
226. Vlasov, G. P.; Bilibin, A. Y.; Skvortsova, N. N.; Kalejs, U.; Kozhevnikova, N. Y.; Aukone, G. In *Peptides 1994. Proceedings of the Twenty-Third European Peptide Symposium*; Maia, H. L. S., Ed.; Escom Science: Leiden, 1995; pp 273–274.
227. Köster, H.; Heyns, K. *Tetrahedron Lett.* **1972**, *16*, 1531–1534.
228. Eichler, J.; Bienert, M.; Stierandova, A.; Lebl, M. *Pept. Res.* **1991**, *4*, 296–307.
229. Frank, R.; Döring, R. *Tetrahedron* **1988**, *44*, 6031–6040.
230. Eichler, J.; Beyermann, M.; Bienert, M. *Collect. Czech. Chem. Commun.* **1989**, *54*, 1746–1752.
231. Frank, R. *Tetrahedron* **1992**, *48*, 9217–9232.
232. Frank, R.; Heikens, W.; Heisterberg-Moutsis, G.; Blöcker, H. *Nucleic Acids Res.* **1983**, *11*, 4365–4377.
233. Ott, J.; Eckstein, F. *Nucleic Acids Res.* **1984**, *12*, 9137–9142.
234. Englebretsen, D. R.; Harding, D. R. K. *Int. J. Pept. Protein Res.* **1994**, *43*, 546–554, and references therein.
235. Neugebauer, W.; Williams, R. E.; Barbier, J.-R.; Brzezinski, R.; Willick, G. *Int. J. Pept. Protein Res.* **1996**, *47*, 269–275.
236. Hansen, P. R.; Holm. A.; Houen, G. *Int. J. Pept. Protein Res.* **1993**, *41*, 237–245.
237. Parr, W.; Grohmann, K.; Hägele, K. *Liebigs Ann. Chem.* **1974**, 655–666.
238. Tam, J. P.; Kent, S. B. H.; Wong, T. W.; Merrifield, R. B. *Synthesis* **1979**, 955–957, and references therein.
239. Matsueda, G. R.; Stewart, J. M. *Peptides* **1981**, *2*, 45–50.
240. Rink, H. *Tetrahedron Lett.* **1987**, *28*, 3787–3790.
241. Albericio, F.; Barany, G. *Int. J. Pept. Protein Res.* **1987**, *30*, 206–216.
242. Sharma, S. K.; Songster, M. F.; Colpitts, T. L.; Hegyes, P.; Barany, G.; Castellino, F. J. *J. Org. Chem.* **1993**, *58*, 3696–3699.
243. Wang, S. S. *J. Am. Chem. Soc.* **1973**, *95*, 1328–1333.
244. Mergler, M.; Tanner, R.; Gosteli, J.; Grogg, P. *Tetrahedron Lett.* **1988**, *29*, 4005–4008.
245. Barlos, K.; Chatzi, O.; Gatos, D.; Stavropoulos, G. *Int. J. Pept. Protein Res.* **1991**, *37*, 513–520.
246. Gait, M. J.; Singh, M.; Sheppard, R. C. *Nucleic Acids Res.* **1980**, *8*, 1081–1096.
247. Matteucci, M. D.; Caruthers, M. H. *J. Am. Chem. Soc.* **1981**, *103*, 3185–3191.
248. Gough, G. R.; Brunden, M. J.; Gilham, P. T. *Tetrahedron Lett.* **1981**, *22*, 4177–4180.
249. Schlatter, J. M.; Mazur, R. H.; Goodmonson, O. *Tetrahedron Lett.* **1977**, *33*, 2851–2852.
250. Jones, D. A., Jr. *Tetrahedron Lett.* **1977**, *33*, 2853–2856.
251. Anwer, M. K.; Spatola, A. *J. Org. Chem.* **1983**, *48*, 3503–3507.
252. Letsinger, R. L.; Kornet, M. J.; Mahadevan, V.; Jerina, D. M. *J. Am. Chem. Soc.* **1964**, *86*, 5163–5165.
253. Felix, A. M.; Merrifield, R. B. *J. Am. Chem. Soc.* **1970**, *92*, 1385–1391.
254. Dixit, D. M.; Leznoff, C. C. *Isr. J. Chem.* **1978**, *17*, 248–252.
255. Burdick, J. D.; Struble, M. E.; Burnier, J. P. *Tetrahedron Lett.* **1993**, *34*, 2589–2592.

256. Hauske, J. R.; Dorff, P. *Tetrahedron Lett.* **1995**, *36,* 1589–1592.
257. Dressman, B. A.; Spangle, L. A.; Kaldor, S. W. *Tetrahedron Lett.* **1996**, *37*, 937–940.
258. Pietta, P. G.; Cavallo, P. F.; Takahashi, K.; Marshall, G. R. *J. Org. Chem.* **1974**, *39*, 44–48.
259. Orlowski, R. C.; Walter, R.; Winker, D. *J. Org. Chem.* **1976**, *41*, 3701–3705.
260. Gaehde, S. A.; Matsueda, G. R. *Int. J. Peptide Protein Res.* **1981**, *18*, 451–458.
261. Blake, J.; Li, C. H. *Proc. Natl. Acad. Sci. USA* **1981**, *78*, 4055–4058.
262. Yamashiro, D.; Li, C. H. *Int. J. Peptide Protein Res.* **1988**, *31*, 322–334.
263. Canne, L. E.; Walker, S. M.; Kent, S. B. H. *Tetrahedron Lett.* **1995**, *36*, 1217–1220.
264. Bernatowicz, M. S.; Daniels, S. B.; Köster, J. *Tetrahedron Lett.* **1989**, *30*, 4645–4648.
265. Stüber, W.; Knolle, J.; Breipohl, G. *Int. J. Peptide Protein Res.* **1989**, *34*, 215–221.
266. Breipohl, G.; Knolle, J.; Stüber, W. *Int. J. Peptide Protein Res.* **1989**, *34*, 262–267.
267. Calmes, M.; Daunis, J.; David, D.; Jacquier, R. *Int. J. Peptide Protein Res.* **1994**, *44*, 58–60.
268. Boojamra, C. G.; Burow, K. M.; Ellman, J. A. *J. Org. Chem.* **1995**, *60*, 5742–5743.
269. Songster, M. F.; Vágner, J.; Barany, G. *Lett. Pept. Sci.* **1996**, *2*, 265–270.
270. Akaji, K.; Kiso, Y.; Carpino, L. A. *J. Chem. Soc., Chem. Commun.* **1990**, 584–586.
271. Kochansky, J.; Wagner, R. M. *Tetrahedron Lett.* **1992**, *33,*, 8007–8010.
272. Sheppard, R. C.; Williams, B. J. *Int. J. Peptide Protein Res.* **1982**, *20*, 451–454.
273. Bernatowicz, M. S.; Kearney, T.; Neves, R. S.; Köster, H. *Tetrahedron Lett.* **1989**, *30*, 4341–4344.
274. Alsina, J.; Rabanal, F.; Giralt, E.; Albericio, F. *Tetrahedron Lett.* **1994**, *35*, 9633–9636.
275. Kaljuste, K.; Undén, A. *Tetrahedron Lett.* **1995**, *36,* 9211–9214.
276. Chao, H.–G.; Bernatowicz, M. S.; Matsueda, G. R. *J. Org. Chem.* **1993**, *58*, 2640–2644.
277. Chao, H.–G.; Bernatowicz, M. S.; Reiss, P. D.; Klimas, C. E.; Matsueda, G. R. *J. Am. Chem. Soc.* **1994**, *116*, 1746–1752.
278. Thompson, L. A.; Ellman, J. A. *Tetrahedron Lett.* **1994**, *35*, 9333–9336.
279. Ramage, R.; Irving, S. L.; McInnes, C. *Tetrahedron Lett.* **1993**, *34*, 6599–6602.
280. Noda, M.; Yamaguchi, M.; Ando, E.; Takeda, K.; Nokihara, K. *J. Org. Chem.* **1994**, *59*, 7968–7975.
281. Sieber, P. *Tetrahedron Lett.* **1987**, *28*, 2107–2110.
282. Chan, W. C.; White, P. D.; Beythlen, J.; Steinauer, R. *J. Chem. Soc., Chem. Commun.* **1995**, 589–592.
283. Chan, W. C.; Mellor, S. L. *J. Chem. Soc., Chem. Commun.* **1995**,1475–1477.
284. Han, Y.; Barany, G. In *Innovation and Perspectives in Solid Phase Synthesis. Peptides, Proteins and Nucleic Acids. Biological and Biomedical Applications*, 1994. Epton, R., Ed.; Mayflower Worldwide Ltd.: Birmingham, England, 1994; pp 525–526.
285. Albericio, F.; Barany, G. *Tetrahedron Lett.* **1991**, *32*, 1015–1018, and references cited therein.
286. Flörsheimer, A.; Riniker, B. In *Peptides 1990, Proceedings of the Twenty-First European Peptide Symposium*; Giralt, E.; Andreu, D., Eds.; Escom Science Publishers: Leiden, 1991; pp. 131–133.
287. Riniker, B.; Flörsheimer, A.; Fretz, H.; Sieber, P.; Kamber, B. *Tetrahedron* **1993**, *49*, 9307–9320.
288. Fréchet, J. M. J.; Nuyens, L. J. *Can J. Chem.* **1976**, *54*, 926–934.
289. van Vliet, A.; Smulders, R. H. P. H.; Rietmann, B. H.; Tesser, G. In *Innovations and Perspectives in Solid Phase Synthesis. Peptides, Polypeptides and Oligonucleotides,1992*, Epton, R., Ed.; Intercept Ltd.: Andover, UK, 1992; pp 475–477.
290. Rietman, B. H.; Smulders, R. H. P. H.; Eggen, I. F.; Van Vliet, A.; Van De Werken, G.; Tesser, G. I. *Int. J. Peptide Protein Res.* **1994**, *44*, 199–206, and other publications from this research group.
291. Zhang, L.; Rapp, W.; Goldammer, C.; Bayer, E. In *Innovation and Perspectives in Solid Phase Synthesis. Peptides, Proteins and Nucleic Acids. Biological and Biomedical*

Applications, 1994. Epton, R., Ed.; Mayflower Worldwide Ltd.: Birmingham, England, 1994; pp. 717–722, and other publications from this research group.

292. Chen, C.; Randall, L. A. A.; Miller, R. B.; Jones, A. D.; Kurth, M. J. *J. Am. Chem. Soc.* **1994**, *116*, 2661–2662 .

293. Plunkett, M. J.; Ellman, J. A. *J. Org. Chem.* **1995**, *60*, 6006–6007.

294. Chenera, B.; Finkelstein, J. A.; Veber, D. F. *J. Am. Chem. Soc.* **1995**, *117*, 11999–12000.

295. Boehm, T. L.; Showalter, H. D. H. *J. Org. Chem.* **1996**, *61*, 6498–6499.

296. Eritja, R.; Robles, J.; Fernandez-Forner, D.; Albericio, F.; Giralt, E.; Pedroso, E. *Tetrahedron Lett.* **1991**, *32*, 1511–1514.

297. Albericio, F.; Giralt, E.; Eritja, R. *Tetrahedron Lett.* **1991**, *32*, 1515–1518.

298. Mutter, M.; Bellof, D. *Helv. Chim. Acta* **1984**, *67*, 2009–2016.

299. Liu, Y.-Z.; Ding, S.-H.; Chu, J.-Y.; Felix, A. M. *Int. J. Pept. Protein Res.* **1990**, *35*, 95–98.

300. Rabanal, F.; Giralt, E.; Albericio, F. *Tetrahedron* **1995**, *51*, 1449–1458.

301. Markiewicz, W. T.; Wyrzykiewicz, T. K. *Nucleic Acids Res.* **1989**, *17*, 7149–7158.

302. Murphy, A. M.; Dagnino, R., Jr.; Vallar, P. L.; Trippe, A. J.; Sherman, S. L.; Lumpkin, R. H.; Tamura, S. Y.; Webb, T. R. *J. Am. Chem. Soc.* **1992**, *114*, 3156–3157.

303. Fehrentz, J.-A.; Paris, M.; Heitz, A.; Velek, J.; Liu, C.-F.; Winternitz, F.; Martinez, J. *Tetrahedron Lett.* **1995**, *36*, 7871–7874.

304. Dinh, T. Q.; Armstrong, R. W. *Tetrahedron Lett.* **1996**, *37*, 1161–1164.

305. Wang, S. S. *J. Org. Chem.* **1976**, *41*, 3258–3261.

306. Rich, D. H.; Gurwara, S. K. *J. Am. Chem. Soc.* **1975**, *97*, 1575–1579.

307. Giralt, E.; Albericio, F.; Pedroso, E.; Granier, C.; Van Rietschoten, J. *Tetrahedron* **1982**, *38*, 1193–1208.

308. Hammer, R. P.; Albericio, F.; Gera, L.; Barany, G. *Int. J. Peptide Protein Res.* **1990**, *36*, 31–45, and references cited therein.

309. Greenberg, M. M.; Gilmore, J. L. *J. Org. Chem.* **1994**, *59*, 746–753.

310. Holmes, C. P.; Jones, D. G. *J. Org. Chem.* **1995**, *60*, 2318–2319.

311. Yoo, D. J.; Greenberg, M. M. *J. Org. Chem.* **1995**, *60*, 3358–3364.

312. Mullen, D. G.; Barany, G. *J. Org. Chem.* **1988**, *53*, 5240–5248.

313. Ramage, R.; Barron, C. A.; Bielecki, S.; Thomas, D. W. *Tetrahedron Lett.* **1987**, *28*, 4105–4108.

314. Farrall, M. J.; Fréchet, J. M. *J. Org. Chem.* **1976**, *41*, 3877–3882.

315. Brugidou, J.; Méry, J. *Peptide Res.* **1994**, *7*, 40–47, and earlier work by these authors cited therein.

316. Asseline, U.; Thuong, N. T. *Tetrahedron Lett.* **1989**, *30*, 2521–2524.

317. Gottikh, M.; Asseline, U.; Thuong, N. T. *Tetrahedron Lett.* **1990**, *31*, 6657–6660.

318. Kumar, P.; Bose, N. K.; Gupta, K.C. *Tetrahedron Lett.* **1991**, *32*, 967–970.

319. Kunz, H.; Dombo, B. *Angew. Chem., Int. Ed. Engl.* **1988**, *27*, 711–713.

320. Lloyd-Williams, P.; Jou, G.; Albericio, F.; Giralt, E. *Tetrahedron Lett.* **1991**, *33*, 4207–4210.

321. Seitz, O.; Kunz, H. *Angew. Chem., Int. Ed. Engl.* **1995**, *34*, 803–805.

322. Guibé, F.; Dangles, O.; Balavoine, G.; Loffet, A. *Tetrahedron Lett.* **1989**, *30*, 2641–2644.

323. Lloyd-Williams, P.; Merzouk, A.; Guibé, F.; Albericio, F.; Giralt, E. *Tetrahedron Lett.* **1994**, *35*, 4437–4440.

324. Kaljuste, K.; Undén, A. *Tetrahedron Lett.* **1996**, *37*, 3031–3034.

325. Lyttle, M. H.; Hudson, D. H.; Cook, R. M. *Nucleic Acids Res.* **1996**, *14*, 2793–2798.

326. Semenov, A. N.; Gordeev, K. Y. *Int. J. Peptide Protein Res.* **1995**, *45*, 303–304.

327. DeGrado, W. F.; Kaiser, E. T. *J. Org. Chem.* **1980**, *45*, 1295–1300.

328. DeGrado, W. F.; Kaiser, E. T. *J. Org. Chem.* **1982**, *47*, 3258–3261.

329 Findeis, M. A.; Kaiser, E. T. *J. Org. Chem.* **1989**, *54*, 3478–3482.

330. Scarr, R. B.; Findeis, M. A. *Peptide Res.* **1990**, *3*, 238–241.
331. Backes, B. J.; Ellman, J. A. *J. Am. Chem. Soc.* **1994**, *116*, 11171–11172.
332. Osborn, N. J.; Robinson, J. A. *Tetrahedron* **1993**, *49*, 2873–2884.
333. Sola, R.; Saguer, P.; David, M.L.; Pascal, R. *J. Chem. Soc., Chem. Commun.* **1993**, 1786–1788.
334. Pascal, R.; Chauvey, D.; Sola, R. *Tetrahedron Lett.* **1994**, *35*, 6291–6294.
335. Kiso, Y.; Fukui, T.; Tanaka, S.; Kimura, T.; Akaji, K. *Tetrahedron Lett.* **1994**, *35*, 3571–3574.
336. Hoffmann, S.; Frank, R. *Tetrahedron Lett.* **1994**, *35*, 7763–7766.
337. Maeji, N. J.; Bray, A. M.; Geysen, H. M. *J. Immunol. Meth.* **1990**, *134*, 23–33.
338. Bray, A. M.; Maeji, N. J.; Geysen, H. M. *Tetrahedron Lett.* **1990**, *31*, 5811–5814.
339. Pinori, M.; DiGregorio, G.; Mascagni, P. In *Innovation and Perspectives in Solid Phase Synthesis. Peptides, Proteins and Nucleic Acids. Biological and Biomedical Applications, 1994.* Epton, R., Ed.; Mayflower Worldwide Ltd.: Birmingham, England, 1994; pp 635–638.

4

Synthesis Tools for Solid-Phase Synthesis

John S. Kiely, Thomas K. Hayes, Michael C. Griffith,
and Yazhong Pei

Before a combinatorial library is constructed on a solid support, many decisions must be made. In this chapter, the processes needed to plan and carry out this construction, and the physical materials needed, such as solid supports, linkers, and building blocks, are described in detail. We address questions important to the conception of the library, including how to decide upon the variables of building block usage—their complexity, number, and method of attachment. Initial chemistry, reaction optimization, library production, mechanisms of cleavage from resin, and the concept of "libraries from libraries" are described.

C ombinatorial chemistry is a response to the need to increase the productivity of the medicinal chemist. The concept of combinatorial libraries seeks to expand the number of molecules prepared in a simultaneous or near-simultaneous manner while maintaining the control and fidelity existing in current solution-phase synthesis methods (*1*). This chapter will focus on describing the conceptual and experimental tools that the chemist uses to construct combinatorial libraries on solid supports. In particular, the focus will be on the processes needed to conceive, plan, and carry out the actual construction of the library and the physical materials needed (solid supports, linkers, and building blocks). Most of the issues addressed here are the same whether the library components are ultimately to be used while still attached to the support (*2*) or in solution after cleavage (*3*). The concerns and choices regarding library formats, deconvolution methods, and analytical techniques are covered by others within this book and will not be addressed here, although these directly impact the choices to be made regarding synthesis tools.

Combinatorial chemistry has resulted in a paradigm shift, with new emphasis placed on solid-phase methods. But the body of chemical knowledge, with the exception of peptide (*4, 5*) and oligonucleotide syntheses (*6, 7*), presupposes that the reaction will be carried out in solution. Methods exist for straightforward reaction monitoring and analyses of the products of solution-phase reactions, but on solid support the chemistry takes place in a more restrictive chemical and mechanical microenvironment, where there are fewer direct analysis–monitoring methods (*8*). The reaction rates and side reactions observed for solid-phase reactions are altered with respect to those observed in solution by the bulk of the polymer support molecule, and because one of the reactants is tethered (*9*).

Even with these limitations, there are advantages in performing reactions on solid phase. These benefits have been well described in the peptide and oligonucleotide literature, which is discussed in the preceding chapter in this book. Primarily, one has the ability to drive the reaction to completion using huge excesses of reagents while retaining a very easy method for removal of the excess and any reaction by-products. The solid support offers a means of easily controlling the need to either segregate or combine the individual members of the library. These advantages drive chemists to adopt solid-phase methodology for the preparation of combinatorial libraries despite the analytical challenges.

Library Selection Criteria

What solid-phase combinatorial library should the chemist build? To answer this question, the chemist must ask a series of questions to define the suitability of the library core structure and the synthetic scheme envisioned to produce it. The answers to these questions will allow the chemist to decide if a reaction scheme yielding a particular core structure is truly amenable to the creation of a library. The questions that need to be asked are as follows:

- Can the desired molecule be assembled in a reasonable number of reactions?

- Are there a sufficient number of building blocks available to reach the desired library size?

- Is there a logical attachment point to the solid support?

- Can the reaction conditions be optimized for the desired variety of substituents or functionalities?

- Is there reason to believe the structure would have desirable properties (i.e., biological activity)?

Once these questions are answered and the relative merits of several different libraries are considered, the chemist can select the first library to be constructed.

The chemist's focus will then turn to the efforts needed to reduce the chemical idea to practice. These efforts are the same as for any solution-phase synthesis. First, the chemist must show in initial experiments that the chemistry works as planned. Second, the reactions must be optimized to some acceptable level with regard to yield and reaction rate. Finally, the product(s) must be made in sufficient quantity for the intended purpose and proven to be of sufficient purity. The following sections will take up these issues in the order that the chemist faces them. Selection of the chemistry based on the criteria just described will to some degree also be influenced by the library format to be adopted for its presentation or deconvolution. Other chapters in this book deal with these issues, and the reader is referred to them for the appropriate discussions.

Initial Synthesis

At the most basic level chemists will need to first demonstrate that the planned synthesis works in their hands. It is likely that this will be carried out in solution using basic building blocks (i.e., those devoid of extraneous substituents). For the building block that will be linked to the support, some relatively large group is used to mimic the bulky attachment to the solid support. For example, a *t*-butyl ester or amide can be used where an amide or ester is to be the planned substrate-to-support linkage. Having demonstrated that the chemistry works under these model conditions, the next step is to transport the synthesis to the solid support and demonstrate that product can be isolated when the reaction is run on the resin. Examples of this are the use of a (4-ethyl)benzyl ester to correspond to a Merrifield resin ester linkage, an α-methyl phenacyl ester to mimic a Wang resin, or a benzhydryl amine as part of an amide bond to mimic a methylbenzhydrylamine (MBHA) resin linkage (Figure 1).

Optimization

A key to preparing a useful library is having a high degree of confidence that all the planned members of the library will be obtained in good yield. To accomplish this, one must perform all optimization work on resin. This seems to be underappreciated by chemists entering the field. Practitioners have learned, painfully, that any optimization done in solution may be of little value when applied to the solid-phase environment.

A second issue is the selection of trial building blocks to be used in the optimization. In addition to some standard molecule, one must use both sterically and electronically demanding building blocks. This is done in order to define the limits of the reaction under a standardized set of conditions. On the basis of the reaction mechanism, one will decide whether sterics or electronics or both are a major concern. When one knows the steric and electronic limits, and the number of available building blocks that are within these limits, the accessible size of the library is defined. With standard quantitative structure–activity relationship (QSAR) parameters for electronic and steric factors, one can do a reasonable job of selecting com-

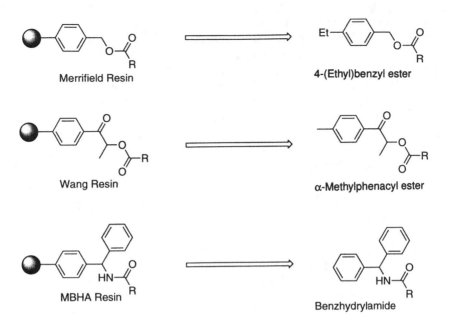

FIGURE 1. Bulky groups can be used in an initial synthesis to mimic resin.

pounds that fall inside the limitations the optimization has set for the reaction of interest.

This can be illustrated by the optimization of the synthesis of a series of 4*N*-(alkylcarboxamido)-3,4-dihydro-1(2*H*)-isoquinolones (*10*). On-resin formation of the ring by condensation of the homophthalic anhydride with the preformed imine was shown to work reasonably well with electron-rich and electron-poor aldehydes, with the exception of 4-hydroxybenzaldehyde, where an unacceptably low yield was encountered (Table I). The cause of the low yield was traced to a problem in the workup procedure, not in the condensation reaction, and was solved by altering the workup procedure. For the formation of the carboxamide/ester, investigation of the possible substituents demonstrated that oxygen nucleophiles and electron-deficient anilines were not sufficiently nucleophilic to undergo the reaction (Table I) (*11*).

From these optimization studies, it was possible to conclude that steric or electronic factors for aromatic aldehydes were not critical in achieving the desired condensation. It was found, however, that only amines that were not α-substituted could be used for the amine portion of the imine (data not shown). For the second reaction, the derivatization of the carboxylic acid of the isoquinolone, it was found that weak amine and oxygen nucleophiles did not undergo condensation well. These data serve to define the scope of the building blocks useful in this reaction sequence. From these limits it was possible to prepare a combinatorial library containing 11 amines at R^1, 38 aldehydes at R^2, and 51 amine nucleophiles at R^3, resulting in a library of 21,736 enantiomeric pairs.

TABLE I. Preparation of Isoquinolone Derivatives

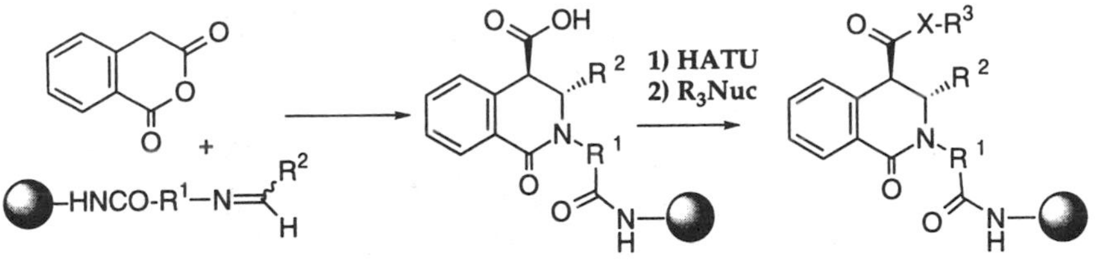

	Yield of Isoquinolone (%) Identity of R^2				*Yield of Carboxamide/Ester*	
Identity of R^1	$-(CH_2)_2-$	$-(CH_2)_3-$	$-(CH_2)_5-$	$-CH_2-$	*R^3Nuc*	*% Conversion*
C_6H_5	89	97	132	68	$(CH_3)_2CHOH$	0
4-HO-C_6H_4	11	11	15	18	$CH_3CH_2CH_2OH$	0
4-CH_3O-C_6H_4	87	58	78	51	$(CH_3)_2CHNH_2$	100
3,5-$(CH_3O)_2C_6H_3$	75	83	82	47	$(CH_3CH_2)_2NH$	100
4-CN-C_6H_4	50	69	74	52		
2-Br-C_6H_4	82	84	83	104	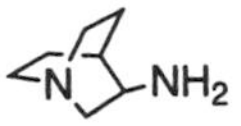	100
					$C_6H_5NH_2$	100
					(2-Et)$C_6H_4NH_2$	100
					4-$NO_2C_6H_4NH_2$	10

Libraries from Libraries

For maximum benefit from the decision to proceed with a particular chemistry, an additional question needs to be asked. Can the library, once prepared, be chemically transformed as a whole to produce a new library possessing altered physicochemical properties? Having expended the energy to create a library for a particular use, one should seek to achieve the greatest possible return on the investment. One means of doing this is by creating a library from an existing library using one or two additional chemical steps (*12*). For a chemist to accomplish this he must have prepared a sufficient quantity of a resin-bound library to divert a portion to a library-from-library synthesis. For the new library to be useful, it is also necessary to carry out a significant alteration in the physicochemical properties of the members of the library so that the new library can be expected to lead to new activities.

The synthesis can involve one of two types of chemical transformation. In a point transformation, the chemist modifies one selected functional group present on all members of the library. An example of the use of this technique is the recently reported transformation of an isolated olefin to an isoxazole ring (Scheme I) (*13*).

Second, in a global transformation all examples of a functional group within the library of a given class, even when there is more than one on a given molecule, are converted to a new group. This is illustrated by the complete reduction of the amides in a tripeptide library to secondary amines yielding a tetraamine library

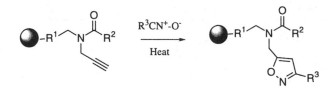

SCHEME I. An example of a point transformation. The olefins generated as part of the original library are converted to isoxazoles.

SCHEME II. Reduction of tripeptides to tetraamines as an example of one possible global transformation of a library.

(Scheme II) (*14*). With these tools, the use of existing, easily assembled building block pools can be maximized.

Individual Synthetic Tools

Having discussed the conceptual and preliminary experimental processes for building a combinatorial library on solid support, we now turn to a consideration of the individual synthetic tools needed to construct the combinatorial library. These tools include the actual solid support(s), the linker between the solid support and the attached building block(s), and the building blocks. Knowledge of the various attributes of each of these tools is important to the selection of the correct ones, and increases the probability of a successful library synthesis.

Solid Supports

The choice of support can be critical to successful solid-phase organic synthesis. The majority of work to date has been performed using either polystyrene cross-linked with 1–2% divinylbenzene (*15*) or a polyethylene glycol–polystyrene block copolymer (PEG–PS) (*16*). A variety of other supports have been developed for peptide synthesis, including polyamides (*17*), polyacrylamides (*18*) and cellulose (*19, 20*), but these have not been used for solid-phase organic synthesis thus far. Two additional supports have seen use in specialized cases: polyethylene functionalized with acrylic acid for synthesis on pins (*21*), and glass surfaces for libraries based on oligonucleotides and modified oligonucleotides (*22, 23*) as well as chip-based arrays (*24*).

Polystyrene cross-linked with divinylbenzene was originally used for solid-phase peptide synthesis, and it remains the most commonly used solid support. Its advantages include high chemical and physical stability, low cost, and relatively high loading capacity (0.3–0.9 mmole/g resin or higher). The main drawback of polystyrene–divinylbenzene is its high hydrophobicity; the use of polar protic solvents results in little resin swelling. Without resin swelling there is limited solvent/reagent accessibility to the reaction sites of the polymer that are on the interior channels of the resin bead.

The PEG–PS copolymers specifically address the limitations of polystyrene by having the reaction sites at the end of polyethylene glycol chains, resulting in a more hydrophilic environment around the reaction sites on the support. This results in high reagent accessibility in solvents such as methanol or water (*25*). For this reason, PEG–PS supports are used when aqueous reaction conditions are required and when resin-bound material will be directly bioassayed. Disadvantages of these copolymers versus polystyrene–divinylbenzene copolymer include lower chemical and physical stability, higher cost, and diminished loading capacity (0.2–0.3 mmole/g resin).

Linkers

A central element of solid-phase synthesis is the means to covalently attach the initial building block to the desired polymeric support. This is accomplished through a linker (*4, 5*). The linker consists of three parts: a resin-anchoring linkage, a building block anchoring/cleavable functional group, and a structural template that influences the stability of that functional group. A linker can consist of a single atom, a functional group, or a larger molecular structure. The linkers must provide bifunctionality whereby the resin and building blocks can be attached to either end of the linker. In addition to serving as a connection site to the resin, the linker performs the role of a protecting group (*26*) for the functional group (on the initial building block) that is used for the attachment. The basic factors that are important in choosing the linker are ease of attachment (including protection of the functional group of the first building block), stability to expected reaction conditions, and ease of cleavage. These principles are the same as those for solution-phase functional group protection. In solid-phase synthesis the linker acts as a protecting group attached to a large polymeric system.

The steric and electronic character of the linker structure and the nature of the chemical bonds within the linker will influence the utility of the linker. Ultimately, a decision concerning the choice of linker system is based primarily on the chemical properties relative to the anticipated reaction conditions and desired method of cleavage. Ideally, after cleavage the linker template should remain attached to the resin and be filtered away with the solid resin. If the linker is cleaved from the resin as a by-product, then this linker by-product should be readily removable (by filtration or evaporation) so as not to interfere with the final purity of the products. Likewise, any reagents used for cleavage should be readily removable. These last two points are important, as purification (extractions or chromatography) is impractical when a large library of compounds is being produced.

Linkers that are currently available for solid-phase organic chemistry have their roots in the solid-phase synthesis of peptides (5) and oligonucleotides (7). The linkers used in these areas rely mainly on ester or amide bonds to attach building blocks to the resin. A large variety of templates have been developed to deal with various difficulties encountered during the preparation of peptides or oligonucleotides. Even with the currently available linkers, new multipurpose and specialty linkers will be required to satisfy the increasingly varied chemistry that practitioners of solid-phase organic synthesis are attempting (1).

Already some variations on these general principles have been introduced to solve problems in combinatorial library development. The linker molecule may at times itself be the initial building block and serve as a temporary linkage point while the penultimate intermediate is constructed. An intramolecular cyclization of this penultimate intermediate then simultaneously forms the product and cleaves it from the support. Alternatively, linkers capable of release of a portion of the material from the resin (e.g., by photochemical means) have been used in screening/deconvolution methods (27, 28). Use of two or more orthogonal linker systems has allowed selective cleavage of the product and/or tagging fragments (29). (With orthogonal linkers the cleavage mechanism of each linker leaves the other unaffected.)

Categorization of linkers is based on their standard cleavage conditions, although many linkers may be fragmented by more than one means. The major classes of linkers release product by intramolecular cyclization, acid treatment, basic conditions (including nucleophiles and reduction), oxidative processes, safety-catch linker systems, and photochemical fragmentation. The following discussion presents a wide array of potential linker systems for carrying out solid-phase synthesis.

Intramolecular Release Linkers

In some cases building blocks can be connected directly to the resin support and then later disconnected by intramolecular cyclization to release the products (Scheme III). A key feature of this strategy is an anchoring functional group attached to the resin that will survive the assembly methods. It must still be sufficiently reactive to effectively release the desired products by intramolecular reaction. The penultimate products that fail to react remain bound to the resin and therefore do not contaminate the products. The preparation of benzodiazepines and hydantoins (30) illustrates of the use of an ester linkage that is cleaved intramolecularly to release the product from the resin. Tetrahydrofuran derivatives can be prepared by electrophilic cyclization of isoxazole substrates (31). A chiral pyrrolidine template can be converted by iodolactonization to yield γ-butyrolactones (32, 33). Formation of a diketopiperazine ring has been shown to release peptide products from a solid support (29). Reduction of an azide to an aminomethyl group results in an intramolecular cyclization to an N-hydroxymethylamide that releases the product compounds (34). The ability to release the final products by facile intramolecular cyclizations is a highly attractive strategy for combinatorial libraries.

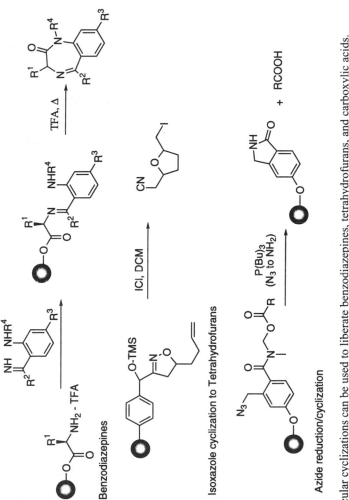

SCHEME III. Intramolecular cyclizations can be used to liberate benzodiazepines, tetrahydrofurans, and carboxylic acids.

Abbreviations: DCM, dichloromethane; ICl, iodine monochloride.

Acid-Cleavable Linkers

The largest class of linkers currently available are those that are cleaved by acid treatment (*35*) (Scheme IV). This class of linkers has been central to solid-phase peptide synthesis (*4*). The significant number of different acid-cleavable linkers can be attributed to the various problems encountered in solid-phase peptide synthesis. For the majority of the acid-labile linkers, treatment with acid results in the formation of a carbocation intermediate. The electronic structure of the linker has a strong effect on the stability of this carbocation, and thus on the ease of cleavage. Often a scavenger such as anisole is used in conjunction with the acid cleavage to minimize carbocation side-reactions. Presuming the products are stable to the conditions, tri-

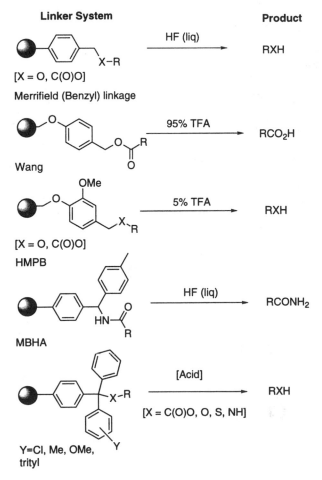

SCHEME IV. Acid-cleavable linkers, the cleavage reactions, and the products of cleavage.

Abbreviation: THP, tetrahydropyran.

Continued on next page

fluoroacetic acid (TFA) and hydrofluoric acid (HF) are normally the acids of choice, as they are easily removed by evaporation. Procedures using trifluoromethanesulfonic acid (TFMSA), hydrobromic acid (HBr), dilute hydrochloric acid, or acetic acids have all been used. The degree of acid sensitivity of a specific linker system may allow synthetic reactions to be conducted in mildly acidic conditions, followed by cleavage with a strong acid.

The classic linker used with Merrifield (hydroxymethyl polystyrene) resin is a benzyl ester or alcohol linkage that can be cleaved with strong acids (HF or TFMSA) (*4*). The Wang linker (4-hydroxymethylphenoxybutyric acid) is structurally related, but release requires less rigorous conditions (95% TFA) (*35*). The HMPB linker (4-hydroxymethyl-3-methoxyphenoxybutyric acid) (*36*) has an additional ortho methoxy group compared to the Wang linker, resulting in an extremely acid-sensitive functionality. The benzhydryl group is very useful as a linker template. MBHA (*p*-methylbenzhydryl) is probably the most popular resin linker of this type and is very extensively used for *N*-Boc-protected peptide synthesis (*4, 35*). MBHA is cleaved with HF or TFMSA, yielding free amides. The family of substituted trityl linker

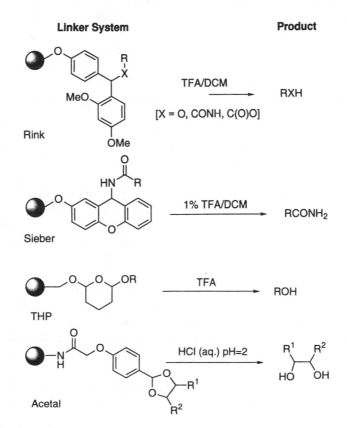

Linker System   **Product**

Rink

TFA/DCM

RXH

[X = O, CONH, C(O)O]

Sieber

1% TFA/DCM

RCONH₂

THP

TFA

ROH

Acetal

HCl (aq.) pH=2

SCHEME IV—*Continued.*

resis is suitable for the attachment of many types of functional groups. The trityl linkages are cleaved with mild acid and can be too unstable for some uses (*35*). The Rink acid linker (4-(2′,4′-dimethoxyphenylhydroxymethyl)-phenoxy resin) is very sensitive to weak acid conditions, while the Rink amide requires much higher concentrations of acid to effect disconnection (*35*). Another group structurally similar to the Rink linker, the Sieber linker, can be cleaved with dilute TFA (*35*). A dihydropyran-based linker can be used to attach alcohols to resin as a tetrahydropyran (THP) moiety. The alcohols are later released by mild acid treatment (*37*). Acetal linkage to a diol has been used in oligonucleotide synthesis with cleavage by treatment with aqueous acid (*38*). These examples indicate the wide range of acid-cleavable linkers that are available for use in solid-phase synthesis.

Linkers Cleavable with Bases, Nucleophiles, and Reducing Agents

Another important category of linker systems are those that can be fragmented by treatment with bases, nucleophiles, or reducing agents (*4, 35*) (Scheme V). Catalytic hydrogenation will detach benzyl (Merrifield) connections, although special techniques are required and this method is not commonly used. A number of the acid-labile linkers mentioned above contain a carboxylic acid group that can be connected to an amine or alcohol to give an ester or amide. An ester linkage to resin can be broken by basic hydrolysis to generate an acid, while reduction with metal hydrides yields alcohol derivatives. Nucleophilic disconnection of an ester leads to different products, depending on the identity of the nucleophile; amines generate amides, hydrazines produce hydrazides, and alkoxides give esters. Again, the structure of the linker template will influence the ease of cleavage. The ester of the 4-hydroxymethylbenzoic acid (HMBA) linker is stable to many acidic conditions, but it can be cleaved readily with bases or nucleophiles. The oxime linker is very reactive toward nucleophiles and can be cleaved from the solid support using amines to prepare amides (*35*). Several base-labile fluorene ring linkers have been studied, with the HMFS (*N*-[(9-hydroxymethyl)-2-fluorenyl] succinamic acid) linker offering stability to tertiary amines but rapid hydrolysis upon piperidine- or morpholine-DMF (dimethylformamide) treatment (*39*). Similarly, building blocks containing amines and alcohols can be attached to carboxyl-functionalized resins as esters or amides. Cleavage of these by any of the methods mentioned above will regenerate the amine or alcohol. Base-catalyzed β-eliminations from a carboxyethylsulfonyl linker (*40*) and nitrophenylethyl (NPE) linker (*41*) have also been described.

Like esters, amide linkages to resin can be hydrolyzed or undergo nucleophilic cleavage. The vigorous conditions needed for these cleavages, while unsuitable for standard peptide synthesis, may not be a problem for the synthesis of many classes of organic molecules. Aldehyde products can be generated from a Weinreb amide-derived linkage system by LiAlH$_4$ reduction (*42*). As with the acid-cleavable linkers, a number of options are available for the class of base-cleavable linkers with the possibility of functional group transformation via nucleophilic displacement or reduction.

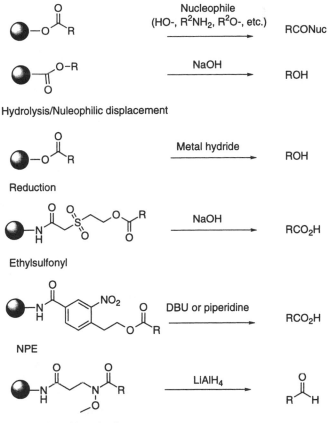

Hydrolysis/Nuleophilic displacement

Reduction

Ethylsulfonyl

NPE

Weinreb amide reduction

SCHEME V. Linkers cleavable with bases, nucleophiles, and reducing agents; the cleavage reactions; and the products of cleavage.

Abbreviation: DBU, 1,8-diazabicyclo[5.4.0]undec-7-ene.

Oxidative-Release Linkers

A few examples of linkages cleaved by oxidative processes have been reported to release products from resin (Scheme VI). As with safety-catch linkers, discussed in the next section, the oxidation normally converts a stable functionality to one that is more readily cleaved. Ceric ammonium nitrate oxidation of alkoxybenzoic acid linkers releases alcohol-containing products (*43*). Oxidation of a phenylhydrazide linker with a copper complex generates an unstable phenyldiimide, resulting in cleavage to an acid (*44*). Iodine-promoted oxidation of a thioether linkage has been used to release compounds from a solid support with subsequent formation of cyclic or symmetric peptide disulfides (*45*). Oxidative-release linkers provide an additional tool by which building blocks can be affixed to a solid support.

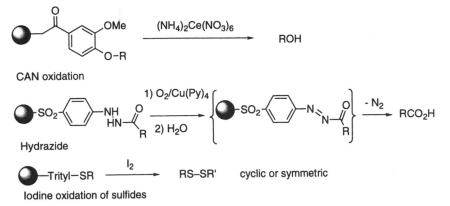

SCHEME VI. Oxidation-cleavable linkers, the cleavage reactions, and the products of cleavage. *Abbreviations:* CAN, ceric ammonium nitrate; Py, pyridine.

Safety-Catch Linker Systems

Linker systems in which a chemical group is transformed from a relatively stable connection to a readily cleavable functionality are termed safety-catch linkers (*46*) (Scheme VII). This type of linker is convenient for conducting a variety of manipulations that do not degrade the linkage, yet make disconnection easier. An example of this class of linkers involves the reduction of a bissulfone benzhydryl group to a bissulfide, which is very acid-sensitive (*47*). Other methylsulfinyl-type linkers, which can undergo one-pot reductive acidolysis, have been introduced (*48*). A linker containing an imidazole side chain that assists intramolecular ester hydrolysis represents a different approach (*49*). The loss of the carbamate moiety has a dramatic effect on the pH sensitivity of the linkage, with cleavage now occurring at neutral pH. Another safety-catch linker system entails the conversion of a phenyl carbamate to a cyclic acylurea (via an isocyanate), generating a linker that can be cleaved by aqueous base (*50*). Safety-catch linker systems allow for chemical reactions to be performed without the possibility of cleavage until after a conversion of the linker to a different functionality.

Photochemical-Release Linkages

The use of chemical bonds that can be broken by exposure to light is convenient because no additional reagents are necessary (Scheme VIII). Limited exposure times and knowledge of the rate of cleavage allow for the release of portions of the products. This partial cleavage of material is important in several deconvolution strategies. The α-methylphenacyl ester, prepared from the brominated Wang resin linker, can be cleaved by hydrolysis or nucleophilic attack but will also undergo photolysis (*51, 52*). Several nitrobenzyl ester linkers that can release product after light exposure have been developed recently (*53*). Photochemical cleavage provides some kinetic control in the release of product. There are some problems with the

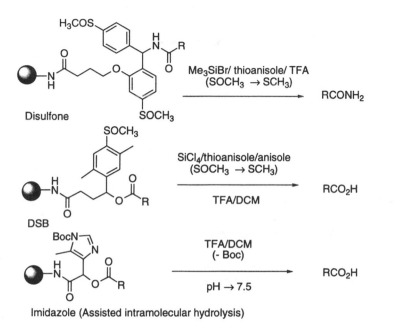

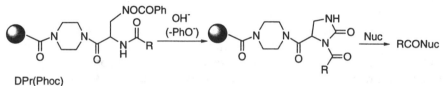

SCHEME VII. Safety-catch linkers, the cleavage reactions, and the products of cleavage. The safety catch is exposed by the first reaction, and only after this reaction can the simple cleavage reaction occur.

Abbreviations: DSB, 4-(2,5-dimethyl-4-methylsulfinylphenyl)-4-hydroxybutanoic acid; Dpr(Phoc), N^3-phenyloxycarbonyl-1-2,3-diaminopropionic acid; Nuc, nucleophile.

generation of reactive aldehydes as by-products of the cleavage, and care must be taken that these do not interfere with the product(s) (*54*).

Miscellaneous Linkers

There are several other linkers that do not fit into the major categories described above but provide additional options for conducting solid-phase chemistry. Included in this group are silicon-based linkers typically cleaved by fluorine, and allyl linkers cleaved by palladium catalysis.

Fluoride Cleavage. Silyl derivatives are in widespread use as protecting groups in organic synthesis (*26*), and therefore it is no surprise that a few such silicon-containing compounds have been adapted for solid-phase synthesis. The silyl linkage

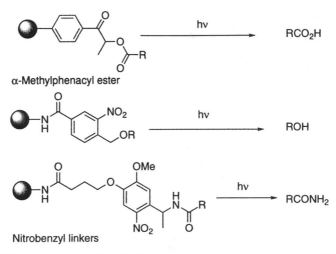

α-Methylphenacyl ester

Nitrobenzyl linkers

SCHEME VIII. Photochemically cleavable linkers and the products of cleavage. The α-methyl-phenacyl ester is derived from a brominated Wang resin.

Abbreviations: h, Planck's constant; ν, frequency of light.

is readily fragmented by treatment with fluorine or in some cases acidolysis (Scheme IX). The preparation and attachment of the silyl linker templates to the solid support are fairly elaborate. Thus, it may be difficult to develop this class of linker for use in practical solid-phase organic synthesis. Peptides were, however, synthesized with the Pbs linker, and fluoridolysis used to cleave the silicon–oxygen bond, releasing the peptide (*55*). Quinone methide is also released from the Pbs linker, and it reacts with the thiophenol scavenger. A similar fluorine-promoted cascade elimination, in which the quinone remains affixed to the resin, has also been published (*55*). The SAC (silyl acid) linker releases product by β-elimination (*56, 57*). Several benzodiazepine derivatives were prepared using an arylsilyl linker system that is cleaved with HF (*58*). Similar arylsilane derivatives that can be cleaved by either TFA, cesium fluoride (CsF), or liquid HF have been described (*59*). There appears to be a great deal of potential for silicon-based linker systems. The current challenge is in devising means to effectively couple a suitable silyl group to the solid support. The difficulty in coupling these groups to the support is in contrast to the ease with which most coupling reactions proceed when using standard linkers.

Allyl-Functionalized Linkers. Several systems that take advantage of the palladium-catalyzed fragmentation of allylic esters have appeared in the literature (*60–63*) (Scheme X). Treatment of the allyl ester linkers with Pd(0) in the presence of a nucleophile releases the acid from the solid support. These systems are stable to many acidic and basic conditions, but allow cleavage under neutral conditions.

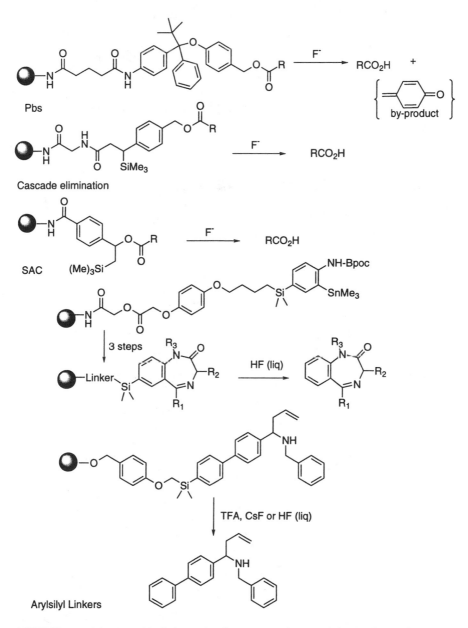

SCHEME IX. Fluoride-cleavable linkers, the cleavage reactions, and the products of cleavage. *Abbreviation:* Bpoc, 2-(4-biphenylyl)isopropyloxycarbonyl.

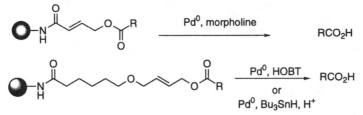

SCHEME X. Palladium-cleavable linkers, the cleavage reactions, and the products of cleavage. *Abbreviation:* HOBT, 1-hydroxybenzotriazole.

Building Blocks for Combinatorial Libraries on Solid Supports

Several criteria have to be considered in selecting building blocks for constructing a chemical library on a solid support. These criteria, which will be discussed separately, include:

- chemical compatibility of the synthetic processes and cleavage conditions;
- commercial availability of the building blocks;
- a large diversity base;
- proper protection of additional functionalities; and
- a proper point for attachment to the solid support.

Chemical Compatibility of the Synthetic Process and Cleavage Conditions

The selection of building blocks for a given library must be considered in conjunction with the design of the reaction sequence for the synthesis of the library. Unlike conventional organic syntheses, solid-phase organic synthesis requires the release of the final products from the solid phase (cleavage). Most of the cleavage conditions have been well developed and are highly chemoselective for the linkage bond(s). It should, however, be kept in mind that the final products must be stable under the cleavage conditions.

Commercial Availability

The building blocks should be readily available from commercial fine chemical suppliers or from a one- or two-step simple and high-yield preparation. Resources for locating suppliers are available in electronic form from Chemical Abstract Services databases or from Molecular Design Limited and Beilstein software packages. It is a common practice to use a large excess of reactants in the liquid media to drive the reaction to completion on the solid phase. As a result, large quantities of building blocks are needed for both methods development and library synthesis. The cost of lengthy preparations of a large number of building blocks will be prohibitive. On the other hand, for a library designed to probe some structure–activity relationship (SAR), it may well be worthwhile to synthesize a particular building block for mapping a diversity point of interest.

Large Diversity Base

As building blocks are the input points for the diversity of a given library, each type of building block used in the library construction should include a wide range of structural variations. These variations include hydrogen-bonding donors and acceptors, positive and negative charges, different chiralities, and hydrophobic and hydrophilic substituents. This will ensure not only that the library contains a large number of components, but also that it maps the maximal space within the diversity spectrum. A common practice when making a library is to incorporate as many of the reasonably available building blocks as is feasible. This method insures as broad a coverage as possible. Conventional medicinal chemistry has demonstrated time after time that minor modifications of a molecule can lead to major changes in its biological properties. Even with practice, it must be recognized that there will be unprobed regions in the diversity spectrum. This approach can be labor-intensive and costly. In addition, the library size can be so large as to complicate the use of some assay systems.

Another approach is to use computer modeling to assess the possible diversity represented by a given library in n-dimensional space, and to incorporate only the minimum number of representative building blocks that map the widest possible diversity. This approach, in theory, will avoid the repetition of diversity points and produce a smaller library. This may not, however, represent the whole diversity spectrum the library could offer, and an important activity may be missed. This technique is still in its infancy, and its effectiveness has yet to be validated.

Proper Protection of Additional Functionalities

Besides the functionality required for their attachment to the solid phase, building blocks may contain other functionalities that require protection. If these additional functionalities provide the attachment points for the incoming building blocks, their protecting groups and the process of deprotecting them must be compatible with the linker. In some cases, the functional group of a given building block can be transformed (by a change of the oxidation state) to another moiety, such as $-NO_2$ to $-NH_2$, providing the attachment point for the next building block. Protecting groups for functionalities that do not participate in the synthetic process should be removed prior to the cleavage of the final product from the solid phase without affecting the linker. If the by-products from the protecting groups have low boiling points and do not disrupt the chemistry of the library, they can also be removed during the cleavage of the final products from the solid phase.

Proper Point for Attachment to the Solid Support

Each library usually has its own unique core structure, which is either preformed or constructed during the library synthesis. In both cases, the attachment point of the core structure to the solid phase will affect the reaction sequence and the format of the library. Therefore, the first building block to be attached to the solid phase should be chosen with ease of synthesis and a simple library format in mind. Normally, the first building block should be at least bifunctional, with one functionality allowing attachment to the solid phase, and the other providing a connection point for the next building block. The nature of this stepwise process creates a situation in which the synthetic design will be linear in nature. As a consequence, solid-phase synthesis does not lend itself readily to convergent synthetic design.

Building blocks that have been used for solid-phase organic synthesis and the core structures of their final products are summarized in Table II.

Conclusions

We have tried to describe both the conceptual and experimental steps to be considered in devising a solid-phase synthesis. The "synthesis tools" that the chemist uses are first mental in nature and then physical. Both types of tools are required equally, and to slight one in preference to the other is certainly a prescription for unnecessary effort in completing a solid-phase synthetic scheme. To begin, one must evaluate a number of chemistries as to their suitability for both solid-phase synthesis and the final function of the molecule(s) to be synthesized. It is logical to expect that these evaluations will not give the same answers if the target is a single entity or is a combinatorial library. However, in either case the process is necessary. After the reaction planning and the evaluations of resins, linkers, and cleavage methods comes the actual exploration of the chemistry, first in terms of true feasibility and second to define conditions for the synthesis that are as close to optimal as possible. It is important to restate that the optimization efforts must be carried out on the solid support. To do otherwise will in all likelihood be a wasted effort. We suggest that the reader consider the points made by Crowley and Rapoport (9); these are as valid today as they were when written more than 20 years ago. The fundamental chemical principles laid out by these authors on the consequences of doing reactions on solid supports are unchanged and must be considered.

By considering the points given here, the practicing chemist can find solid-phase synthesis a useful technique to add to her or his arsenal for creating the desired molecule(s). Additional information on solid-phase synthesis has appeared in several recent reviews (78–80).

TABLE II. Building Blocks and Their Reported Uses

Building Blocks	Reactions	Products	Ref.
Amino acids	Coupling and deprotection	Peptides	64
Amino acids and alkylated peptides	Coupling, deprotection, and alkylation	N-alkyl halides	12
α-Bromoacetic acid and amines	Coupling and displacement	Peptoids	65
α-Bromoacetic acid and amines	Coupling, displacement, and cyclization–cleavage	Diketopiperazines and diketomorpholines	66
Amino acids and aldehydes	Coupling, deprotection, reductive amination, and cyclization–cleavage	Diketopiperazines	67
Alkynes or alkenes and nitroalkanes	Coupling, deprotection, displacement, and [3+2] cycloaddition	Isoxazoles or isoxazolines	13
Amino acids, aldehydes, and alkenes	Coupling, deprotection, imine formation, and [3+2] cycloaddition	Pyrrolidines	68
Amino acids, aldehydes, and acid chlorides	Coupling, deprotection, imine formation, and [2+2] cycloaddition	Azetidin-2-ones (β-lactams)	69
Amino acids and isocyanates	Coupling, deprotection, and acylation	Ureas and hydantoins	30
2-Aminophenones, amino acids, and alkyl halides	Coupling, deprotection, cyclization, and alkylation	Benzodiazepines	70
(2-Aminoaryl)stannane, acid chlorides, amino acids, and alkyl halides	Coupling, Stille coupling, deprotection, coupling, cyclization, and alkylation	Benzodiazepines	58
Amino acids, aldehydes, homophthalic anhydrides, and amines	Coupling, deprotection, imine formation, acylation–cyclization, and coupling	Isoquinolinones	11
trans-4-Bromo-2-butenoic acid, amines, and 2-iodoaroyl chlorides	Coupling, displacement, acylation, and cyclization/ Heck reaction	Isoquinolinones	71
β-Ketoesters, aldehydes, and 1,3-diketones	Condensation and condensation–cyclization (Hantzsch reaction)	1,4-Dihydropyridines	72
Dimethoxyphenylalanine and carboxylic acids	Coupling, acylation, Bischler–Napieralski cyclization, and reduction	Di- and tetra-hydroisoquinolines	73

Continued on next page

TABLE II—*Continued*

Building Blocks	Reactions	Products	Ref.
Amino acids, 1,2-diones, aldehydes, and ammonium acetate	Coupling and condensation–cyclization	Imidazoles	74
Cysteine, aldehydes, and acid chlorides	Coupling, deprotection, condensation, and acylation	Thiazolidines	75
Amino acids, aldehydes, and α- or β-mercapto-carboxylic acids	Coupling, deprotection, and condensation–cyclization	Thiazolidinones, 4-metathiazanones	76
β-Nitro-4-aminophenol and acid chlorides	Coupling, acylation, and reduction	Phenol derivatives	77

References

1. Terrett, N. K.; Gardner, M.; Gordon, D. W.; Kobylecki, R. J.; Steele, J. *Tetrahedron* **1995**, *51*, 8135–8173.
2. Lam, K. S.; Salmon, S. E.; Hersh, E. M.; Hruby, V. J.; Kazmierski, W. M.; Knapp, R. J. *Nature (London)* **1991**, *354*, 82–84.
3. Houghten, R. A.; Pinilla, C.; Blondelle, S. E.; Appel, J. R.; Dooley, C. T.; Cuervo, J. H. *Nature (London)* **1991**, *354*, 84–86.
4. Atherton, E.; Sheppard, R. C. *Solid Phase Peptide Synthesis: A Practical Approach*; IRL: Oxford, 1989.
5. *Solid-Phase Peptide Synthesis*; Barany, G.; Merrifield, R. B., Eds.; Academic: New York, 1979; Vol. 2.
6. *Oligonucleotides and Analogues: A Practical Approach*; Eckstein, F., Ed.; Oxford University: New York, 1991.
7. Beaucage, S. L.; Iyer, R. P. *Tetrahedron* **1992**, *48*, 2223–2311.
8. Fitch, W. L.; Detre, G.; Holmes, C. P. *J. Org. Chem.* **1994**, *59*, 7955–7956.
9. Crowley, J. I.; Rapoport, H. *Acc. Chem. Res.* **1976**, *9*, 135–144.
10. Cushman, M.; Madaj, E. J. *J. Org. Chem.* **1986**, *52*, 907–915.
11. Griffith, M. C.; Dooley, C. T.; Houghten, R. A.; Kiely, J. S. In *Exploiting Molecular Diversity and Solid-Phase Synthesis*; Chaiken, I.; Janda, K. D., Eds.; American Chemical Society: Washington, D. C., 1996; pp 50–57.
12. Ostresh, J. M.; Husar, G. M.; Blondelle, S. E.; Dorner, B.; Weber, P. A.; Houghten, R. A. *Proc. Natl. Acad. Sci. U.S.A.* **1994**, *91*, 11138–11142.
13. Pei, Y.; Moos, W. H. *Tetrahedron Lett.* **1994**, *35*, 5825–5828.
14. Cuervo, J. H.; Weitl, F.; Ostresh, J. M.; Hamashin, V. T.; Hannah, A. L.; Houghten, R. A. In *European Peptide Symposium*; Maia, H. L. S., Ed.; ESCOM Science: Braga, Portugal, 1994; pp 465–466.
15. Merrifield, R. B. J. *J. Am. Chem. Soc.* **1963**, *85*, 2149–2154.
16. Bayer, E. *Angew. Chem. Int. Ed. Engl.* **1991**, *30*, 113–129.
17. Dryland, A.; Sheppard, R. C. *J. Chem. Soc., Perkin Trans. 1* **1986**, 125–137.
18. Ashardy, E.; Atherton, E.; Gait, M. J.; Lee, K.; Sheppard, R. C. *J. Chem. Soc., Chem. Commun.* **1979**, 423–425.
19. Frank, R.; Doring, R. *Tetrahedron Lett.* **1988**, *19*, 6031–6040.

20. Lebl, M.; Stierandova, A.; Eichler, J.; Patek, M.; Pokorny, V.; Jehnicka, J.; Mudra, P.; Zenisek, K.; Kalousek, J. In *Innovation and Perspectives in Solid Phase Synthesis; 2nd Intl. Symposium*; Epton, R., Ed.; Intercept Ltd: Andover, 1992; pp 251–257.

21. Geysen, H. M.; Meleon, R. H.; Barteling, S. J. *Proc. Natl. Acad. Sci. U.S.A.* **1984**, *81*, 3998–4002.

22. Wyatt, J. R.; Vickers, T. A.; Roberson, J. L.; Buckheit, R. W.; Klimkait, T.; DeBaets, E.; Davis, P. W.; Rayner, B.; Imbach, J. L.; Ecker, D. J. *Proc. Natl. Acad. Sci. U.S.A.* **1994**, *91*, 1356–1360.

23. Hebert, N.; Davis, P. W.; DeBaets, E. L.; Acevedo, O. *Tetrahedron Lett.* **1994**, *35*, 9509–9512.

24. Fodor, S. P. A.; Read, J. L.; Pirrung, M. C.; Stryer, L.; Lu, A. T.; Solas, D. *Science (Washington, DC)* **1991**, *251*, 767–773.

25. Virgilio, A. A.; Ellman, J. A. *J. Am. Chem. Soc.* **1994**, *116*, 11580–11581.

26. Green, T. W.; Wuts, P. G. M. *Protective Groups in Organic Synthesis; Second Ed.*; John Wiley & Sons: New York, 1991.

27. Lebl, M.; Patek, M.; Krchnak, V.; Hruby, V. J.; Salmon, S. S.; Lam, K. S. *Int. J. Pept. Protein Res.* **1993**, *41*, 201–203.

28. Ohlmeyer, M. H. J.; Swanson, R. N.; Dillard, L. W.; Reader, J. C.; Asouline, G.; Kobayashi, R.; Wigler, M.; Still, W. C. *Proc. Natl. Acad. Sci. U.S.A.* **1993**, *90*, 10922–10926.

29. Kocis, P.; Krchnak, V.; Lebl, M. *Tetrahedron Lett.* **1993**, *34*, 7251–7252.

30. DeWitt, S. H.; Kiely, J. S.; Stankovic, C. J.; Schroeder, M. C.; Cody, D. M. R.; Pavia, M. R. *Proc. Natl. Acad. Sci. U.S.A.* **1993**, *90*, 6909–6913.

31. Beebe, L.; Schore, N. E. *J. Am. Chem. Soc.* **1992**, *114*, 10061–10062.

32. Moon, H.-S.; Schore, N. E.; Kurth, M. J. *J. Org. Chem.* **1992**, *57*, 6088–6089.

33. Moon, H.-S.; Schore, N. E.; Kurth, M. J. *Tetrahedron Lett.* **1994**, *35*, 8915–8918.

34. Osborn, N. J.; Robinson, J. A. *Tetrahedron* **1993**, *49*, 2873–2884.

35. *Novabiochem Combinatorial Chemistry Catalog and Solid Phase Organic Chemistry Handbook*; Calbiochem-Novabiochem Corp.: La Jolla, CA, 1996; Vol. February 1996.

36. McMurray, J. S.; Lewis, C. A. *Tetrahedron Lett.* **1993**, *34*, 8059–8062.

37. Thompson, L. A.; Ellman, J. A. *Tetrahedron Lett.* **1994**, *35*, 9333–9336.

38. Palom, Y.; Alazzouzi, E.; Gordillo, F. *Tetrahedron Lett.* **1993**, *34*, 2195–2198.

39. Rabanal, F.; Giralt, E.; Albericio, F. *Tetrahedron* **1995**, *51*, 1449–1458.

40. Katti, S. B.; Misra, P. K.; Hag, W.; Mathur, K. B. *J. Chem. Soc., Chem. Commun.* **1992**, 843–844.

41. Albericio, F.; Giralt, E.; Eritja, R. *Tetrahedron Lett.* **1991**, *32*, 1515–1518.

42. Fehrentz, J.-A.; Paris, M.; Heitz, A.; Velek, J.; Liu, C.-F.; Winternitz, F.; Martinez, J. *Tetrahedron Lett.* **1995**, *36*, 7872–7874.

43. Burbaum, J. J.; Ohlmeyer, M. H. J.; Reader, J. C.; Henderson, I.; Dillard, L. W.; Li, G.; Randle, T.; Sigal, N. H.; Chelsky, D.; Baldwin, J. J. *Proc. Natl. Acad. Sci. U.S.A.* **1995**, *92*, 6027–6031.

44. Semenov, A. N.; Gordeev, K. Y. *Int. J. Pept. Protein Res.* **1995**, *45*, 303–304.

45. Rietman, B. H.; Smulders, R. H. P. H.; Eggen, I. F.; Vliet, A. V.; Werken, G. V. D.; Tesser, G. I. *Int. J. Pept. Protein Res.* **1994**, *44*, 199–206.

46. Kenner, G. W.; McDermott, J.R.; Sheppard, M. R. C. *J. Chem. Soc., Chem. Commun.* **1971**, 636–637.

47. Patek, M.; Lebl, M. *Tetrahedron Lett.* **1991**, *32*, 3891–3894.

48. Kiso, Y.; Fukui, T.; Tanaka, S.; Kimura, T.; Akaji, K. *Tetrahedron Lett.* **1994**, *21*, 3571–3574.

49. Hoffman,S.; Frank, R. *Tetrahedron Lett.* **1994**, *35*, 7763–7766.

50. Sola, R.; Saguer, P.; David, M.-L.; Pascal, R. *J. Chem. Soc., Chem. Commun.* **1993**, 1786–1788.

51. Wang, S.-S. *J. Org. Chem.* **1976**, *41*, 3258–3261.
52. Tjoeng, F.-S.; Heavner, G. A. *J. Org. Chem.* **1983**, *48*, 355–359.
53. Holmes, C. P.; Jones, D. G. *J. Org. Chem.* **1995**, *60*, 2318–2319.
54. Sucholeiki, I. *Tetrahedron Lett.* **1994**, *35*, 7307–7310.
55. Mullen, D. G.; Barany, G. *J. Org. Chem.* **1988**, *53*, 5240–5248.
56. Chao, H.-G.; Bernatowicz, M. S.; Matsueda, G. R. *J. Org. Chem.* **1993**, *58*, 2640–2644.
57. Chao, H.-G.; Bernatowicz, M. S.; Reiss, P. D.; Limas, C. E.; Matsueda, G. R. *J. Am. Chem. Soc.* **1994**, *116*, 1746–1752.
58. Plunkett, M. J.; Ellman, J. A. *J. Org. Chem.* **1995**, *60*, 6006–6007.
59. Chenera, B.; Finkelstein, J. A.; Veber, D. F. *J. Am. Chem. Soc.* **1995**, *117*, 11999–12000.
60. Birr, C.; Nguyen-Trong, H.; Becker, G.; Muller, T.; Schramm, M.; Kunz, H.; Dombo, B.; Kosch, W. *Peptides* **1990**, 127–128.
61. Blankemeyer-Menge, B.; Frank, R. *Tetrahedron Lett.* **1988**, *29*, 5871–5874.
62. Kunz, H.; Dombo, B.; Kosch, W. *Peptides* **1988**, 154–156.
63. Guibe, F.; Dangles, O.; Balavoine, G. *Tetrahedron Lett.* **1989**, *30*, 2641–2644.
64. Pinilla, C.; Appel, J.; Blondelle, S.; Dooley, C.; Dorner, B.; Eichler, J.; Ostresh, J.; Houghten, R. A. *Biopolymers (Pept. Sci.)* **1995**, *37*, 221–270.
65. Zuckermann, R. N.; Kerr, J. M.; Kent, S. B. H.; Moos, W. H. *J. Am. Chem. Soc.* **1992**, *114*, 10646–10649.
66. Scott, B. O.; Siegmund, A. C.; Marlowe, C. K.; Pei, Y.; Spear, K. L. *Molecular Diversity* **1995**, *1*, 125–129.
67. Gordon, D. W.; Steele, J. *Bioorg. Med. Chem. Lett.* **1995**, *5*, 47–52.
68. Murphy, M. M.; Schullek, J. R.; Gordon, E. M.; Gallop, M. A. *J. Am. Chem. Soc.* **1995**, *117*, 7029–7030.
69. Ruhland, B.; Bhandari, A.; Gordon, E. M.; Gallop, M. A. *J. Am. Chem. Soc.* **1996**, *118*, 253–255.
70. Bunin, B. A.; Ellman, J. A. *J. Am. Chem. Soc.* **1992**, *114*, 10997–10998.
71. Goff, D. A.; Zuckermann, R. N. *J. Org. Chem.* **1995**, *60*, 5748–5749.
72. Gordeev, M. F.; Patel, D. V.; Gordon, E. M. *J. Org. Chem.* **1996**, *61*, 924–928.
73. Meutermans, W. D. F.; Alewood, P. F. *Tetrahedron Lett.* **1995**, *36*, 7709–7712.
74. Sarshar, S.; Mjalli, A. M. M.; Siev, D. *Tetrahedron Lett.* **1996**, *37*, 835–838.
75. Patek, M.; Drake, B.; Lebl, M. *Tetrahedron Lett.* **1995**, *36*, 2227–2230.
76. Holmes, C. P.; Chinn, J. P.; Look, G. C.; Gordon, E. M.; Gallop, M. A. *J. Org. Chem.* **1995**, *60*, 7328–7333.
77. Meyers, H. V.; Dilley, G. J.; Durgin, T. L.; Powers, T. S.; Winssinger, N. A.; Zhu, H.; Pavia, M. R. *Molecular Diversity* **1995**, *1*, 13–20.
78. Ellman, J. *Chem. Rev.* **1996**, *96*, 555–600.
79. Fruchtel, J. S., Jung, G. *Angew. Chem. Int. Ed. Engl.* **1996**, *29*, 132–143.
80. Hermkens, P. H. H., Ottenheijm, H. C. J., Rees, D. *Tetrahedron* **1996**, *52*, 4527–4554.

On-Resin Analysis in Combinatorial Chemistry

Michael J. Shapiro, Mengfen Lin, and Bing Yan*

The practice of combinatorial chemistry has revived interest in solid-phase synthesis and the analysis of reactions on polymer supports. IR, NMR, and mass spectroscopy has been used to monitor reaction progress and determine the structure of products. Single-bead IR techniques are more practical for reaction monitoring, whereas improved NMR techniques allow more complete structure determinations.

The need to isolate many discrete compounds in combinatorial chemistry makes solid-phase synthesis very attractive. Furthermore, with solid-phase synthesis, reactions can be driven to completion with large excesses of reagents. This chemistry demands the development of new detection methods, both to follow and optimize reactions, and to determine the structure of products, and in this chapter we discuss the ways that various detection methods have been developed and tailored to these new requirements.

Fourier-Transform Infrared (FTIR) Spectroscopy

Macro-FTIR or FT-Raman Techniques

FTIR spectroscopy has historically been accepted as a very powerful technique for determining the structures of chemicals. As with any other FT technique, many

---

*Address correspondence concerning IR spectroscopy to Bing Yan and NMR spectroscopy to Michael Shapiro.

spectral measurements are made over time, averaged, and the average transformed into the more customary frequency domain. The technique typically results in a far higher signal-to-noise ratio than with data collection in the continuous-wave (CW) mode, and therefore permits spectra to be obtained on smaller samples. The nondestructive nature of this technique, coupled with its extreme sensitivity, enables its effective use for both reaction monitoring and product analysis, especially when the chemistry involved is a functional group change.

Several FTIR techniques are valuable tools for monitoring reactions carried out on solid supports (1–3). Conventional FTIR techniques (macro-FTIR, using KBr pellets) have been used to monitor solid-phase reactions since the early 1970s. FTIR (4, 5) and FT-Raman (6) techniques have been used to detect the formation and conformation of peptide bonds in compounds on solid supports, on the basis of the analysis of amide I and amide III bands. Pioneers in solid-phase organic synthesis frequently relied on FTIR spectroscopy to determine the outcome of reactions (7–12). For example, unidirectional Dieckmann cyclizations (8) and the multistep synthesis of a solid support for the synthesis of oligosaccharides, poly[p-(1-propen-3-ol-1-yl)styrene], were both monitored in this way (9).

The advent of combinatorial chemistry has revived interest in solid-phase organic synthesis. FTIR has been used to monitor various solid-phase organic syntheses (13–16) and is once again an indispensable technique, which is now available to average laboratories. IR spectra can be obtained with about 10 mg of beads (a 1.0-mg sample contains approximately 10,000 beads with a diameter of 50 μm), which are treated in the same way as "regular" solid samples and then pressed as a KBr pellet for FTIR measurements. A more detailed description of the experimental procedure for KBr–IR spectroscopy has been given by Crowley and Rapoport (8). Spectra can also be obtained using swollen beads held between NaCl windows as described by Narita et al. (17).

To better follow reaction progress, the chemist needs an analytical tool that has the ease and speed of thin-layer chromatographic (TLC) analysis. The KBr pellet IR approach is not satisfactory for combinatorial chemistry based on resin beads for several reasons. First, although it can provide information on the final product, the KBr method consumes the sample. Second, the scale of solid-phase combinatorial synthesis is small, and the amount of sample required for the KBr method is usually too large to allow this method to be used for quality control at every reaction step. It is crucial to gain access to this kind of information rapidly in the early development phase of new solid-phase reactions. Third, the KBr method is useless for dealing with a collection of beads carrying compound libraries (mixtures). In compound libraries like those assembled by "split-and-pool" approaches, each bead carries one compound. To address these critical issues, a single-bead FTIR microspectroscopy (SBFTIRM) technique was developed (18). This method increases sensitivity by using a greatly reduced sampling area, and thus extracts information regarding functional group changes in a compound that is attached to a single resin bead.

Single-Bead FTIR Microspectroscopy (SBFTIRM)

The Method of SBFTIRM

To measure a spectrum from a single bead, an IR microscope accessory (commercially available from IR instrument manufacturers) is required. This accessory includes a regular transmission objective and an attenuated total reflection (ATR) objective (*see* the discussion later in this section). At any time point during a reaction, a tiny drop (reminiscent of a TLC sample) of resin suspension is taken out of the reaction mixture and subjected to washing and vacuum drying. A few beads are transferred onto a well-polished NaCl window. The small size of the sample can be seen in Figure 1, which compares the size of the beads to the width of a human hair.

By using the view mode and the *X–Y* platform of the microscope, the incident radiation is focused on a bead-free area of the NaCl window that is the size of the selected bead to collect the background spectrum, and also on a single resin bead to collect the bead spectrum. Transmission mode is used for these measurements. Typically, 32 or 64 scans are enough for a high-quality spectrum, with a spectrum recording time of one minute. A spectrum of a flattened bead is superior to that from a globular bead (*19*), possibly because the spectral distortion and broadening that is a result of the long and irregular path length of radiation passing through a

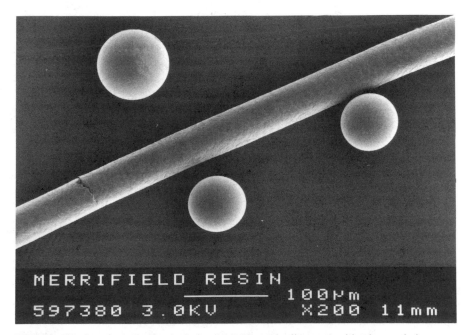

FIGURE 1. Size comparison of a single bead (<100 μm in diameter) with a human hair.

globular bead can be minimized by flattening the bead. A bead can be easily flattened by pressing it between two NaCl windows.

Single beads appear to be representative of the whole population of beads in a reaction mixture, as IR spectra obtained from each of ~30 different beads on which several different chemical reactions had been carried out indicated that all the beads from a given reaction showed the same degree of product formation.

Spectra can be collected from the partial surface of a bead by replacing the regular objective with an attenuated total reflection (ATR) objective. This objective uses an internal-reflection element with a higher refractive index than the sample to reduce the effective path length, thus restricting the sampling volume to a small area from the surface of a bead with a penetration depth of ~2 μm. Figure 2 shows the sampling volumes for ATR microspectroscopy and the regular transmission FTIR microspectroscopy method. The dark area shows the cross-section of the sampling volume. In the transmission mode, the IR beam passes through the whole bead, and the sampling volume is the whole bead. In the ATR mode, an ATR objective is in contact with a bead. The IR beam penetrates into the relevant part of the surface of the bead and is reflected with attenuated intensity. The minimal detectable signal from the partial surface of a single bead corresponds to about 130 fmol (1.3×10^{-13} mol) based on the detection of carbonyls (*20*). This amount is ~1/3850 of the total compound on a single bead with ~100 μm diameter and a 1.0 μmol g^{-1} loading. This result demonstrates that SBFTIRM is an extremely sensitive technique.

Advantages of SBFTIRM

Soon after SBFTIRM was introduced, it became a routine tool in our laboratories as well as many others. The practical value of this technique is that it gives a qualitative yes-or-no answer, a quantitative estimation of the reaction yield, and allows the time course of a reaction course to be followed (*see* the section "Following Reaction Time Courses"). Triple bonds and carbonyls serve as sensitive markers for chemical transformations on beads, but a wide range of functional transformations can be detected, including, more recently, heterocycle synthesis (*21*).

Yes-or-No Answers: SBFTIRM as the TLC Equivalent. When work is performed in solution, the progress of reactions is typically followed with TLC. SBFTIRM is an equivalent method for following solid-phase reactions, as shown in Figure 3 for the multistep synthesis described in the next two paragraphs.

The reactions shown were carried out on Rink amide resin **1** (Figure 3A). The success of the first step, which involves the creation of an amide linkage to form **2**, was followed by observing the formation of a band at 1665 cm^{-1} (compare Figure 3C with Figure 3B). Additional evidence for progress of the reaction is the appearance of the nitro frequency at 1540 cm^{-1}. The next step, the formation of **3**, is a nucleophilic replacement of a fluorine by a group containing a methyl ester. The success of this conversion is indicated by the observation of an ester carbonyl frequency at 1738 cm^{-1} (Figure 3D). The yields of these reaction steps cannot be directly determined

ATR   **Transmission**

FIGURE 2. Two-dimensional drawing of the sampling volumes (in black) in two SBFTIRM experiments. With an ATR objective (left), only a small area of the bead is sampled, and the penetration is shallow. In transmission mode (right), the whole bead (here shown flattened) is sampled.

from these IR spectra, but as the newly formed peaks are so distinct, the reaction time course can be monitored by taking a single-bead IR spectrum.

Quantitative Answers. When the signal of a functional group from the starting material disappears in a reaction step, an estimation of the transformation yield can be obtained. For example, for the reaction shown in Figure 4A, the hydroxymethyl polystyrene resin **4** (*see* Figure 4B) was converted into ester **5** (*see* Figure 4C). The yield was estimated to be >95% based on the complete loss of intensity from the –OH peak at 3450 cm^{-1}. Acid chloride **5** was converted to amide **6**, as shown in Figure 4D. This transformation can be followed by observing the appearance of the carbon–carbon triple bond and terminal alkyne stretches at 2120 cm^{-1} and 3288 cm^{-1}, as well as the disappearance of the acid chloride carbonyl at 1800 cm^{-1} and the observation of an amide carbonyl at 1670 cm^{-1}. These changes indicated that the conversion is nearly quantitative (>95%).

Following Reaction Time Courses. Solid-phase reactions are expected to take longer than their solution counterparts. Without good analysis tools, it is difficult to determine how much time is required for a reaction to reach the steady state. A method needs to be quick, sensitive and versatile to provide information on solid-phase reaction rates. SBFTIRM (*18, 20*) readily allows the time course of a reaction to be monitored (Figure 5). The rate of solid-supported reactions was found to be faster than expected for several of the reactions studied.

The rates of reaction on the surface of the bead (Figure 5B) and in the bead interior (Figure 5C) have been compared for several reactions, and no significant differences in reaction rates were found. The amount of compound sampled is calculated to be about 120 fmol for the surface experiment and about 500 pmol for the whole-bead experiments.

FTIR of Pins

Multipin synthesis (*22*) is a powerful technology for the parallel synthesis of discrete compounds. It also serves as a platform for rapid optimization of reactions

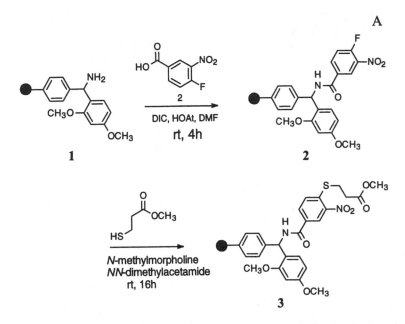

FIGURE 3. Single-bead FTIR spectra can give yes-or-no answers, indicating that reactions have occurred. (A) The multistep synthesis to be followed.

Abbreviation: rt, room temperature.

Continued on next page

intended for the synthesis of either peptide (*22, 23*) or nonpeptide (*24, 25*) compounds on solid supports. Analytical methods that do not require the cleavage of the compound from the pins are FTIR microspectroscopy using an ATR objective and regular FTIR using a macro-ATR accessory.

Figure 6 shows IR spectra recorded by these two methods for an uncoated polyethylene pin and a pin that has been coated with a reactive linker, 2-hydroxyethyl methacrylate (HEMA). Both methods yield high-quality spectra. An example of the use of this method was the observation of Fmoc deprotection using the Split-Pea method (*26*).

Future Perspectives

IR techniques have not been fully explored as a structural elucidation tool in the past, as they have been overshadowed by NMR and mass spectrometry techniques. Macro-FTIR techniques will continue to be used in relatively large scale synthesis because of the accessibility of the instrumentation in the average laboratory, but techniques used to monitor solid-phase reactions by FTIR or FT-Raman microspectroscopy are developing rapidly. SBFTIRM can identify and characterize changes in

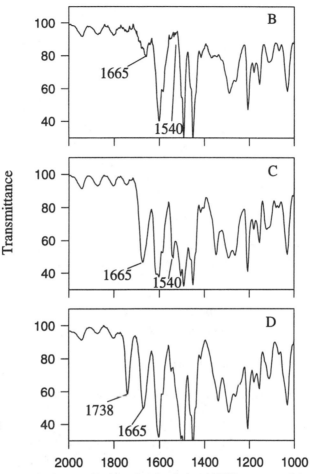

FIGURE 3—*Continued.* The panels show the single-bead FTIR spectra of the starting Rink amide resin **1** (B), the first product **2** (C), and the final product **3** (D). See the text for details.

many functional groups, and it is becoming one of the most sensitive on-bead analysis tools.

NMR Spectroscopy

Although FTIR techniques are quite sensitive and versatile, and mass spectroscopy provides confirmation of molecular weight, NMR spectroscopy is still the only method that can determine the detailed structure of a compound covalently bound to a resin support. Two techniques can be used to obtain NMR data for compounds

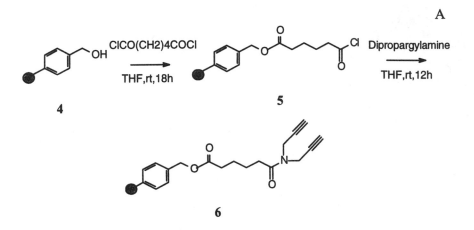

FIGURE 4. Single-bead IR spectroscopy can be used to monitor reaction products to give quantitative answers. (A) The reaction under study.

Continued on next page

bound to a resin: standard NMR using gel-phase samples (prepared as indicated in the section "Sample Preparation") and magic-angle spinning (MAS) NMR (*27*).

Standard NMR of Gel-Phase Samples

Standard NMR of gel-phase samples (gel-phase NMR) is generally limited to the detection of heteronuclei such as ^{13}C, ^{19}F, and ^{31}P, where there is a great deal of chemical shift dispersion. Gel-phase ^{13}C NMR is an established NMR technique dating back to 1971 (*28*). This method was first used to investigate the properties of swollen cross-linked polymers. In 1980 Epton et al., using a poly(acryloylmorpholine)-based matrix, showed that peptide synthesis could be followed by ^{13}C NMR (*29*). This matrix was chosen because it swells in highly polar solvents, allowing maximal access to the polymer surface. Epton et al. were able to follow the deprotection of a Boc group as well as the growth of the peptide chain. Leibfritz demonstrated that ^{13}C NMR can be used to determine the structure of peptides on polyethylene supports (*30*). Giralt et al. also investigated the general application of this method to monitor both growth of the peptide chain and the deprotection steps in the solid-phase synthesis of peptides (*31*). Narrow signals are obtained for carbons that are further away from the polymer backbone, and the NMR signals get broader as the peptide chain grows. The ^{13}C T_1 relaxation time correlates with the mobility of the peptide, which in turn can be used to gauge reaction time and hence to optimize reaction yield. The compound on the solvent-swollen resin exhibited relatively narrow lines in the ^{13}C spectrum. A typical carbon gel-phase spectrum is shown in Figure 7.

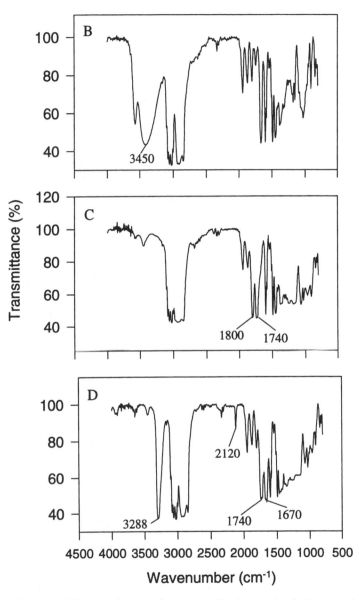

FIGURE 4.—*Continued.* The panels show the spectra for the starting hydroxymethyl resin **4** (B), the initial product **5** after reaction for 12 h (C), and the final product **6** after reaction for 18 h (D). See the text for details.

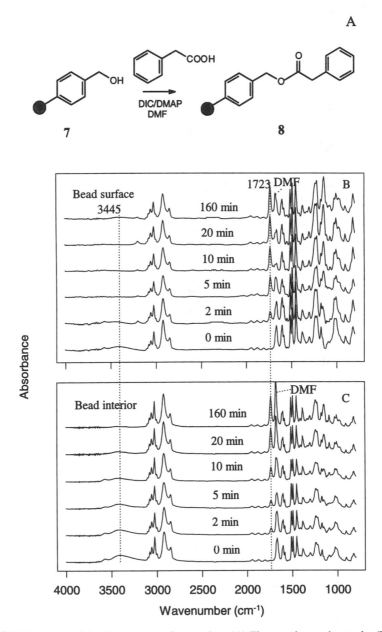

FIGURE 5. IR spectra of the time course of a reaction. (A) The reaction under study. (B) Spectra taken from the partial surface by FTIR microspectroscopy with an ATR objective. (C) Spectra taken from the whole bead using FTIR microspectroscopy in transmission mode. The sampling volumes used with the two different methods are shown in Figure 2. Dotted lines highlight the IR absorbance bands that either disappear (3445 cm^{-1}, attributable to the disappearing hydroxy group) or appear (1738 cm^{-1}, attributable to the emerging carbonyl group).

The first time that gel-phase NMR was used to follow organic solid-phase synthesis was in 1980, when Manatt et al. detected the degree of chloromethylation in a cross-linked polymer (*32*).

Jones and colleagues showed that ^{13}C NMR data could be obtained from a wide variety of organic substrates (*33*). Although they used 2% cross-linked resin, which generally gives rise to broader signals due to decreased mobility of the compounds, the spectral quality was acceptable. Interestingly, although they did not perform the experiment, they suggested the use of MAS NMR to enhance the data. The use of MAS was, however, to surface only some 12 years later! Application of gel-phase NMR to polymer-bound steroids has also been reported (*34*).

Sample Preparation

Gel-phase NMR is a reliable, nondestructive, simple technique that involves taking a spectrum in a standard NMR tube with a standard high-resolution NMR system. Between 50 and 100 mg of resin is placed in a standard 5-mm NMR tube. Sufficient solvent, usually 0.2–0.3 mL, is placed over the resin to cover the active region of the NMR sample. The sample can be vortexed or sonicated to produce a "homogeneous" gel. Data are accumulated using standard solution-state conditions. As the spectra that are obtained are usually not of the quality of standard solution data, the free induction decay (fid) can be worked up with considerable line broadening (4 Hz) to enhance the signal-to-noise ratio.

Real-Time Monitoring of Reactions

^{13}C NMR. As typical ^{13}C gel-phase NMR involves taking thousands of transients, it is impractical for the monitoring of reactions in real time. A method that is faster was, however, recently described (*35, 36*). This method uses an insert in the NMR tube to restrict the resin to the active region of the tube, so that only 20 mg of resin is needed, and uses ^{13}C labeling to monitor the reaction and further enhance sensitivity. Key atoms of the molecules are ^{13}C-labeled to give the best data without interference from solvent and the unenriched peaks from unlabeled atoms. The entire procedure can usually be completed in 30 min. After the removal of a small amount of resin, the resin is washed 3–4 times with the reaction solvent and then several times with methanol. The resin is filtered and then dried under high vacuum for 10–15 min. The appropriate solvent is added to the dried resin to make a slurry, which is placed into the NMR tube using a syringe. Data is collected for 64 transients on a 300-MHz instrument.

^{1}H NMR. It would be desirable to obtain ^{1}H NMR data in this manner, but the proton spectra under gel-phase conditions are generally broad and featureless. However, an example of high-resolution ^{1}H gel-phase NMR has been reported for an octapeptide on a polyacrylic resin swelled in dimethyl sulfoxide (DMSO) (*37*). The resolution of the spectrum was enhanced using a deconvolution method that allowed the observation of several of the peptide resonances.

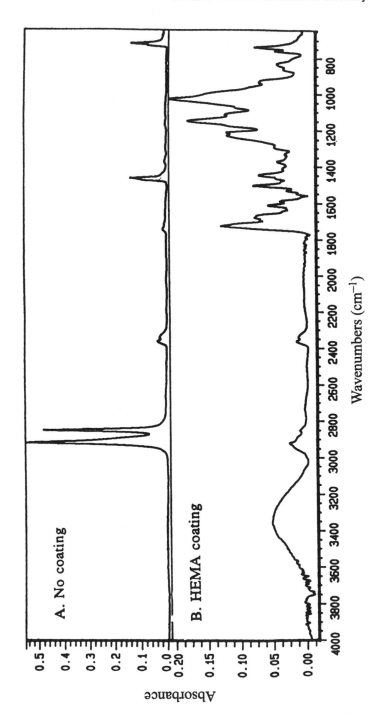

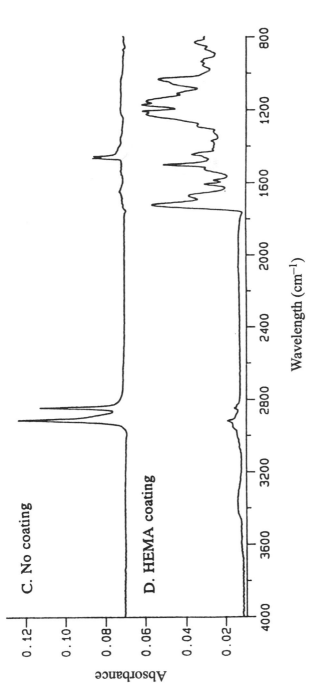

FIGURE 6. FTIR spectra of the surface of pinheads (crowns) collected using an ATR objective (A and B) or by a regular FTIR spectrometer coupled to a SplitPea accessory (C and D). The spectra in (A) and (C) are of an uncoated polyethylene pinhead, and those in (B) and (D) are of a grafted pinhead with 2-hydroxyethylmethacrylate (HEMA), a reactive linker support. Spectral accumulation time was ~1 min.

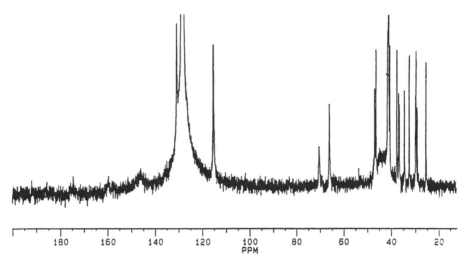

FIGURE 7. A typical standard NMR spectrum with a ^{13}C gel-phase sample.

^{19}F Gel-Phase NMR. Based on their early success with ^{13}C gel-phase results, Manatt et al. applied the same methodology to ^{19}F-containing compounds and obtained high-quality ^{19}F gel-phase data after incorporating ^{19}F into a protecting group on a peptide (*32*). They also discussed methods for the quantitative analysis of reaction progress in peptide synthesis. We have used similar techniques to monitor a substitution nucleophilic aromatic (SNAr) nucleophilic displacement reaction (Figure 8). Here fluorine is the leaving group, so ^{19}F NMR can be used to observe reaction progress. A standard Bruker quadranuclear probe was used for the experiment. Shown in Figure 8 are spectra for two time points (15 and 30 min) during the course of the reaction. The broad peak in the spectrum arises from fluorine-containing components of the probe and can be used as an external standard. The disappearance of the ^{19}F signal allows the reaction progress to be followed.

^{31}P Gel-Phase NMR. If phosphorus-containing reagents are used, reactions can be followed by gel-phase ^{31}P NMR, as shown first by Giralt and colleagues (*31*). This has proven to be a highly sensitive method of following reactions on solid phase (*39, 40*). Spectra can be obtained in just 10 min with 50 mg of resin in exactly the same manner as described for ^{13}C gel-phase NMR.

Magic-Angle Spinning (MAS) NMR

Gel-phase heteronuclear NMR is far from the perfect technique for monitoring polymer-bound reactions as it requires relatively large quantities of resin to obtain data, and proton NMR is of little value in solid-phase synthesis as the spectra have broad peaks. A variant of gel-phase NMR is MAS NMR, which involves spinning

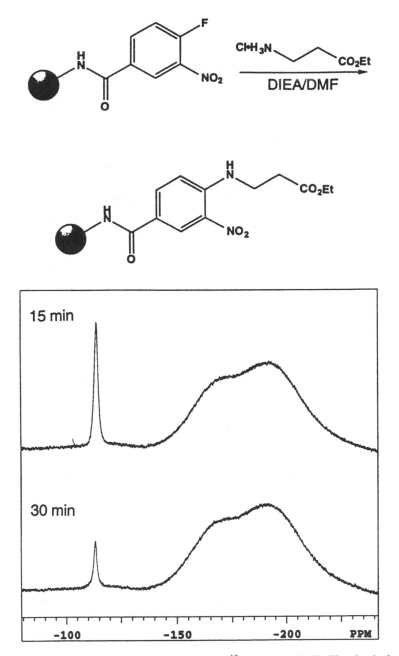

FIGURE 8. The monitoring of an SNAr reaction using ^{19}F gel-phase NMR. Fluorine is the leaving group in this nucleophilic displacement, so reaction progress can be followed by monitoring the loss of the narrow ^{19}F peak.

the sample at a relatively high speed at the "magic angle" relative to the magnetic field. The utility of ^{1}H MAS NMR in characterizing cross-linked networks of polystyrene gels was demonstrated in the early 1980s (41), but ^{1}H MAS NMR was first used for structure determination in combinatorial chemistry only recently (42, 43). We and others demonstrated that MAS NMR is a useful technique by which "solution quality" NMR data can be obtained for resin-supported molecules. An example of the quality of data that can be obtained is shown in Figure 9.

Unlike gel-phase NMR, MAS requires special equipment. MAS spectra can be obtained using either a Varian Nano probe or a Bruker high-resolution MAS accessory as shown in the schematic in Figure 10. Any standard MAS probe can be used to collect data, but the resolution will probably not be as good.

All solvent–resin combinations do not yield high-quality ^{1}H MAS NMR spectra. A useful starting point can be found in Table I (44). The table shows that the choice of resin remains the dominant factor in the quality of the data, with the choice of swelling solvent playing a secondary role. Good solvents include CD_2Cl_2, $CDCl_3$, C_6D_6, DMF-d_6, and DMSO-d_6. TentaGel resins generally give rise to the best quality data. It must be noted, however, that the identity of the best solvent not only relies on the nature of the resin but is also dependent on the attached compound (M. J. Shapiro, unpublished results). For instance, charged and neutral compounds would require different solvent conditions. The quality of the spectra can be enhanced by using presaturation (elimination of the resin signal by applying a small decoupling pulse to the signal prior to data collection), or by using spin-echo techniques (see Figure 11), which have the effect of reducing the intensity of broader signals.

^{13}C MAS NMR

In the late 1980s Fréchet showed that ^{13}C MAS data could readily be obtained on swollen gels with a standard MAS probe (45, 46), and that the data were superior to those collected using gel-phase NMR. ^{13}C MAS NMR was first used in combinatorial chemistry to follow a potential reaction complication, the production of two similar compounds (43). The resolution of the spectra was extremely good, and the two isomers could be easily distinguished. These spectra also demonstrated for the first time that structure determination on resin is important for understanding the course of chemical reactions, as the identity of the resin affected the products of the reac-

TABLE I. Spectral Quality for Various Resin–Solvent Combinations

Resin	CD_2Cl_2	DMF-d_6	DMSO-d_6	Benzene-d_6
Fmoc–Asp(OtBu)^a–Wang	10.0	8.7	ND	13.9
Fmoc–Asp(OtBu)–NovaSyn TGA	8.4	10.0	8.8	25.1
Fmoc–Asp(OtBu)–NovaSyn TGT	4.1	5.9	5.8	19.4
Boc–Asp(OBzl)^b–PAM	12.1	10.7	16.8	15.2

^{a}Spectral quality indicated by line widths of t-butyl groups (Hz).
^{b}Spectral quality indicated by line widths of methylene group of the benzyl moiety (Hz).

Abbreviation: ND, no data.

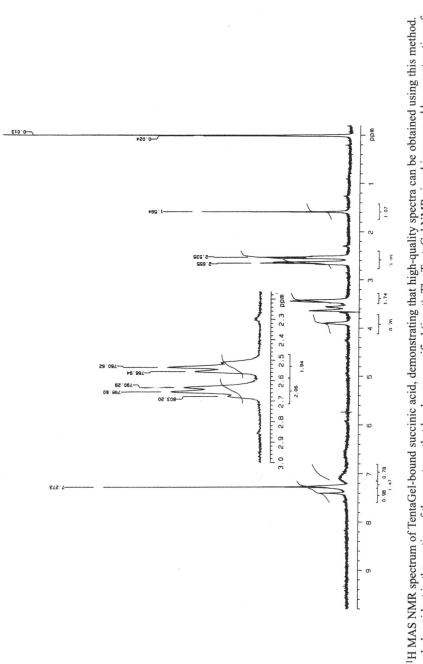

FIGURE 9. A ^{1}H MAS NMR spectrum of TentaGel-bound succinic acid, demonstrating that high-quality spectra can be obtained using this method. This is particularly evident in the portion of the spectrum that has been magnified (inset). The TentaGel NMR signal is removed by presaturation of the signals arising from the polyethylene glycol moiety.

A

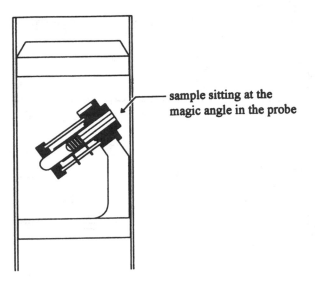

sample sitting at the
magic angle in the probe

B

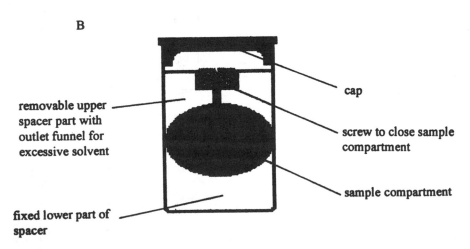

cap

removable upper
spacer part with
outlet funnel for
excessive solvent

screw to close sample
compartment

sample compartment

fixed lower part of
spacer

FIGURE 10. Schematics of the sample compartments of (A) a Varian Nano probe and (B) the Bruker high-resolution MAS accessory. Special features of these instruments are labeled.

A

B

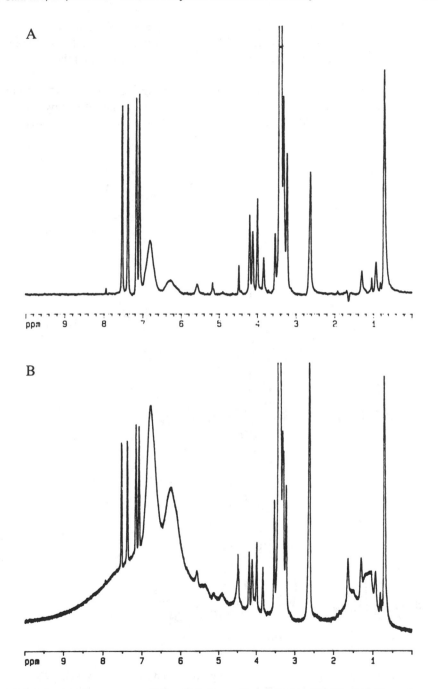

FIGURE 11. ^{1}H MAS spectrum of TentaGel–Fmoc–Ile swelled in CD_2Cl_2. (A) Normal spectrum. (B) Spin-echo spectrum.

tion. As shown in Figure 12, the ratio of products is clearly different after the reaction of norbornane carboxylic acid on Merrifield and Wang resins to produce the ester, with Wang resin yielding a 60:40 ratio of exo/endo product and the Merrifield resin an 80:20 ratio.

The improvement in spectra obtained when using ^{13}C MAS NMR to follow reactions can be seen in the comparison of a deprotection step monitored by standard gel-phase NMR or ^{13}C MAS NMR (Figure 13). Although the sample sizes are

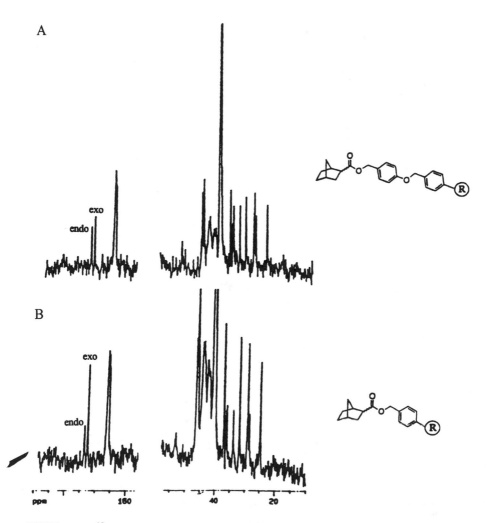

FIGURE 12. The ^{13}C MAS spectrum for a norbornyl compound on benzene-swollen Wang resin (A) and Merrifield resin (B). The spectra demonstrate that the esterification reaction has yielded a higher ratio of exo to endo product on the Merrifield resin (80:20) than on the Wang resin (60:40). The peaks corresponding to the two products are indicated.

roughly equivalent for the two experiments, there is a difference of almost two orders of magnitude in the time required for data collection (overnight for the gel-phase data and 20 min for the MAS). The spectral quality of the MAS data is exceptional.

The high quality of the data obtained in MAS NMR of swollen resins means that it is possible to use the same high-resolution NMR techniques as would be used if the samples were solutions (Figures 14–16) (*47*). Figure 14 shows the direct-observe CH-correlated data for the exo/endo norborane mixture. On the right, for comparison, is the solution data for norbornyl carboxylic acid. As can be seen, the correspondence between the MAS data and the solution NMR spectrum is excellent.

A more complete assessment of a structure can be obtained using a combination of data from heteronuclear multiple quantum coherence (HMQC) spectroscopy and total correlation spectroscopy (TOCSY), as shown for Wang–Fmoc–Lys–Boc in Figure 15. We have demonstrated that these typical 2D NMR structure determination tools can be used with swollen resins without problems even though we are performing the experiment using MAS. The combination of these data sets allows complete assignment of the ^{1}H and ^{13}C spectra and therefore confirmation of the structure.

As chemists, we like to be able to observe coupling constants directly so that we can better evaluate structure. For samples such as those coupled to Wang resin, the coupling constants are not generally visible, as the line widths approach or are greater than the *J*-values. Therefore, we cannot distinguish the two methyls in

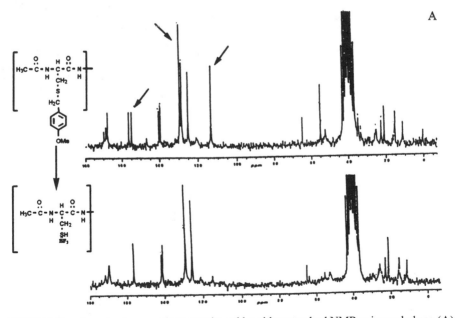

FIGURE 13. Two deprotection reactions monitored by either standard NMR using gel phase (A) or ^{13}C MAS NMR (B). Arrows indicate disappearing peaks.

Continued on next page

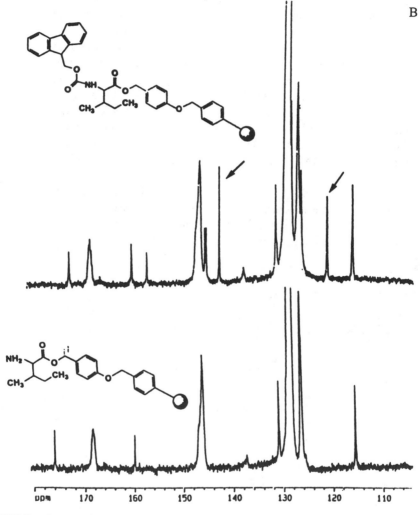

FIGURE 13—*Continued.*

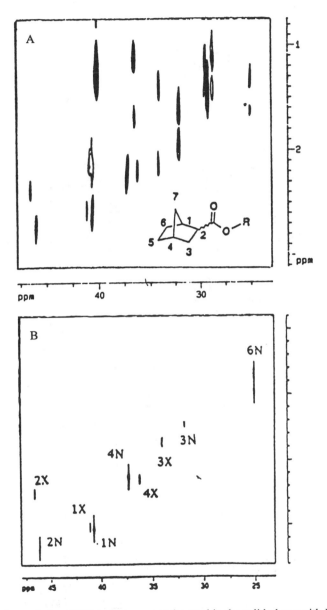

FIGURE 14. High-resolution NMR techniques can be used in the solid phase with MAS NMR. The figure shows direct-observe CH-correlated NMR spectra of (A) norborane on benzene-d_6-swollen resin, obtained using MAS conditions and (B) norbornyl carboxylic acid dissolved in benzene-d_6, obtained in solution conditions. The similarity of the two spectra shows that MAS brings NMR of solid-phase samples almost to the level of resolution achieved with solution-phase compounds.

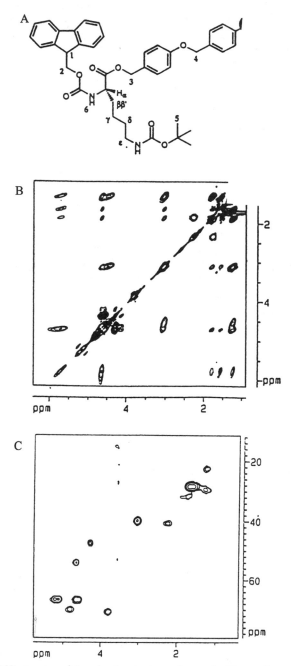

FIGURE 15. 2D NMR structure determination tools can be used with swollen resins by incorporating MAS techniques. The figure shows MAS NMR data obtained for benzene-swollen Wang–Fmoc–Lys–Boc [structure in (A)] using TOCSY (B) and HMQC (C).

TABLE II. Conditions for MAS NMR Experiments

NMR Experiment	*Conditions and Comments*
1D ^{1}H	16 scans with 1–3 mg of resin
1D ^{13}C	10 min for 50 mg of resin
CPMG spin echo	16–32 scans of 60 ms each
2D *J*-resolved	16 scans with a maximum of 64 increments
COSY, TOCSY, NOESY	8–32 scans using a 2048 × 512 data matrix
HMQC	32 scans using a 2048 × 256 data matrix and 50 mg of resin takes about 20 min.
HMBC	50 mg of resin takes about 2 h. Line width for protons should be <10 Hz for good results.

Abbreviations: 1D, one-dimensional; 2D, two-dimensional; CPMG, Carr–Purcell–Meibloom–Gill; COSY, correlation spectroscopy; NOESY, nuclear Overhauser enhancement spectroscopy; HMBC, heteronuclear multiple bond correlation.

Ile–Wang in the 1D MAS proton spectrum. Using the *J*-resolved H–H correlated 2D NMR spectrum, however, the doublet and triplet nature of the two methyl groups is revealed and the assignment can be made (Figure 16).

As an aid to performing the MAS NMR experiments described, a set of typical conditions is given in Table II.

Future Perspectives

It is now possible to generate an NMR spectrum of a single resin bead (*48*). Although this demonstrates the ultimate sensitivity of the MAS technique, it does not yet seem practical to use these techniques routinely, as it requires specific labeling and is very time consuming. Automation in sample switching and data collection will be increasingly important in the efficient use of MAS NMR in combinatorial chemistry because of the large number of samples that can be generated. As an illustration of the potential of this method, an unattended Bruker instrument now has the capability to perform MAS NMR on 30 samples.

Mass Spectrometry

The use of mass spectrometry to evaluate combinatorial chemistry is covered in Chapter 8. A discussion of one application of mass spectroscopy does, however, seem appropriate here. Mass spectrometry is an ideal method to monitor reactions as it is time- and material-efficient. While not using truly on-resin techniques, the following two examples represent an on-bead analysis. Mass spectrometry is considerably more sensitive than either NMR or macro-FTIR, and, as with SBFTIRM, the analysis can be performed on a single bead. Two very similar reports have appeared describing the use of matrix-assisted laser desorption/ionization time-of-flight (MALDI-TOF) mass spectroscopy for analysis of on-resin products. In both

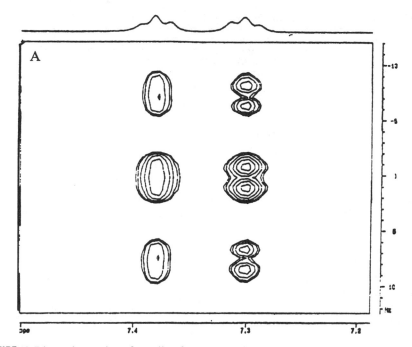

FIGURE 16. Direct observation of coupling for compounds on resins. Shown are illustrative HH J-resolved MAS spectra for the aromatic region of Fmoc in CDCl₃-swollen, Fmoc-protected TentaGel resin (A) and the methyl region of CD₂Cl₂-swollen Fmoc–Ile–Wang resin (B).

Continued on next page

cases a single bead (Rink resin) is removed from the reaction mixture (not a trivial task) and placed into a chamber that is then perfused with gaseous TFA to cleave small amounts of compound from the polymer (*49–51*). The bead is irradiated with a laser beam at 337 nm, producing the molecular ion from the cleaved compound. Only a small amount, ~1%, of the material is desorbed off the bead, so in principle the bead can be used for further testing. This method allows structural information to be obtained directly without the use of encoding strategies. The application of MALDI-TOF techniques to photocleavable linkers will add additional utility.

Acknowledgments

We acknowledge Paul Keifer of Varian Associates, Manfred Spraul of Bruker Instruments, and B. Fitch and G. Detre of Affymax for providing us with copies of their data. We also acknowledge the co-workers listed in references 18–21, 43, and 47 for contributions to this work, and P. Grosenstein and G. Argentieri for the photomicrographs of beads.

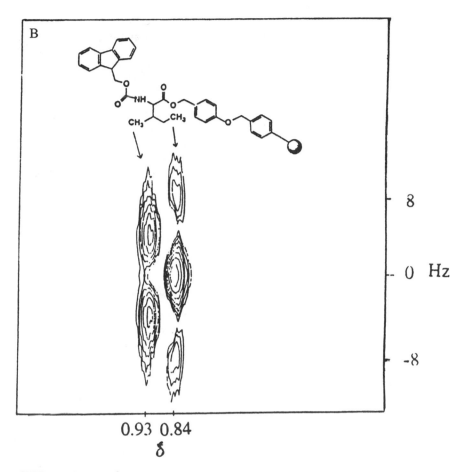

FIGURE 16—*Continued.*

References

1. Fréchet, J. M. *Tetrahedron* **1981**, *37*, 663–683.
2. Thompson, L. A.; Ellman, J. A. *Chem. Rev.* **1996**, *96*, 555–600.
3. Fruchtel, J. S.; Jung, G. *Angew. Chem. Int. Ed. Engl.* **1996**, *35*, 17–42.
4. Narita, M.; Honda, S.; Umeyama, H.; Ogura, T. *Bull. Chem. Soc. Jpn.* **1988**, *61*, 1201–1206.
5. Narita, M.; Isokawa, S.; Honda, S.; Umeyama, H.; Kakei, H.; Obana, S. *Bull. Chem. Soc. Jpn.* **1989**, *62*, 773–779.
6. Larsen, B. D.; Christensen, D. H.; Holm, A.; Zillmer, R.; Nielsen, O. F. *J. Am. Chem. Soc.* **1993**, *115*, 6247–6253.
7. Crowley, J. I.; Rapoport, H. *J. Am. Chem. Soc.* **1970**, *92*, 6363–6365.
8. Crowley, J. I.; Rapoport, H. *J. Org. Chem.* **1980**, *45*, 3215–3227.
9. Fréchet, J. M.; Schuerch, C. *J. Am. Chem. Soc.* **1971**, *93*, 492–496.

10. Leznoff, C. C.; Greenberg, S. *Can. J. Chem.* **1976**, *54*, 3824–3829.
11. Scott, L. T.; Rebek, J.; Ovsyanko, L.; Sims, C. L. *J. Am. Chem. Soc.* **1977**, *99*, 625–626.
12. Goldwasser, J. M.; Leznoff, C. C. *Can. J. Chem.* **1978**, *56*, 1562–1568.
13. DeWitt, S. H.; Kiely, J. S.; Stankovic, C. J.; Schroeder, M. C.; Reynolds Cody, D. M.; Pavia, M. R. *Proc. Natl. Acad. Sci. U.S.A.* **1993**, *90*, 6909–6913.
14. Chen, C.; Randall, L. A. A.; Miller, R. B.; Jones, A. D.; Kurth, M. J. *J. Am. Chem. Soc.* **1994**, *116*, 2661–2662.
15. Young, J. K.; Nelson, J. C.; Moore, J. S. *J. Am. Chem. Soc.* **1994**, *116*, 10841–10842.
16. Hauske, J. R.; Dorff, P. *Tetrahedron Lett.* **1995**, *36*, 1589–1592.
17. Narita, M.; Tomotake, Y.; Isokawa, S.; Matsuzawa, T.; Miyauchi, T. *Macromolecules* **1984**, *17*, 1903–1906.
18. Yan, B.; Kumaravel, G.; Anjaria, H.; Wu, A.; Petter, R.; Jewell, C. F., Jr.; Wareing, J. R. *J. Org. Chem.* **1995**, *60*, 5736–5738.
19. Yan, B.; Kumaravel, G. *Tetrahedron* **1996**, *52*, 843–848.
20. Yan, B.; Fell, J. B.; Kumaravel, G. *J. Org. Chem.* **1996**, *61*, 7467–7472.
21. Yan, B.; Gstach, H. *Tetrahedron Lett.* **1996**, *37*, 8325–8328.
22. Geysen, H. M.; Rodda, S. J.; Mason, T. J.; Tribbick, G.; Schoofs, P. G. *J. Immunol. Methods* **1987**, *102*, 259–274.
23. Bray, A. M.; Jhingran, A. G.; Valerio, R. M.; Maeji, N. J. *J. Org. Chem.* **1994**, *59*, 2197–2203.
24. Bunin, B. A.; Plunkett, M. J.; Ellman, J. A. *Proc. Natl. Acad. Sci. U.S.A.* **1994**, *91*, 4708–4712.
25. Virgilio, A. A.; Ellman, J. A. *J. Am. Chem. Soc.* **1994**, *116*, 11580–11581.
26. Gremlich, H.-U.; Berets, S. L. *Appl. Spectrosc.* **1996**, *50*, 532–536.
27. Fyfe, C. A. *Solid State NMR for Chemists*; CFC Press: Guelph, Ontario, 1983.
28. Sterlicht, H.; Kenyon, G. L.; Packer, E. L.; Sinclair, J. *J. Am. Chem. Soc.* **1971**, *93*, 199–208.
29. Epton, R.; Goddard, P.; Ivin, K. J. *Polymer* **1980**, *21*, 1367–1371.
30. Leibfritz, D.; Mayr, W.; Oekonomopulos, R.; Jung, J. *Tetrahedron* **1978**, *34*, 2045–2050.
31. Giralt, E.; Rizo, J.; Pedroso, E. *Tetrahedron* **1984**, *40*, 4141–4152.
32. Manatt, S. L.; Amsden, S. F.; Bettison, C. A.; Frazer, W. T.; Gudman, J. T.; Lenk, B. E.; Lubetich, J. F.; McNelly, E. A.; Smith, S. C.; Templeton, D. J.; Pinnell, R. P. *Tetrahedron Lett.* **1980**, *21*, 1397–1400.
33. Jones, A. J.; Leznoff, C. C.; Svirskaya P. I. *Org. Magn. Reson.* **1982** , *18*, 236–240.
34. Blossey, E. C.; Cannon, R. G.; Ford, W. T.; Periyasamy, M.; Mohanraj, S. *J. Org. Chem.* **1990**, *55*, 4664–4668.
35. Look, G. C.; Holmes, C. P.; Chinn, J. P.; Gallop, M. A. *J. Org. Chem.* **1994**, *59*, 7588–7590.
36. Look, G. C.; Murphy, M. M.; Campbell, D. A.; Gallop, M. A. *Tetrahedron Lett.* **1995**, *36*, 2937–2940.
37. Mazure, M.; Calas, B.; Cave, A.; Parello, J. C. *R. Acad. Sc. Paris* **1986**, *303*, 553–556.
38. Bardella, F.; Eritja, R.; Pedroso, E.; Giralt., E. *Bioorg. Med. Chem. Lett.* **1993**, *3*, 2793–2796.
39. Johnson, C. R.; Zhang, B. *Tetrahedron Lett.* **1995**, *36*, 9253–9256.
40. Tian, Z.; Gu, C.; Roeske, R. W.; Zhou, M.; Van Etten, R. L. *Int. J. Pept. Protein Res.* **1993**, *42*, 155–158.
41. Schneider, B.; Doskočilová, D.; Dubal, J. *Polymer* **1985**, *26*, 253–259.
42. Fitch, W. L.; Detre, G.; Holmes, C. P.; Schoolery, J. N.; Keifer, P. *J. Org. Chem.* **1994**, *59*, 7955–7956.
43. Anderson, R. C.; Jarema, M. A.; Shapiro, M. J.; Stokes, J. P.; Ziliox, M. *J. Org. Chem.* **1995**, *60*, 2560–2651.
44. Keifer, P. *J. Org. Chem.* **1996**, *61*, 1558–1559.

45. Stöver, H. D. H.; Fréchet, J. M. J. *Macromolecules* **1989**, *22*, 1574–1576.
46. Stöver, H. D. H.; Fréchet, J. M. J. *Macromolecules* **1991**, *24*, 883–888.
47. Anderson, R. C.; Shapiro, M. J.; Stokes, J. P. *Tetrahedron Lett.* **1995**, *36*, 5311–5314.
48. Sarkar, S.; Garigipati, R. S.; Adams, J. L.; Keifer, P. A. *J. Am. Chem. Soc.* **1996**, *118*, 2305–2306.
49. Bernatoicz, M. S.; Daniels, S. B.; Koster, H. *Tetrahedron Lett.* **1989**, *30*, 4645–4648.
50. Zambias, R. A; Boulton, D. A.; Griffin, P. R. *Tetrahedron Lett.* **1994**, 35, 4283–4286.
51. Egner, B. J.; Langley, G. J.; Bradley, M. *J. Org. Chem.* **1995**, *60*, 2652–2653.

Deconvolution Tools for Solid-Phase Synthesis

John J. Baldwin and Roland Dolle

Combinatorial libraries are being widely used to discover novel bioactive lead compounds and perform initial optimization steps on them. This approach can rapidly supply a significant database on which to establish a medicinal chemistry program. But without the ability to rapidly determine the structure of all compounds contained in a library, that library is of less value. In earlier deconvolution methods the synthetic reactions were reconstructed step by step, but chemical and electronic tags are increasingly being used as a less cumbersome alternative. The improvement of these tagging methods should result in the ability to encode larger libraries, providing inferred structures in a convenient and facile manner.

A library is usually built around a common structural motif and may possess as few as a dozen or up to millions of compound members. At the moment, the preferred methods for building libraries are parallel synthesis, split synthesis, or a combination of the two. Parallel synthesis is in essence an automated form of the traditional "one at a time" approach. It can use either solution or solid-phase chemistry and can be truly combinatorial; alternatively, individual compounds can be selected from a virtual library, a computer-generated library that contains the full range of combinatorial possibilities for synthesis. Libraries prepared by split synthesis are always combinatorial. In other words, a range of synthons (or reactants) are used at each step, in principle giving a library in which every possible compound from every possible combination of serial steps using these synthons is represented.

When the parallel approach is used, the structure of the product can be inferred from knowledge of the order of addition of reagents. In contrast, split synthesis and the related strategy of mixture synthesis produce a complex mixture of compounds, the structures of which are more difficult to determine. Tools designed to assist in structure determination include highly sensitive analytical methods (nuclear magnetic resonance (NMR) or mass spectroscopy), deconvolution using bioanalysis to guide resynthesis of interesting members, and encoding (introducing readily identifiable chemical modalities in parallel with the steps required for chemical synthesis). This chapter will cover the deconvolution and encoding methods that have been developed to determine the structure of members of combinatorial libraries.

Deconvolution Tools for Mixture Synthesis

The mixture synthesis approach to library generation has been applied principally to peptide libraries built on solid supports. In this synthetic strategy, one or more positions in a peptide chain are defined, and the remaining positions contain all members of a selected group of amino acids (*1–7*). For example, in a simple case of an 8000-membered peptide library, a mixture of all 20 natural amino acids is coupled to a solid support, such as a pin. After deprotection, a second set of 20 amino acids is coupled to the first set, yielding a mixture of all possible dipeptides (a total of 400). Twenty such pins can then be individually coupled to a single amino acid to give 20 sets of 400 tripeptides, with the third amino acid of each tripeptide being known. The most active tripeptide in this 8000-member set is usually identified biologically by the iterative resynthesis method of deconvolution.

The first step in iterative resynthesis is to identify which of the 20 sublibraries (contained on the 20 pins) is the most active. This determines which amino acid is preferred at position 3. The next step is to prepare a second tripeptide library in which the first position is random, and the last position is always the amino acid that is preferred at position 3. In this step it is the second position that is used to split the library into 20 sublibraries. Testing these sublibraries defines the preferred amino acid at position 2, and it is then a simple matter to synthesize and test the 20 possible tripeptides that have the correct preferred amino acid at both position 2 and position 3. Thus, the daunting problem of how to test the binding of each member of a library of 8000 tripeptides has been reduced to a test of two sets of 20 sublibraries (one for position 3, one for position 2) and one set of 20 individual tripeptides (*8, 9*).

Deconvolution Tools for Split Synthesis

In split synthesis, multiple copies of a single compound are built onto a single solid support particle in a series of reactions. The method was pioneered by Furka and by Lam (*10, 11*) for peptide synthesis but has since been extended to the preparation of nonpeptide libraries: it is now the preferred method for generating large combinatorial collections. In the first synthetic step, each of the synthons is added to its own pool of resin. After reaction, these pools are combined, mixed, and redivided (split) into a second set of pools. Each of these pools is then reacted with a different second-stage synthon. This process of reaction, pooling, mixing, and redividing is continued until the library is complete. Each bead then carries only one compound, but the mixture of beads that forms the library carries all possible compounds that could be made from the set of synthons used.

Iterative Resynthesis

To determine which of the library members are biologically active, researchers have used deconvolution by iterative resynthesis, microanalysis, or encoding. The method of iterative resynthesis for split synthesis is similar to that used in mixture synthesis. In the simplest case, the final sublibraries formed on the last step of the synthesis are not pooled but are tested as separate mixtures. The determination of which pool is most active then defines which synthon is preferred in the last synthetic step. The synthesis is then repeated to the penultimate step, and the unmixed pools are then reacted with the "best" synthon from the last step and tested separately. This reductive process is repeated until the structure of the most active library member is defined.

Microanalytical Methods

As each resin bead holds only one compound, an alternative strategy, in principle, is to use highly sensitive microanalytical methods such as mass spectrometry to determine the structure of the compound directly from the bead. A typical resin bead used in split synthesis can release ~200 pmol of compound for biotesting and structural analysis. Some success has been achieved with this approach, but it is not yet clear whether this method is generally useful and reliable for large libraries of non-oligomeric compounds. Innovative approaches under development to increase the usefulness of this strategy include the judicious selection of pool members to minimize molecular weight redundancy, and the use of capping reagents so that truncated sequences can be analyzed as well as final products (*12, 13*). In simple cases, on-bead analysis is possible using the magic-angle spinning (MAS) NMR technique. Complete assignment of structure has been made using MAS heteronuclear multiple-quantum coherence (HMQC) and total-correlation spectroscopy (TOCSY) (*14–16*).

Encoding Tools

Encoding techniques provide an alternative to the iterative and recursive methods of deconvolution. Simply stated, in encoding strategies chemical synthesis is performed on a solid support such as a bead, and simultaneously with the synthetic step a "tag" is coupled to the bead. Different readily identifiable tags are employed for each different reaction used to construct the library. This technique thus records the synthetic steps the bead has been subjected to, allowing one to identify individual compounds on single beads and documenting the steps that lead to that particular compound.

Figure 1 shows an example in which reaction x is carried out introducing synthon X on a bead. In a separate tagging step, tag A is attached to the bead. This is followed by reaction y, generating compound X-Y, and the introduction of tag B. Subsequent reaction z and the introduction of Z followed by tag C yields an "encoded bead". Decoding the sequence A-B-C from that bead informs the chemist or biologist that a chemical substance having the structure X-Y-Z is present on the bead. One can consider A-B-C as a bar code for compound X-Y-Z (or, more accurately, a bar code for the reaction events x, y, and z). As indicated in Figure 1, tags can be added as a "growing chain" (Figure 1A) or in a nonsequential fashion (Figure 1B).

The advantages of encoding techniques are threefold. First, encoding techniques avoid the need for traditional physical and spectroscopic characterization of the compounds. This is important as only tiny amounts of compounds (<1 nmol) are generated on single beads, and thus it may not be possible to provide for biological analysis and determine the chemical structures reliably by NMR or mass spectroscopy. Second, encoding is well suited to split-pool synthesis, where the synthesis often involves several steps, and large (>100,000-compound) libraries are produced. Deconvolution using iterative or recursive methods is often difficult and unreliable with these large libraries because of the tremendous labor that would be involved synthetically and the additivity effects observed in bioanalysis (which can result in a mixture of weakly active library members giving the same result in a screen as a single member of greater potency). Third, classical deconvolution methods typically yield only the most active structure in a library. In contrast, encoding allows one to identify any and all structures of interest from a library. Decoding multiple beads provides a more complete structure–activity analysis of the library. This information is invaluable for the design and synthesis of follow-up libraries or individual compounds.

At present the four types of encoding tags used are:

1. DNA strands (*17, 18*).

2. Peptide strands (*19–21*).

3. Detachable molecular tags based on electrophoric aromatics (*22, 23*) and aliphatic dialkylamines (*24*).

4. Radio-frequency transponders (25, 26).

DNA Strands

Encoding with DNA strands was the very first tagging method to be developed, and it was used for the indexing of peptides on solid supports (*17, 18*). Oligonucleotide tag synthesis and peptide synthesis occur at separate sites on the bead. Upon completion of the solid-phase synthesis, an on-bead binding assay is performed. The beads of interest are decoded by a polymerase chain reaction (PCR) amplification of the tag followed by DNA sequencing. If multiple beads are to be decoded, an individual PCR reaction is required for each bead. As each individual compound will be

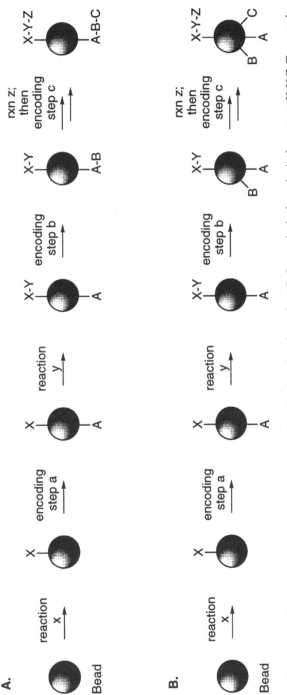

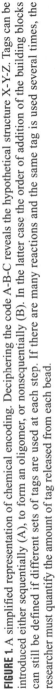

FIGURE 1. A simplified representation of chemical encoding. Deciphering the code A-B-C reveals the hypothetical structure X-Y-Z. Tags can be introduced either sequentially (A), to form an oligomer, or nonsequentially (B). In the latter case the order of addition of the building blocks can still be defined if different sets of tags are used at each step. If there are many reactions and the same tag is used several times, the researcher must quantify the amount of tag released from each bead.

represented by an oligonucleotide of unique sequence, there is no ambiguity regarding peptide structure.

A peptide library encoded with DNA strands would be constructed by first functionalizing solid-support beads with two different linkers: an oxybutyrate linker containing an acid-labile 4,4'-dimethoxytrityl- (DMT-) protected alcohol and a Knorr-carboxamide linker containing a base-sensitive 9-fluorenylmethoxycarbonyl- (Fmoc-) protected amine. These orthogonally protected linkers differentiate the sites for DNA tagging and compound synthesis on the bead (Figure 2). The DMT group is removed and the beads are subjected to automated oligonucleotide synthesis incorporating five or more nucleotide triplets (codons) representing a common PCR priming sequence. The Fmoc protecting group on the Knorr linker is removed, exposing the free amino group, and the beads are divided equally into a number of reaction vessels to begin split synthesis, with unique pairs of nucleotides encoding for each amino acid. Upon completion of peptide synthesis, the beads are pooled and subjected en masse to ~35 cycles of automated oligonucleotide synthesis to attach additional PCR priming sites and an ancillary oligonucleotide sequence homologous to the primer used for sequencing the template. For a library of seven amino acid peptides, a 69-mer oligonucleotide is required (Figure 3). A library of 1,000,000 heptapeptides prepared in this way was screened against a fluorescent antipeptide antibody, and ligands possessing nanomolar affinity for the artificial receptor were identified (17, 18).

The development of this technique marked the beginning of the tagging paradigm for encoded split synthesis. One great advantage of the DNA strand technology is the ability to amplify the tag to high levels. Only minute amounts of the DNA template are needed for PCR amplification, so these tags should be compatible with microscopic solid supports and miniaturized reaction vessels. Both natural and unnatural amino acids may be incorporated into such libraries, overcoming a major limitation of the bacteriophage peptide libraries, which can only display the natural L-amino acids. The DNA tagging method does, however, have potential shortcomings. First, the peptides cannot be removed from the beads, so only "on-bead" assays can be performed. The strand of DNA could interfere with the binding assay, leading to anomalous results when the decoded peptide is resynthesized and tested. Second, the DNA strands are fragile polymers. They are destroyed by heat, strong aqueous and organic acids or bases, Lewis acids, oxidants, and many other common reagents and reaction conditions that are used routinely in organic synthesis. Their utility for encoding libraries of nonpeptidic, small organic molecules therefore appears to be limited.

Peptide Strands

An alternative to DNA encoding is to use sequenceable peptide strands as the tag (19). Decoding of the tag is by amino acid microsequencing, which uses Edman degradation followed by HPLC analysis of the phenylthiohydantoin amino acid derivatives (Figure 4). As with DNA encoding, the peptide tags can be synthesized independently of the ligand using distinct protection schemes.

The first of two variations on this theme starts with beads containing the acid-labile Knorr linker (Figure 5) (*21*). The beads are treated with a lysine linker that has two functionalities protected by two distinct protecting groups. The base-labile Fmoc is the site for ligand synthesis, while the very acid labile *N*-[[2-(3,5-dimethoxyphenyl)prop-2-yl]oxy] (Ddz) is the site for encoding synthesis. Upon completion of the library synthesis, the beads are treated with strong acid, liberating the tethered ligand and coding peptide as one unit. The ligand–peptide complex is assayed for biological activity. As peptides are inherently biologically active, the coding strand may interact with the receptor or enzyme, giving rise to false positives. This phenomenon was not, however, observed in a peptide library containing non-natural amino acids (*21*).

The second variation can only be used to encode libraries of peptides, and in this case the tag and ligand are identical. This method involves the use of an "exclusive-release" linker, which has three sites for ligand attachment (*20*). The three attachment sites differ in their sensitivity to pH, and hence partial release of a peptide is possible. The structure of the linkers is shown in Figure 6, and the screening method (a two-step arraying process) that takes advantage of this unique linker is outlined in Figure 7. In the arraying process, ~500 beads from a library are placed in each well of a 96-well assay plate, and the beads are treated with dilute acid (pH 4.5 buffer). The acid cleaves one of the three linkages on the beads (Figure 6), and one-third of the bound peptide is released. These peptides are filtered into a derivative plate for assay. When a well with an active derivative is identified, the beads corresponding to that active well are redistributed in a new assay plate as single beads. The beads are treated with dilute base and the peptides that are released (an additional one-third) are transferred to a second derivative plate for assay. The active well, now containing only a single peptide, is identified, and the corresponding bead from the second assay plate is retrieved. The remaining portion of peptide still bound to the bead is subjected to Edman microsequencing for structure identification.

Using this method, a pentapeptide library was prepared (*20*) in which glycine was incorporated as the first residue followed by the random incorporation of 19 amino acids (excluding cysteine) into the remaining 4 positions. This produced a library of 19^4 or 130,321 peptides of different sequence, from which peptide ligands for a monoclonal antibody were successfully identified.

The use of amino acids for encoding requires that >1 pmol of peptide tag be present on the bead, because of the limits of Edman microsequencing. Care must be taken in that the peptide tags may racemize or react with common reagents used in small-molecule library synthesis, rendering them unreadable.

Molecular Tags

Molecular tags represent a significant advance over the previously described oligomeric strands. Unlike the DNA and peptide tags, which suffer from chemical sensitivity and/or detection limitations, these tags are chemically robust and are readily analyzed by sensitive chromatographic techniques. The two known types of

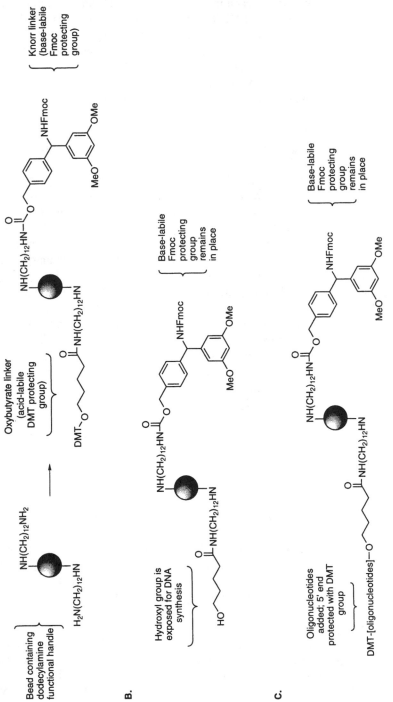

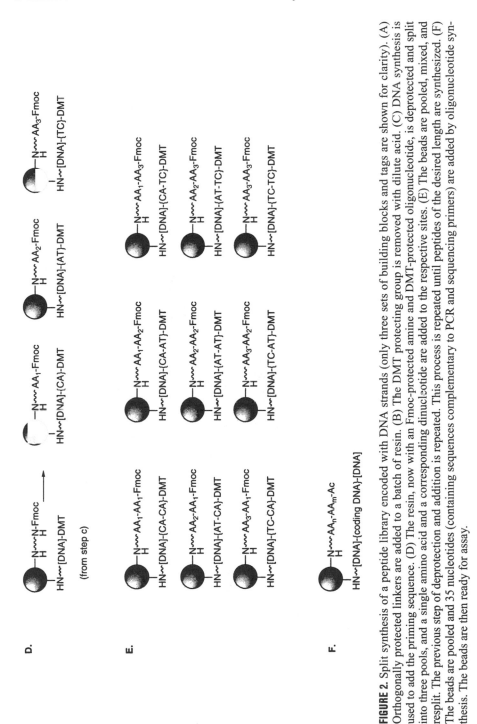

FIGURE 2. Split synthesis of a peptide library encoded with DNA strands (only three sets of building blocks and tags are shown for clarity). (A) Orthogonally protected linkers are added to a batch of resin. (B) The DMT protecting group is removed with dilute acid. (C) DNA synthesis is used to add the priming sequence. (D) The resin, now with an Fmoc-protected amine and DMT-protected oligonucleotide, is deprotected and split into three pools, and a single amino acid and a corresponding dinucleotide are added to the respective sites. (E) The beads are pooled, mixed, and resplit. The previous step of deprotection and addition is repeated. This process is repeated until peptides of the desired length are synthesized. (F) The beads are pooled and 35 nucleotides (containing sequences complementary to PCR and sequencing primers) are added by oligonucleotide synthesis. The beads are then ready for assay.

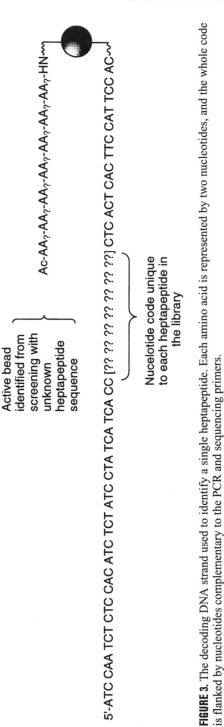

FIGURE 3. The decoding DNA strand used to identify a single heptapeptide. Each amino acid is represented by two nucleotides, and the whole code is flanked by nucleotides complementary to the PCR and sequencing primers.

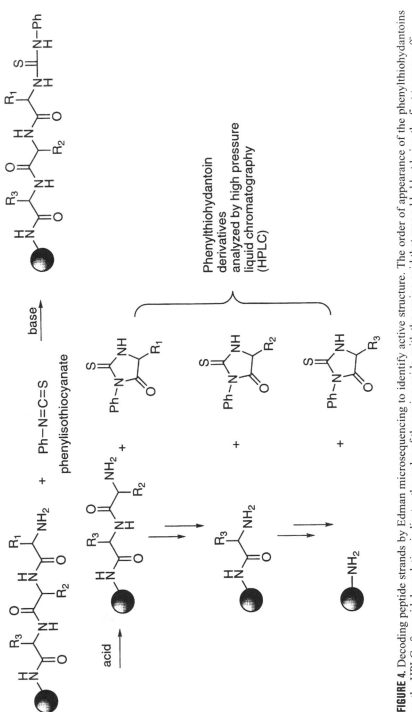

FIGURE 4. Decoding peptide strands by Edman microsequencing to identify active structure. The order of appearance of the phenylthiohydantoins on the HPLC after acid degradation indicates the order of the amino acids, with the amino acid that was added last being the first to come off.

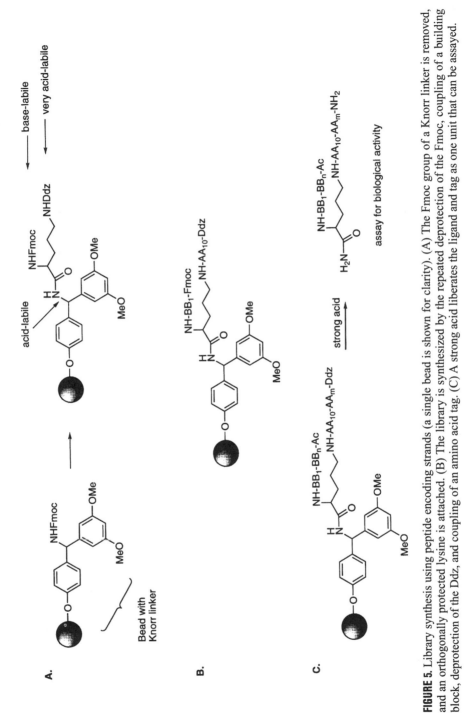

FIGURE 5. Library synthesis using peptide encoding strands (a single bead is shown for clarity). (A) The Fmoc group of a Knorr linker is removed, and an orthogonally protected lysine is attached. (B) The library is synthesized by the repeated deprotection of the Fmoc, coupling of a building block, deprotection of the Ddz, and coupling of an amino acid tag. (C) A strong acid liberates the ligand and tag as one unit that can be assayed.

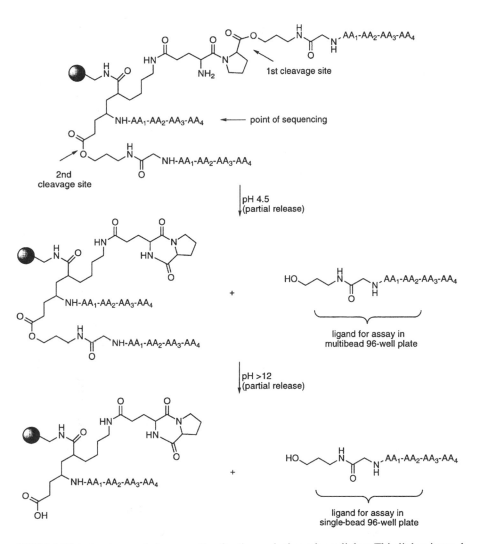

FIGURE 6. The structure and cleavage sites for the exclusive-release linker. This linker is used to synthesize three identical peptides at three discrete sites, and each peptide can be released selectively under different conditions. This is necessary for the decoding method described in Figure 7.

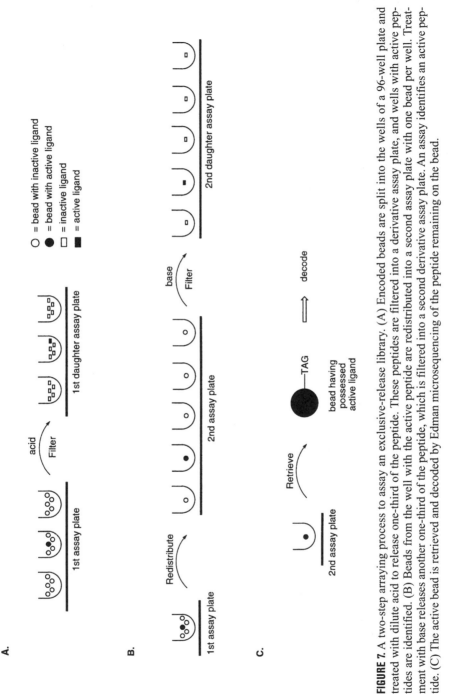

FIGURE 7. A two-step arraying process to assay an exclusive-release library. (A) Encoded beads are split into the wells of a 96-well plate and treated with dilute acid to release one-third of the peptide. These peptides are filtered into a derivative assay plate, and wells with active peptides are identified. (B) Beads from the well with the active peptide are redistributed into a second assay plate with one bead per well. Treatment with base releases another one-third of the peptide, which is filtered into a second derivative assay plate. An assay identifies an active peptide. (C) The active bead is retrieved and decoded by Edman microsequencing of the peptide remaining on the bead.

molecular tags are electrophoric aromatics (*22, 23*) and fluorescent dialkylsulfon-amides (*24*).

Electrophoric Tags. Electrophoric tags are halogen-substituted phenoxyalkyl alco-hols (Figure 8). The attachment of the tag occurs by metal-catalyzed carbene inser-tion of a diazoketone precursor directly into the polystyrene bead matrix. (This is in contrast to the original photolabile tags, which were attached to the resin by amide bond formation (*22*).) Treatment with ceric ammonium nitrate (CAN), an oxidant, detaches the tag. The released tag alcohols are derivatized to their more volatile trimethylsilyl ethers and analyzed by gas chromatography using an electron-capture detector (ECD). The sensitivity for tag detection is at or below the femtomolar range (actual detection limit for the ECD is 10^{-18} M). Thus, only trace quantities of tags are required to encode a library member.

Direct insertion of tags negates the need for preforming a differentially func-tionalized bead. The tagging process is compatible with photo-, acid-, and base-labile linkers commonly used for ligand synthesis. As they are chemically unreac-tive, they are also compatible with a whole host of organic reactions, such as classical carbon–carbon bond-forming reactions, including cycloadditions and aldol condensations. Only those reactions using strong oxidants must be excluded from the synthetic repertoire.

The greatest benefit of direct tag insertion is that it permits binary encryption of libraries constructed by the split synthesis technique (Figure 9). With this method, which is analogous to the binary code used in computing, only the presence or absence of tag need be determined. Using a binary code, n encoding bits can encode for 2^n building blocks. In practice, the null binary bit (no tags) is not used to ensure a measure of quality control during tag synthesis. The number of building blocks that can be encoded by a set of tags is thus given by $2^n - 1$: three tags encode 7 building blocks, four tags encode 15 building blocks, and so on. Some forty tags have been synthesized, allowing for the theoretical encryption of over one trillion library members.

Biological evaluation of the library is carried out using the two-step arraying process as described previously for the exclusive-release linker (Figure 7). Instead of using acid for partial release, longwave UV light (365 nm) is used. Brief expo-sure of a mixture of beads in a 96-well assay plate to UV causes ~50% of the ligand on each bead to be released. Upon single-bead rearray, the beads are again exposed to UV, releasing the remaining portion of ligand. The tags stay attached to the beads during photolysis. Only beads of interest are retrieved from the second assay plate and their tags detached by oxidation.

As "proof of principle" applications of these tags, a 6727-membered amino-acid-based library and a 1143-membered benzopyran library were constructed (*27, 28*). Both libraries contained a portion of sulfonamides, a well-known pharma-cophore for carbonic anhydrase. Biological evaluation of the libraries led to the dis-covery of potent inhibitors of this enzyme.

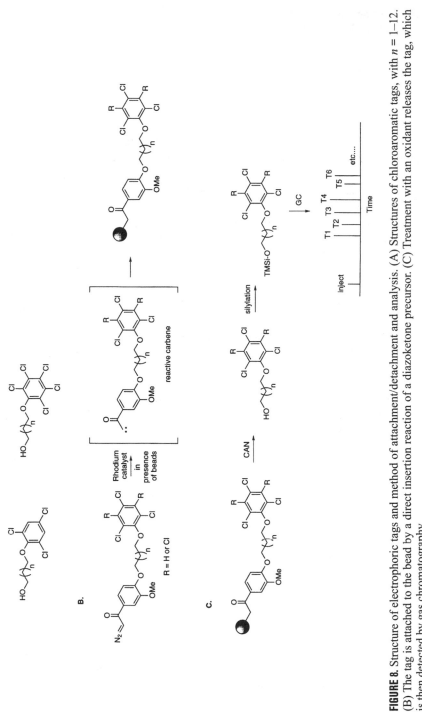

FIGURE 8. Structure of electrophoric tags and method of attachment/detachment and analysis. (A) Structures of chloroaromatic tags, with $n = 1$–12. (B) The tag is attached to the bead by a direct insertion reaction of a diazoketone precursor. (C) Treatment with an oxidant releases the tag, which is then detected by gas chromatography.

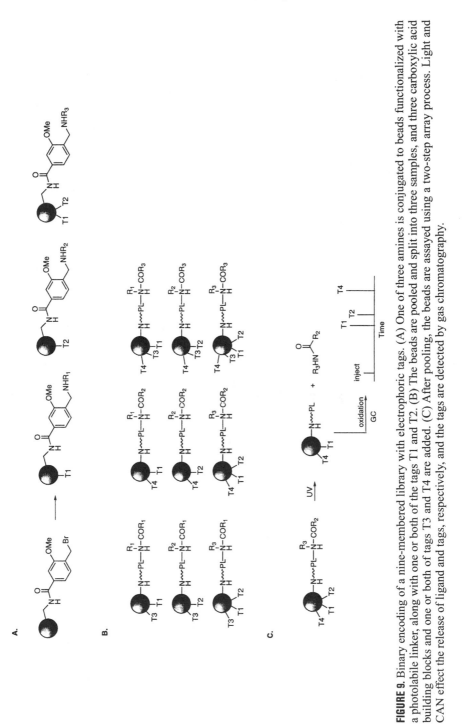

FIGURE 9. Binary encoding of a nine-membered library with electrophoric tags. (A) One of three amines is conjugated to beads functionalized with a photolabile linker, along with one or both of the tags T1 and T2. (B) The beads are pooled and split into three samples, and three carboxylic acid building blocks and one or both of tags T3 and T4 are added. (C) After pooling, the beads are assayed using a two-step array process. Light and CAN effect the release of ligand and tags, respectively, and the tags are detected by gas chromatography.

Dialkylamine Tags. The use of dialkylamines as binary encoding tags has recently been described (*24*). In contrast to the direct insertion method used for electrophoric tags, differentially functionalized resin is necessary to provide separate sites for tag addition and ligand synthesis (Figure 10). Tag monomers, *N*-[(dialkylcarbamoyl)-methyl]glycines, are added sequentially as ligand synthesis proceeds (Figure 11). Ligand can be cleaved from the solid support leaving the tags intact. Tags (secondary amines) are removed from the beads upon hydrolysis with 50% aqueous hydrochloric acid and converted to their corresponding highly fluorescent dansyl sulfonamide derivatives. Subpicomolar amounts of the dansylated secondary amine tags are easily resolved by reverse-phase HPLC.

As for electrophoric tags, the secondary amine tags are chemically robust and compatible with a wide range of polymer-supported chemistries. Encoded split synthesis of β-lactams, 4-thiazolidinones, and pyrrolidines have been reported (*24*).

Radio-Frequency Tags

Two methods of encoding with radio-frequency transponders have been developed (*25, 26*). In both methods a tiny glass-encapsulated microchip is incorporated into a mesh capsule loaded with polymer beads (Figure 12).

In the first method, the chip emits a binary code that is scanned each time a synthetic building block is reacted with the beads in the capsule (*26*). The identity of both the capsule of beads and the building block it is reacted with are registered and uploaded to a computer, capturing the capsule's reaction history. Once the library synthesis is complete, the ligand is released for biological evaluation and its structure identified by scanning the transponder trapped inside the capsule and com-

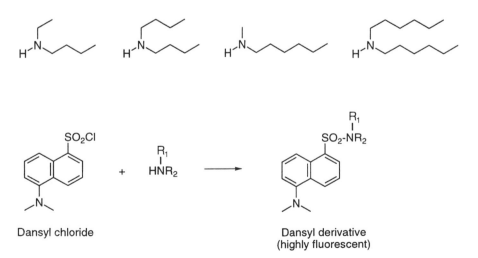

Dansyl chloride

Dansyl derivative
(highly fluorescent)

FIGURE 10. Structures of some representative dialkylamine tags and a dansyl derivative.

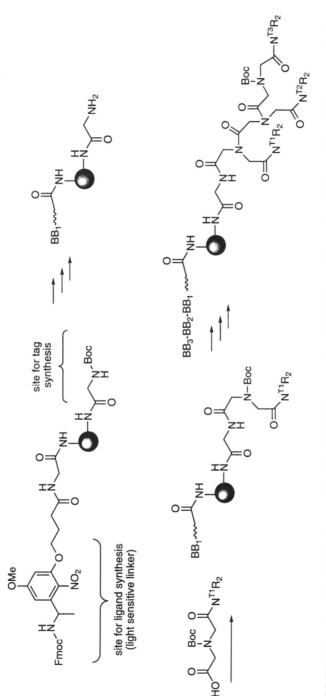

FIGURE 11. Encoding with dialkylamine tags. The ligand is built upon a photolabile linker, and the tag is added to a differentially protected amine. Each tag is a secondary amine. After removal of the ligand by light treatment, the tags are removed by acid hydrolysis and detected by HPLC.

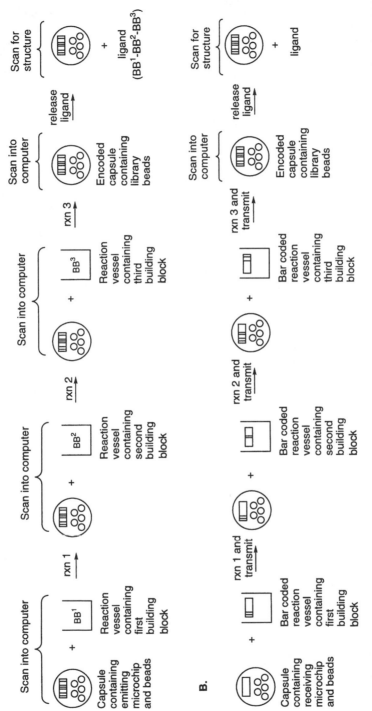

FIGURE 12. Encoding a three-step reaction sequence using radio-frequency tags. (A) Encoding with an emitting transponder. Each capsule, whenever it undergoes a reaction, transmits its identity to a computer that also records the nature of the reaction. (B) Encoding with a recording transponder. The transponder in each capsule of beads receives a signal whenever the beads take part in a reaction. At the end of the synthesis, this accumulated code can be read off in one step.

paring it with the reaction histories recorded earlier. The encoding of a three-step reaction sequence is illustrated in Figure 12A.

The second method makes use of a chip that does not emit a bar code, but rather receives and records binary information sent by a transmitter each time a building block is added to the solid support (Figure 12B) (*25*). A computer generates a set of bar codes for each building block used in a given library. As each capsule is reacted with a particular building block, the building block's bar code is transmitted to the microchip in the capsule. At the end of the synthesis, the information recorded on the chip is scanned and uploaded to a computer. In this instance, only one scan of the capsule is required (after completion of the library synthesis), rather than the scan after every step that is necessary with the first method.

As these microchips are encased in glass they are supposedly solvent-, reagent-, and temperature-resistant. Encoding with transponders would allow any reaction compatible with solid support and linker to be used. Chemical synthesis is not required to introduce the tags. This eliminates the requirement for orthogonally protected tags and ligands and reduces the overall number of chemical reactions performed in constructing the library.

Two libraries have been constructed by split synthesis using the transponders. One was a 24-member peptide library (*25*), and the other was a 125-member library of cinnamic acids (*26*). A protein tyrosine phosphatase inhibitor with a K_i (reversible inhibition constant) of 44 nM was identified from the latter library.

Summary

The synthesis of combinatorial chemical libraries can be achieved with relative ease. With a small measure of optimization, most organic reactions can be adapted to solid-phase synthesis and library preparation (*29*). The key to efficient use of libraries lies with the methods for elucidating the structures of the compounds they contain. Early methods for deconvoluting libraries have largely been superseded by encoding techniques. Encoding with chemically robust tags and microchips currently offer the most reliable methods for encryption. Continued advances in encoding, in the refinement in analytical detection methods, and in synthesis automation are expected as the importance of chemical libraries for discovering new drugs is realized.

References

1. Geysen, H. M.; Rodda ,S. J.; Mason T. J. *Mol. Immunol.* **1986**, *223*, 709–715.
2. Geysen, H. M.; Rodda, S. J.; Mason J.; Tribick, G.; Schoofs, P. G. *J. Immunol. Methods* **1987**, *102*, 259–274.
3. Brag, A. M.; Maeji, N. J.; Geysen, H. M. *Tetrahedron Lett.* **1990**, *31*, 5811–5814.
4. Geysen, H. M.; Mason, T. *Bioorg. Med. Chem. Lett.* **1993**, *3*, 397–404.
5. Houghten, R. A.; Pinilla, C.; Blondelle, S. E.; Appel, J. R.; Dooley, C. T.; Cuervo, J. H. *Nature (London)* **1991**, *354*, 84–86.

6. Houghten, R. A.; Appel, J. R.; Blondelle, S. E.; Cuervo, J. H.; Dooley, C. T.; Pinilla, C. *Biotechniques* **1992**, *13*, 412–421.
7. Houghten, R. A.; Dooley, C. T. *Bioorg. Med. Chem. Lett.* **1993**, *3*,405–412.
8. Pinilla, C.; Appel, J. R.; Blanc, P.;Houghten, R. A. *Biotechniques* **1992**, *13*, 901–905.
9. Dooley, C. T.; Houghten, R. A. *Life Sci.* **1993**, *52*, 1509–1517.
10. Furka, A.; Sebestyen, F.; Asgedom, M.; Dibo, G. *Int J. Pept. Protein Res.* **1991**, *37*, 487–493.
11. Lam, K. S.; Salmon, S. E.; Hersh, E. M.; Hruby, V. J.; Kazmierski, W. M.; Knapp, R. J. *Nature (London)* **1991**, *354*, 82–86.
12. Dunayevskiv, Y.; Vouros, P.; Carell, T., Wintner, E. A.; Rebek, J. *J. Anal. Chem.* **1995**, *67*, 2906–2915.
13. Carell, T.; Wintner, E. A.; Sutherland, A. J.; Rebek, J.; Dunayeveskiy Y. M.; Vouros, P. *Chem. Biol.* **1995**, *2*, 171–183.
14. Fitch, W. L.; Detre, G.; Holmes, C. P.; Shoolery, J. N.; Keifer, P. A. *J. Org. Chem.* **1994**, *59*, 7955–7956.
15. Anderson, R. C.; Jarema, M. A.; Shapiro, M. J.; Stokes, J. P.; Zilioz, M. *J. Org. Chem.* **1995**, *60*, 2650–2651.
16. Anderson, R. C.; Stokes, J. P.; Shapiro, M. J. *Tetrahedron Lett.* **1995**, *36*, 5311–5314.
17. Brenner, S.; Lerner, R. A. *Proc. Natl. Acad. Sci. U.S.A.* **1992**, *89*, 5381–5383.
18. Nielsen, J.; Brenner, S.; Janda, K. D. *J. Am. Chem. Soc.* **1993**, *115*, 9812–9813.
19. Nikolaiev, V.; Stierandova, A.; Krchnak, V.; Seligmann, B.; Lam, K. S.; Salmon, S. E.; Label, M. *Pept. Res.* **1993**, *6*, 161–170.
20. Salmon, S. E.; Lam. K. S.; Label, M.; Kandola, A.; Khattri, P. S.; Wade, S.; Patek, M.; Kocis, P.; Krchnak, V.; Thorpe, D.; Felder, S. *Proc. Natl. Acad. Sci. U.S.A.* **1993**, *90*, 11708–11712.
21. Kerr, J. M.; Banville, S. C.; Zuckerman, R. N. *J. Am. Chem. Soc.* **1993**, *115*, 2529–2531.
22. Ohlmeyer, M. H. J.; Swanson, L.; Dillard, L. W.; Reader, J. C.; Asouline, G.; Kobayaski, R.; Wigler, M.; Still, W. C. *Proc. Natl. Acad. Sci. U.S.A.* **1993**, *90*, 10922–10926.
23. Nestler, H. P.; Bartlett, R. A.; Still, W. C. *J. Org. Chem.* **1994**, *59*, 4723–4724.
24. Ni, Z.-J.; Maclean, D.; Holmes, C. P.; Murphy, M. M.; Ruhland, B.; Jacobs, J. W.; Gordon, E. M.; Gallop, M. A. *J. Med. Chem.* **1996**, *39*, 1601–1608.
25. Nicolaou, K. C.; Xiao, X. Y.; Parandoosh, Z.; Senyei, A.; Nova, M. P. *Angew. Chem. Int. Ed. Engl.* **1995**, *34*, 2289–2291.
26. Moran, E. J.; Sarshar, S.; Cargill, J. F.; Shahbaz, M. M.; Lio, A.; Mjalli, A. M. M.; Armstrong, R. W. *J. Am. Chem. Soc.* **1995**, *117*, 10787–10788.
27. Chabala, J. C.; Baldwin, J. J.; Burbaum, J. J.; Chelsky, D.; Dillard, L. W.; Henderson, I.; Li, G.; Ohlmeyer, M. H. J.; Randle, T. L.; Reader, J. C.; Rokosz, L.; Sigal, N. H. *Perspect. Drug Disc. Des.* **1994**, *2*, 305–318.
28. Burbaum, J. J.; Ohlmeyer, M. H. J.; Reader, J. C.; Henderson, I.; Dillard, L. W.; Li, G.; Randle, T. L.; Sigal, N. H.; Chelsky, D.; Baldwin, J. J. *Proc. Natl. Acad. Sci. U.S.A.* **1995**, *92*, 6027–6031.
29. Ellman, J.A. *Chem. Rev.* **1996**, *96*, 555–600.

Solution-Phase Strategies

7

Synthesis Tools for Solution-Phase Synthesis

Ted L. Underiner and John R. Peterson

Recent advances in the field of small-molecule combinatorial chemistry offer new hope that the enormous costs and time associated with pre-clinical drug discovery and development may be reduced (1–3). A chemist using classical techniques may produce 50 or so compounds per year; using combinatorial techniques, he or she may instead produce hundreds or thousands of compounds (4). Synthesis tools for combinatorial chemistry can be broadly grouped into solid-phase and solution-phase methods. Here, we focus on the solution phase. We discuss the factors affecting the choice of single-compound or mixture synthesis, outline various examples of each, and conclude with a brief discussion of the practical tools used in solution-phase chemistry.

Solid- or Solution-Phase Synthesis?

The selection of an appropriate combinatorial chemistry strategy depends, in part, on whether one needs to discover a lead or optimize one that has already been found. Some other factors that need be considered are the synthetic methods required to make molecules of the appropriate molecular structure, the format of the screen for activity (and whether it can tolerate mixtures of compounds), and how pure the products need to be.

The advantages and disadvantages of solid- and solution-phase synthetic methods have been reviewed previously (5–9). One advantage of solution-phase chemistry is that there is no need to develop a strategy for either the attachment of the substrate to the solid support, or its subsequent cleavage. This simplification, coupled with the maturity of the field of solution-phase chemistry, results in significantly shorter reaction development times. Solution-phase techniques are, however, limited to short reaction sequences that either use reagents in stoichiometric quantities, or use reagents that can be separated (e.g., by extraction or filtration) from the final products. An extreme example is the use of polymer-bound reagents, which are easily separable from the reagents in solution (*10–12*). In contrast, solid-phase synthetic routes need not be limited to stoichiometric reagents or short reaction sequences, because reaction workup and product purification occur simultaneously and simply when solvents and reagents are washed away from the resin to which the modified substrates are attached.

Individual Compounds or Mixtures?

Mixtures of compounds are typically easier to prepare and may require fewer resources for screening than individual compounds. Indeed, a number of pharmaceutical companies routinely mix pure compounds (clips) in their drug-discovery screening programs as a means of increasing assay throughput and controlling costs. Before initiating a mixture synthesis there are, however, a number of issues surrounding the synthesis, screening, and identification of compounds that should be considered.

Synthetic Implications

When solid-phase chemistry is used to produce mixtures of compounds (*1, 2, 5–9*), reactions are driven to completion with excess reagents. With solution-phase chemistry, however, the chemist must be careful to ensure that all compounds have reacted completely and by the desired reaction mechanism. Otherwise, differing amounts of reaction products may be obtained because of the competing kinetics of individual reactions. This problem can be especially acute if a stoichiometric excess of one or more reactants is required to drive reactions to completion (*13, 14*). An additional problem is the analytical challenge of monitoring the reaction progress of large mixtures. There has therefore been some effort to prepare libraries of individual compounds by parallel synthetic techniques (*15–22*). The use of automation for dispensing reagents and quenching reactions can make these processes highly efficient and cost-effective.

Screening Implications

The results from the assaying of pooled compounds are often uncertain, and this uncertainty may be attributed to a number of factors, including a lack of sensitivity or precision in the assay, the large number and variable potency of the elements in the pool, and the narrow concentration range of test substrate that the assay can tolerate (*23*). Consider a hypothetical example in which identification of a lead compound with a 50% inhibitory concentration (IC_{50}) < 1 μM is desired (*7*). In this example, the solubility of the compound mixture and the background noise of the assay limit the total screening concentration to < 50 µM. (With the given background noise, a higher concentration of compounds would result in all or most samples testing positive.) If the components of the mixture are to be present at concentrations of greater than 1 µM (the potency threshold), the mixture must consist of no more than 50 equimolar components. To screen larger mixtures, one must raise the potency threshold (e.g., $IC_{50} < 0.1$ µM), increase sample solubility, or decrease the background noise of the assay. This example assumes that only one compound in the mixture is active in the assay.

The differing demands of lead discovery and lead optimization must be taken into account here. It is more likely that similar compounds (whether peptides or small molecules) will be found in a pool from a lead-optimization library than in one from a lead-discovery library. A tenet of medicinal chemistry is that similar compounds will probably exhibit similar biological activities. Thus, especially when screening lead-optimization libraries, it must be considered that the most active pool may not contain the most active compound(s) (7), but rather a mixture of several moderately active compounds.

Workers at Pfizer have recently screened mixtures of 20, 100, 992, or 8836 compounds (24). The efficient detection of valid "hits" in mixtures with the larger numbers of compounds was, indeed, reported to be problematic (e.g., the activity disappeared upon fractionation), whereas the mixtures with fewer compounds yielded the most productive data.

Compound Identification

Researchers using solid-phase chemistry have used molecular tags as a means of identifying active components in compound mixtures (1, 2, 7, 8), but such tagging techniques have not yet become popular in solution-phase libraries. Methods for identifying active component(s) from solution-phase mixture libraries rely on iterative synthesis (14) or "indexing" (13, 23) (see Chapter 9 for more details). In iterative synthesis, various subsets of the products are synthesized and evaluated until the active compound(s) is identified. For example, if a mixture of 50 products were found to be active in an assay, the 50 compounds are synthesized as two groups of 25 compounds. To determine which group the active compound resides in, the two groups are biologically tested. This process is repeated on additional subsets of compounds until the active product is isolated. Use of the second method has been reported by Pirrung and Chen (23), who use an "indexed" combinatorial library (Figure 1). In this method, two libraries are prepared as pools, using n components of type A, and m components of type B. Library 1 consists of m pools, where the first pool is generated by combining A_1 through A_n with B_1, and the mth pool is generated by combining A_1 through A_n with B_m. Library 2 consists of n pools, where the first pool is generated by combining B_1 through B_m with A_1, and the nth pool is generated by combining B_1 through B_m with A_n. All the pools are screened, and, in the ideal case, the active component, A_xB_y, is identified by the indices of the active pools (i.e., the active compound is at the intersection of the active row and the active

$$A + B \longrightarrow AB$$

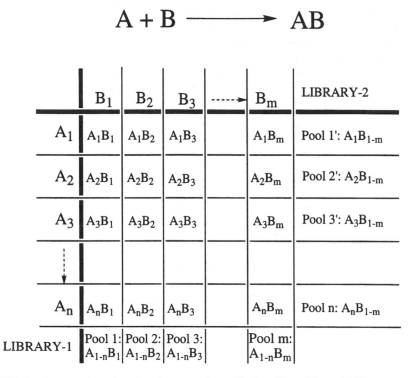

FIGURE 1. Product matrix and pool assignments for an "indexed" combinatorial library.

column in the reaction matrix). For example, if pool 2 and pool 3' are active in the library depicted in Figure 1, then the activity would be attributed to compound A_3B_2. An example of this strategy is described on the following pages.

The Relevance of Purity

Although pure compounds are necessary for biological quantitative structure–activity relationship (QSAR) studies and lead optimization, the fact that the screening of mixtures has been effective for identifying lead structures argues that the traditional requirement for pure compounds in a screen is unnecessary.

Recognizing this point, and the need to contain research and development costs, the combinatorial chemistry field now favors an altered sequence of compound processing. Compounds are first made, then evaluated biologically, and only later purified and characterized. This is a paradigm shift, as chemists have traditionally purified every compound and done biological evaluation last. In the new method, however, the focus is on compound production, with purification only of biologically active compounds. Given the number of compounds that can be produced and evaluated by combinatorial methods, the improvements in practicality and cost:benefit ratios from using the new approach should be enormous.

Libraries of Mixtures Produced by Solution-Phase Synthesis

Whether preparing mixtures or individual compounds, ideal reactions for solution-phase synthesis use starting materials that are readily available or easily synthesized, are kinetically and thermodynamically favorable, are tolerant of other functionalities, require the same general conditions for a broad range of reactants, and yield either pure products or those that can be easily purified by extractive workups (*17, 20, 21, 25*). For these reasons, many research groups have selected reactions such as condensations of nucleophiles (e.g., amines, hydrazines, alcohols, and thiols) with a variety of acylating reagents (e.g., activated carboxylic acids, esters, sulfonyl chlorides, isocyanates, and chloroformates), addition reactions of nucleophiles to epoxides, and condensation reactions of activated methylene compounds with aldehydes.

We now turn to the reactions themselves, highlighting several examples that illustrate the solution-phase combinatorial chemistry techniques and strategies just described. We first examine mixture libraries.

Smith et al. (*13*) reported the generation of a 1600-compound indexed library of amides and esters by reacting 40 alcohols and amines (N_{1-40}, Scheme I) with 40 acid chlorides (A_{1-40}, Scheme I). Two libraries were prepared. In each pool of the first library, a single acid chloride was reacted with an equimolar mixture of the nucleophiles, while in each pool of the second library, a single nucleophile was reacted with an equimolar mixture of the acid chlorides. Thus, each library contained 40 pools each with 40 components. The libraries underwent pharmacological screening, and the detection of activity in a pool identified either the carboxylic acid or the amino–alcohol portion of the putatively active amide or ester. Two weak leads were identified using a neurokinin (NK3) assay (**1**) and metalloprotease-1 (MMP-1) assay (**2**).

The high reactivity of the acid chlorides ensured that reactions with a range of nucleophiles with widely differing reactivities went to completion. Two pools were analyzed by GC–MS, and more than 25 of the 40 expected products were found to be present in each mixture.

A similar experiment was carried out by Pirrung and Chen (*23*), who generation an indexed library of 54 carbamates by individually reacting 9 alcohols with an equimolar mixture of 6 isocyanates, and then individually reacting 6 isocyanates with an equimolar mixture of 9 alcohols (Figure 2). The 15 product mixtures were screened against electric eel acetylcholinesterase, and their activities were used as "indices" to the rows and columns of a two-dimensional matrix (as depicted in Figure 1). Carbamate **3** was identified as the most potent inhibitor, and its identity and activity were confirmed by independent synthesis.

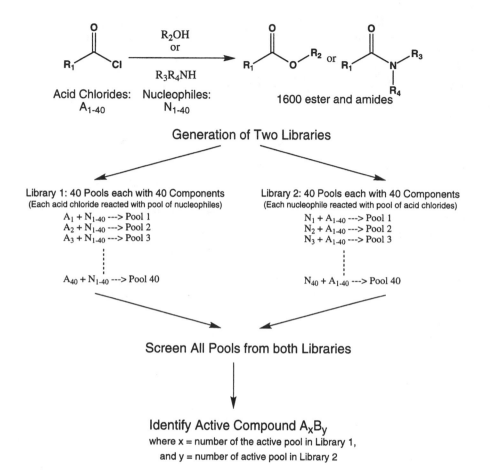

Generation of Two Libraries

Library 1: 40 Pools each with 40 Components
(Each acid chloride reacted with pool of nucleophiles)

$A_1 + N_{1-40}$ ---> Pool 1
$A_2 + N_{1-40}$ ---> Pool 2
$A_3 + N_{1-40}$ ---> Pool 3

$A_{40} + N_{1-40}$ ---> Pool 40

Library 2: 40 Pools each with 40 Components
(Each nucleophile reacted with pool of acid chlorides)

$N_1 + A_{1-40}$ ---> Pool 1
$N_2 + A_{1-40}$ ---> Pool 2
$N_3 + A_{1-40}$ ---> Pool 3

$N_{40} + A_{1-40}$ ---> Pool 40

Screen All Pools from both Libraries

Identify Active Compound A_xB_y
where x = number of the active pool in Library 1,
and y = number of active pool in Library 2

SCHEME I. Synthesis and screening strategy for an ester–amide indexed combinatorial library.

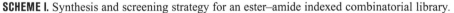

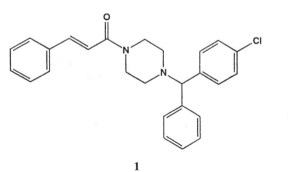

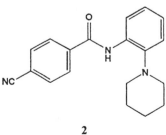

1   2

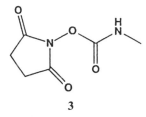

3

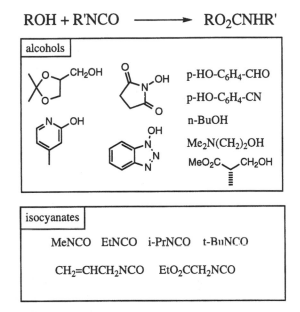

$$ROH + R'NCO \longrightarrow RO_2CNHR'$$

FIGURE 2. Preparation and components of a carbamate library.

Rebek and colleagues constructed a combinatorial library by attaching an array of building blocks to multifunctional core structures (*14, 26, 27*). Tetracarboxylic acids **4** and **5** were selected as the core structures because of their complimentary architecture. Attaching building blocks to the xanthene core, **4**, gave a disklike structure, whereas attachment to the cubane core, **5**, resulted in a more spherical arrangement. Tetracarboxylic acid **4** was reacted with pools of 4, 7, 12, or 21 amines (amino acid derivatives and heterocycles) to generate libraries that should, theoretically, contain 136, 1225, 10,440, or 97,461 compounds. Because of the higher symmetry of the cubane core, the corresponding reactions of **5** with these pools of amines should, theoretically, generate product mixtures with 36, 245, 1860, or 16,611 components. The amines were carefully selected to have comparable reactivities toward the core structures. To minimize competition effects between the amines for the acid chloride groups, four moles of an amine mixture were used for every mole of tetracarboxylic acid.

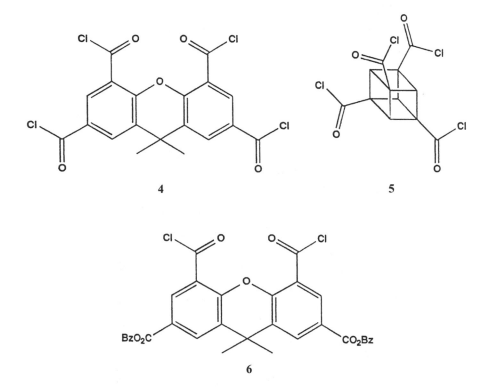

4

5

6

A hint that the libraries were becoming more diverse as the number of amines in the mixture increased was revealed by high-pressure liquid chromatographic (HPLC) analysis of the product mixtures. As the number of amines reacted with **4** increased, the HPLC chromatograms of the product mixtures had increasingly complex traces, and the trace of the most complex mixture contained no well-resolved peaks at all. In a similar study (*26*), a model library derived from coupling compound **6** with 10 representative amines was analyzed by fast atom bombardment (FAB) mass spectrometry after fractionation by HPLC. Out of a theoretical total of 55 molecular ion peaks, 42 (75%) were observed.

An iterative synthesis and screening strategy was devised for identifying active compounds from product mixtures derived from coupling compound **4** with 18 amines (*26*). In this example, compounds were screened for their ability to inhibit the trypsin-catalyzed hydrolysis of *N*-benzoyl-DL-arginine-*p*-nitroanilide. Identification of active compounds involved an iterative process of sublibrary construction and activity testing. All possible amides were initially prepared in 6 sublibraries, and each sublibrary was a mixture resulting from the reaction of **4** with 15 of the 18 amines. From this analysis, 9 amines were identified as having the greatest inhibitory activity. Nine sublibraries were then prepared using pools missing one of

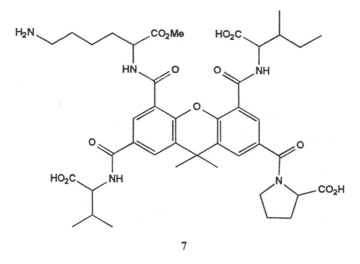

7

these 9 amines. Screening the 9 sublibraries led to the identification of the 5 most important amines. This still represents over 300 compounds—far too many for individual synthesis and screening. Furthermore, preparation of a single derivative of **4** in which each of the four acid chloride functionalities was converted to a unique amide would require a laborious protection–deprotection or purification strategy. Therefore, screening of mixtures was the most expedient route. Four amines were ultimately determined to be necessary for activity; that is, omission of any one of these four amines resulted in sublibraries with significantly less inhibitory activity. Screening mixtures of products derived from these four amines narrowed the possibilities to two structural isomers; subsequently, these were independently prepared, and a novel trypsin inhibitor, **7**, was identified. The authors note that **7** is neither the only nor the most effective inhibitory compound present in the original library. Still, this synthesis–screening strategy led to the identification of a potent lead compound that can be further optimized.

Solution-Phase Synthesis of Individual-Component Libraries

There are few documented reports of the use of parallel solution-phase synthesis to generate large libraries of individual components, although many research groups in both academia and industry have presented, or alluded to, work on this subject (*6, 15–22, 28*).

A one-step procedure for preparing a modest library of 20 discrete 2-aminothiazoles by modification of the conventional Hantzsch synthesis was reported by Watson and colleagues (*15*). Reaction of primary thioureas (**8**) with α-bromoketones (**9**) in *N,N*-dimethylformamide (DMF) at 70 °C yielded relatively pure 2-

aminothiazoles (**10**), as determined by a combination of spectroscopic techniques (Figure 3). The reaction tolerated the presence of both acidic and basic functionalities on the reactants without recourse to protection, and the products did not require purification. It was reported that a library of 2500 compounds was prepared by this procedure using commercially available substrates.

Several groups have reported using the Ugi reaction for the solution-phase synthesis of combinatorial libraries. The Ugi reaction is a one-pot, four-component condensation reaction that generates α-acylaminoamides (**11**) by coupling a carboxylic acid with an amine, an aldehyde, and an isonitrile as depicted in Scheme II (*29, 30*). The utility of this method is limited somewhat by the lack of commercially available isonitriles. Although routes exist to convert amines to isonitriles, the noted reactivity, toxicity, and odor of isonitriles complicate their use (*31*).

This drawback has been overcome by substituting a single compound for all the isonitriles. 1-Isocyanocyclohexene is used in the Ugi reaction to prepare inter-

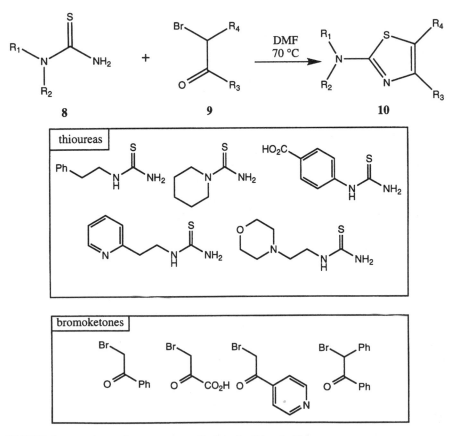

FIGURE 3. Preparation and components of a 2-aminothiazole library.

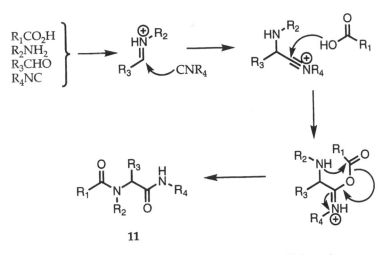

SCHEME II. Proposed mechanism of the four-component, one-pot, Ugi reaction.

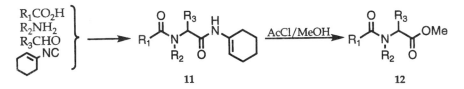

SCHEME III. Methanolysis of Ugi products.

mediate **11**, which upon acid-catalyzed methanolysis gives methyl esters (**12**, Scheme III). Thus, the contribution of the isocyanide component has been converted to a single carbon atom (*31*).

Weber et al. (*32*) used this reaction to prepare 20 libraries, each containing 20 individual Ugi products, which were screened for thrombin-inhibitory activity. Each library sequentially evolved on the basis of a genetic algorithm that used assay results, comparisons of predicted and calculated activities, and other design attributes to select appropriate starting materials for synthesis. The most active compounds (**13** and **14**) had IC_{50} values of 1.4 µM and 0.22 µM, respectively. Note that compound **14** is a side-product that did not incorporate the carboxylic acid functionality because of the low nucleophilicity of 4-aminobenzamidine.

A similar condensation, known as the Passerini reaction (*30*), involves reaction of an isocyanide with a carboxylic acid and an aldehyde or ketone to generate an α-acyloxyamide (**15**). The mechanism for this reaction is shown in Scheme IV.

Another example of a multiple-component reaction is ArQule's alkylation of acylhydrazines (**16**). These substrates are generated in situ by condensation of an ester with an *N,N*-dialkylhydrazine (Scheme V) (*18, 19, 28*). Thus, heating an iso-

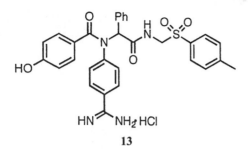
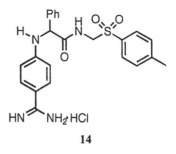

13   14

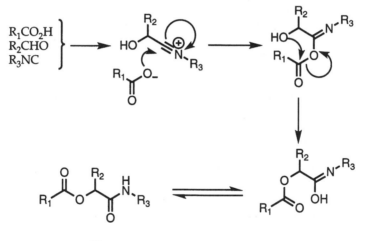

15

SCHEME IV. Proposed mechanism for the Passerini reaction.

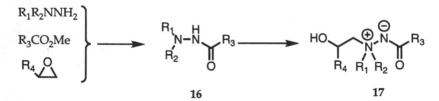

16   17

SCHEME V. Preparation of hydroxyaminimides.

propanol solution of an ester, a hydrazine, and an epoxide generates hydroxyamin-imides (**17**). This method was used to prepare aminide **18**, which has been identified as a human immunodeficiency virus 1 (HIV-1) protease inhibitor (*18, 19*).

ArQule has also presented work on methods of preparing libraries based on azlactone chemistry to give products of type **19** (Scheme VI) (*18, 33*).

Panlabs has reported a semiautomated method for making libraries based on multifunctional scaffolds (diamines and amino ketones) to which two or more phar-macophoric residues are attached (*16, 17*). Using traditional organic chemistry syn-thetic techniques, the first group was attached to the core structure to generate bulk quantities of an intermediate that was subsequently derivatized by a semiautomated method to yield multiple, individual reaction products.

As shown in Scheme VII, acylation of 4-piperidone monohydrate (**20**) gave intermediates **21**, which were readily purified by acidic and basic extractions. Reductively coupling these intermediates with amines (in the presence of NaCNBH₃) led to a library of 8000 4-aminopiperidines (**22**).

Preparation of piperazine derivatives (**25–30**) required a more laborious pro-cedure and involved protection and deprotection, as illustrated in Scheme VIII. Con-densation of *N*-Boc-piperazine (**23**-Boc) with activated carboxylic acids gave the corresponding *N*-Boc-amides, which upon treatment with HCl/dioxane generated the aminoamides (**24**). These compounds were then reacted with a variety of reagents to prepare the products (**25–30**) depicted in Scheme VIII. By shuffling the sequence of reactions, a variety of products (7500 overall) not depicted in Scheme

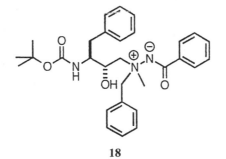

18

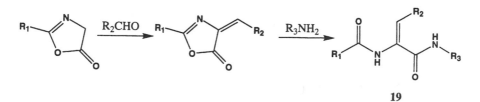

19

SCHEME VI. Tandem condensation–addition reaction of azlactones.

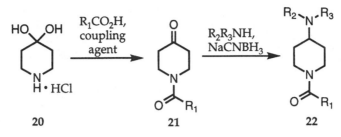

SCHEME VII. Preparation of a 4-aminopiperidine library.

VIII (including bis-amines, bis-sulfonamides, bis-ureas and urea-sulfonamides) were also prepared.

Aliphatic amines are better nucleophiles than aromatic amines, and this difference has been exploited in the preparation of a series of derivatives of 4-aminobenzylamine by a two-step procedure (17). As shown in Scheme IX, coupling 4-aminobenzylamine (**31**) with activated acids yielded 4-aminobenzylamides (**32**). Subsequent reactions at the aromatic nitrogen gave over 6000 bis-adducts (**33–35**).

Amidation chemistry was used to prepare libraries based on two peptidomimetic scaffolds (**36** and **40**; Scheme X) (20, 21). These templates contained three positions that were individually functionalized with a variety of amines and acylating agents. Diacid **36** was converted in situ to the corresponding cyclic anhydride, and sequentially reacted with amines to initially generate a monoamide (**37**) and then a diamide (**38**). Removal of the Boc protecting group followed by acylation yielded the final products (**39**) (20). In each step of the sequence, reagents, starting materials, and by-products were removed by liquid–liquid or liquid–solid extraction. Thus, in the case of **36**, a 3 × 3 × 3 matrix (i.e., 3 R_1NH_2 × 3 R_2NH_2 × 3 R_3CO_2H) yielded 39 unique components (27 final products plus 12 intermediates). In a similar manner, scaffold **40** was used in a 3 × 3 × 3 matrix to make triamides **41**; analysis of the products by HPLC and NMR indicated purity levels exceeding 90% (21). In addition, **40** was used in a 6 × 8 × 20 matrix to prepare a 1158-component library. Overall yields ranged from 10 to 71% (10–148 mg of each individual product).

A library of 9000 aminophenols (**42**) was prepared by the Mannich reaction of phenols, secondary amines, and formaldehyde (Scheme XI) (33, 34). (These reactions have also been conducted using solid-phase chemistry (33).) The reactions were carried out on a 20 mg scale, and the purity of the products was reported to be greater than 80%.

Practical Tools for Preparation of a Combinatorial Library

From suppliers of specialized glassware, reagents, and mechanical devices, to manufacturers of custom and "off-the-shelf" robots for dispensing reagents and for puri-

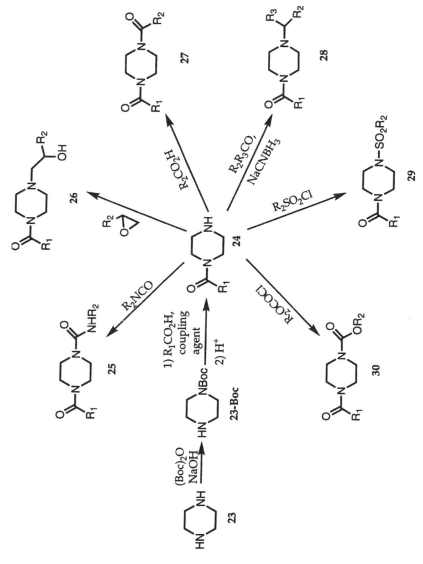

SCHEME VIII. Preparation of a piperazine-based library.

Abbreviation: Boc, *tert*-butyloxycarbonyl.

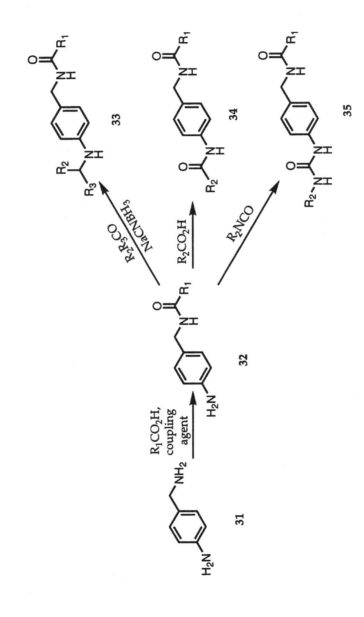

SCHEME IX. Preparation of a 4-aminobenzylamine-based library.

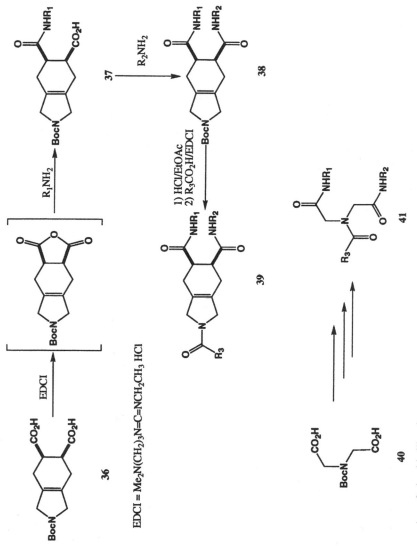

SCHEME X. Preparation of triamide libraries.

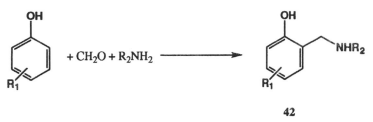

SCHEME XI. Preparation of an aminophenol library.

fying reaction products, a cottage industry has evolved to provide tools that allow combinatorial chemists to more efficiently prepare large libraries of compounds. The number and variety of tools utilized will be different for each combinatorial chemistry application, but some required elements include software for tracking products, vessels and equipment in which to carry out reactions and store products, equipment to strip solvents from solutions of compounds, and automated equipment for preparing and isolating compounds. A noncomprehensive list of some equipment and suppliers is given in Table I.

Data Management

The minimal data that must be kept on every compound includes product structure, method of synthesis, and physical location (whether microtiter-plate well or storeroom shelf). This is no different from the data kept for compounds prepared by traditional methods, but traditional record-keeping techniques cannot keep pace with production in a typical parallel synthesis program. Indeed, more time could be spent keeping and tracking records than preparing the compounds themselves!

What is required is an electronic database. The available software (such as MDL's Project Library, and Tripos's Legion and Selector, among others) allows the assembly and incorporation of two-dimensional (2-D) structures as a searchable field in a database. Assigning product ID numbers for each compound allows for an orderly storage of archive samples and the tracking of each product through the biological evaluation process. Including the method of synthesis will facilitate the resynthesis and purification of active compounds.

Reactors and Vessels for Compound Storage

Depending on the type of combinatorial chemistry approach (i.e., mixture or individual-compound synthesis) and its scale, reactions can be carried out in microtiter plates, individual vessels (e.g., microvials, scintillation vials, or capped test tubes), or in traditional round-bottom flasks. Agitation of heterogeneous reaction mixtures can be carried out by placing trays of reaction vessels onto orbital shakers, reciprocal shakers, or vibrators.

To carry out reactions under conditions other than ambient temperature and pressure, heating and cooling blocks and inert atmosphere and pressure manifolds

TABLE I. Equipment and Suppliers for Solution-Phase Combinatorial Chemistry

Equipment	Use	Supplier
Mechanical micropipetter	Dispensing small solutions of reagents and products	Gilson
Mechanical repeat pipetter	Dispensing solutions of reagents and products	Eppendorf
Solvent dispenser	Dispensing solvents and solutions of reagents	Brinkmann
Mechanical pipetter	Dispensing solutions of reagents and extracting products	Drummond
Orbital shaker	Agitating reactions	Lab-Line
Reciprocal shaker	Agitating reactions	Lab-Line
Heating block	Running reactions at high temperature	Laboratory Devices
Circulating heating and cooling bath	Heating and cooling reactions from −20 to 100 °C (when used with a channeled reaction block)	Brinkmann
Liquid addition and aliquot robot	Quenching and extracting reactions and dispensing solutions of compounds into storage vessels	Bohdan Automation
Multichannel liquid-handling robot	Dispensing solutions of compounds into microtiter plates	Bohdan Automation
Variable multichannel liquid-handling robot	Dispensing solutions of compounds into microtiter plates	Packard
Centrifugal evaporator	Concentrating and drying solutions of compounds	Savant
Multiple-task robots	Mix, heat, cool; solvent and reagent transfer	
	Automated RAM Synthesizer	Bohdan Automation
	Nautilus Synthesizer	Argonaut Technologies
	BenchMark Synthesizer	Advanced ChemTech

NOTE: Similar equipment may be available through other suppliers.

have been engineered to accommodate arrays of reaction vessels. Units typically consist of an aluminum block that holds an array of reaction vessels. The desired temperature is maintained by either placing the block on a thermostatically controlled hot plate, or by circulating hot or cold fluid (from an externally controlled, constant-temperature, circulating bath) through channels bored through the aluminum block. Commercially available electric block heaters can accommodate a variety of vials and test tubes. When placed on orbital shakers, these apparatuses can mix and heat reactions at temperatures above 200 °C. Using components from microscale chemistry sets (such as threaded test tubes, connectors, and air-cooled columns), reactions can be conducted under reflux conditions.

Isolated reaction products are typically distributed into microtiter plates (to expedite screening) and storage vials (for archival purposes). Microtiter plates come in a variety of sizes and are made using a variety of materials, not all of which are compatible with organic solvents. The use of polypropylene plates with well depths appropriate for the necessary dilution volumes has become standard in the industry.

Product Concentration

Methods for the evaporation of solvents from microtiter plates or trays of vials include vacuum-assisted centrifugal evaporation, vacuum-assisted orbital evaporation, and air-assisted evaporation. These methods involve placing a tray of samples on a rotor or orbital shaker that is enclosed in a sealed chamber. Solvent is removed by applying either a vacuum or positive air-flow to the chamber. For removing volatile solvents and co-reactants, any of these methods should be effective. The efficient removal of less volatile solvents (such as DMSO, DMF, or n-BuOH) may, however, require the use of heat as well as a vacuum.

Automation

There are many vendors (Advanced ChemTech, Bodhan Automation, Gilson, and Zymark, to name a few) of automated chemical synthesizers. These machines combine elements described in previous sections (e.g., thermally regulated reaction vessels and mixers) with solution-handling technology. The ability to robotically transfer solutions allows one to automate reagent allocation, product extraction, and product dispensing (e.g., into microtiter plates). Most vendors offer a range of automated workstations capable of performing single or multiple operations. A series of individual, dedicated workstations arranged in an assembly-line fashion, each performing one task such as reagent preparation, reaction initiation, liquid–liquid extraction, or product distribution can achieve higher throughput and be more flexible than a single multipurpose robot. Of course, for some applications, these advantages may be irrelevant, and a multipurpose robot may be preferable.

To illustrate the labor savings involved in using automation, we can compare the construction of a 4-aminopiperidine library of individual compounds (as described in Scheme VII) by a semiautomated approach with construction by manual synthesis (*17*). Bulk quantities (e.g., 10 mmol) of *N*-acyl-4-piperidones (**21**) were prepared using traditional techniques. Using various acid chlorides, individual 1 M methanolic solutions of ten different amides (**21**) were prepared, and a 1 mL aliquot of each amide solution dispensed either robotically or manually (e.g., with a mechanical, repeat pipetter) into 10 reaction vessels arranged in columns of a 10 × 8 array. Similarly, individual 1 M methanolic solutions of eight different ketones were prepared and a 1 mL aliquot of each ketone solution dispensed (either robotically or manually) into eight reaction vessels arranged in rows of the 10×8 array. Finally, 0.5 mL of a 2 M methanolic HOAc solution and then 0.5 mL of a 2 M NaC-NBH_3 methanolic solution were added (either robotically or manually) to all 80 vessels, and the array was agitated on an orbital shaker for 48 h. The reactions were quenched and worked up by adding 1 mL of a 2 M aqueous HCl solution, agitating the resulting mixtures, and then adding 0.5 mL of a 7 M aqueous NaOH solution. The resulting amines (**22**) were extracted from this mixture by adding 4 mL of chloroform to all 80 reaction vessels, agitating, and separating the organic layer from the aqueous portion. This could be accomplished manually using a mechanical pipetter equipped with a 10 mL pipette. Alternatively, a robot could be calibrated or equipped with a sensor to remove an appropriate volume of the desired layer. In either case, the solutions were dispensed into storage vials and diluted, and an appropriate volume from each vial was transferred to a well of a microtiter plate. A micropipette could be used for this operation, but a robot would be ideally suited for the repetitive transfer of solutions from one storage format to another.

In this example, there are 9 transfers of solvent or reagent to each of 80 reaction vials, a total of 720 operations. Other reactions may require fewer steps, but the advantages of automation are obvious. Once a robot is programmed to perform a series of steps, it can be used to prepare many separate batches of compounds.

Future Directions

The need to develop faster, more economical methods for drug discovery (including the steps of lead discovery, lead optimization, and bringing the drug to market) will fuel the development of the methods for solution-phase synthesis of combinatorial libraries. Advances will include the development not only of high-yielding synthetic reactions to access a vast array of structures, but also of computational tools for designing and evaluating libraries and better automated methods to prepare libraries more efficiently. In particular, the development of automated methods to prepare, analyze, and purify library products will allow combinatorial chemistry to become a more effective lead optimization tool in the hands of the medicinal chemist, as well as a powerful tool in process research, materials and polymer science, and other disciplines.

References

1. Gallop, M. A.; Barrett, R. W.; Dower, W. J.; Fodor, S. P. A.; Gordan, E. M. *J. Med. Chem.* **1994**, *37*, 1233–1251.
2. Gordan, E. M.; Barrett, R. W.; Dower, W. J.; Fodor, S. P. A.; Gallop, M. A. *J. Med. Chem.* **1994**, *37*, 1385–1401.
3. Ecker, D. J.; Crooke, S. T. *Biotechnology* **1995**, *13*, 351–360.
4. Mitscher, L. A. *Chemtracts – Organic Chemistry* **1995**, *8*, 19–21.
5. Ellman, J. A. *Chemtracts – Organic Chemistry* **1995**, *8*, 1–4.

6. Storer, R. *Drug Disc. Today* **1996**, *1*, 248–254.
7. Terrett, N. K.; Gardner, M.; Gordon, D. W.; Kobylecki, R. J.; Steele, J. *Tetrahedron* **1995**, *51*, 8135–8173.
8. Thompson, L. A.; Ellman J. A. *Chem. Rev.* **1996**, *96*, 555–600.
9. Balkenhohl, F.; von dem Bussche-Hünnefeld, C.; Lansky, A.; Zechel, C. *Angew. Chem. Int. Ed. Engl.* **1996**, *35*, 2288–2337.
10. Bergbreiter, D. E. In *Polymer Reagents and Catalysts*; Ford, W. T., Ed.; ACS Symposium Series 308; American Chemical Society: Washington, DC, 1986.
11. Hodge, P. In *Synthesis and Separations Using Functionalized Polymers*; Hodge, P., Sherrington, D. C., Eds; John Wiley and Sons: New York, 1988; pp 43–144.
12. Hodge, P. In *Polymer-Supported Reactions in Organic Synthesis*; Hodge, P., Sherrington, D. C., Eds; Wiley-Interscience: Chichester, 1980.
13. Smith, P. W.; Lai, J. Y. Q.; Whittington, A. R.; Cox, B.; Houston, J. G.; Stylli, C. H.; Banks, M. N.; Tiller, P. R. *Bioorg. Med. Chem. Lett.* **1994**, *4*, 2821–2824.
14. Carell, T.; Wintner, E. A.; Bashier-Hashemi, A.; Rebek, J., Jr. *Angew. Chem., Int. Ed. Engl.* **1994**, *33*, 2059–2061.
15. Bailey, N.; Dean, A. W.; Judd, D. B., Middlemiss, D.; Storer, R.; Watson, S. P. *Bioorg. Med. Chem. Lett.* **1996**, *6*, 1409–1414.
16. Peterson, J. R. Presented at Exploiting Molecular Diversity: Small Molecule Libraries for Drug Discovery, La Jolla, CA, January 1995.
17. Garr, C. D.; Peterson, J. R.; Schultz, L.; Oliver, A. R.; Underiner, T. L.; Cramer, R. D.; Ferguson, A. M.; Lawless M. S.; Patterson D. E. *J. Biomol. Screen.* **1996**, *1*, 179–186.
18. Hogan, J. C. Presented at the IBC Conference on Synthetic Chemical Libraries in Drug Discovery, London, UK, October, 1995.
19. Peisach, E.; Casebier, D.; Gallion, S. L.; Furth, P.; Petsko, G. A.; Hogan, J. C., Jr.; Ringe, D. *Science (Washington, D.C.)* **1995**, *269*, 66–69.
20. Boger, D. L.; Tarby, C. M.; Myers, P. L.; Caporale, L. H. *J. Am. Chem. Soc.* **1996**, *118*, 2109–2110.
21. Cheng, S.; Comer, D. D.; Williams, J. P.; Myers, P. L.; Boger, D. L. *J. Am. Chem. Soc.* **1996**, *118*, 2567–2573.
22. Selway, C. N.; Terrett, N. K. *Biorg. Med. Chem. Lett.* **1996**, *4*, 645–654.
23. Pirrung, M. C.; Chen, J. *J. Am. Chem. Soc.* **1995**, *117*, 1240–1245.
24. Terrett, N. Presented at the University of Exeter Combinatorial Synthesis Symposium, Exeter, UK, July 1995.
25. Curran, D. P.; Hadida, S. *J. Am. Chem. Soc.* **1996**, *118*, 2531–2532.
26. Carell, T.; Wintner, E. A.; Bashier-Hashemi, A.; Rebek, J., Jr. *Angew. Chem., Int. Ed. Engl.* **1994**, *33*, 2061–2064.
27. Shipps, G. W., Jr.; Spitz, U. P.; Rebek, J., Jr. *Biorg. Med. Chem. Lett.* **1996**, *4*, 655–657.
28. Gallion, S. L. Presented at Combinatorial Chemistry '96—Track B, San Diego, CA, March 1995.
29. Ugi, I.; Domling, A.; Horl. W. *Endeavour* **1994**, *18*, 115–122.
30. Gokel, G.; Ludke, G.; Ugi, I. In *Isonitrile Chemistry*; Ugi, I., Ed.; Academic: New York, 1971; pp 145–199.
31. Keating, T. A.; Armstrong, R. W. *J. Am. Chem. Soc.* **1995**, *117*, 7842–7843.
32. Weber, L.; Wallbaum, S.; Broger, C.; Gubernator, K. *Angew. Chem., Int. Ed. Engl.* **1995**, *34*, 2280–2282.
33. Hogan, J. C., Jr. *Nature (London)* **1996**, *384* (Suppl), 17–19.
34. Hogan, J. C., Jr. Presented at Exploiting Molecular Diversity: Small Molecule Libraries for Drug Discovery, San Diego, CA 1996.

Analytical Tools for Solution-Phase Synthesis

Christopher E. Kibbey

This chapter focuses on analytical tools suitable for qualitative and quantitative analyses of solution-phase combinatorial libraries, particularly approaches for interfacing information-rich detectors with high-efficiency separation techniques to provide the combinatorial chemist with the speed of analysis required for automated synthesis. I briefly describe three high-resolution separation techniques, then review the various interfaces developed to couple chromatographic methods with Fourier transform infrared, nuclear magnetic resonance, and mass spectrometric detectors for qualitative analyses. I conclude with a discussion of sample quantitation approaches and a summary of the data-handling requirements for library analysis.

Analysis is essential for tasks ranging from reaction monitoring and optimization to the identification of a compound that is active in a biological assay. The solution-phase chemist must contend with the fact that the three analytical methods commonly used by organic chemists, infrared and nuclear magnetic resonance spectroscopy and mass spectrometry, all detect solvent as well as solute. I will describe various techniques that have been developed to eliminate solvent from these analyses, sometimes at the expense of sensitivity. A further complication is that the compounds involved are often in complex mixtures. I will therefore begin my discussion by considering high-performance liquid chromatography (HPLC), the high-resolution separation technique most familiar to synthetic organic chemists, and the one most commonly used in conjunction with spectroscopic or spectrometric analytical techniques.

High-Resolution Separations of Organic Molecules

A number of books (*1, 2*) and review articles (*3*) describing the theory and application of HPLC have been written, and I will therefore discuss the techniques of HPLC only briefly in this chapter, highlighting the attributes of conventional HPLC and two recent techniques that separate compounds with greater resolution.

HPLC relies on the partition of compounds between two phases; in conventional HPLC those phases are the solid support and the mobile phase, with the latter moved by hydraulic pressure. Separations by HPLC are generally carried out using stainless-steel columns (with internal diameters (i.d.) of 0.5–4.6 mm) packed with stationary porous particles with diameters of 3–10 μm, although in a given column they are of a uniform size (Figure 1A). The separation mode can be either normal-phase, in which case the solid support is more polar than the mobile phase (also known as the liquid eluant), or reversed-phase, in which case the solid support is less polar than the liquid eluant. The polarity of the silica-based solid support is decreased for reversed-phase HPLC by chemically bonding a hydrophobic organic molecule onto the surface of the support particles. As solutes pass through the column in the moving eluant stream, they are "sorbed" on and off the stationary solid support. Solute mixtures are separated into their constituent components on the basis of differences in their partition coefficients between the mobile and stationary phases. The hypothetical maximum of separation is not achieved because of a number of band-broadening processes, including those related to the laminar-flow profile of the mobile phase in HPLC (which results from the slower flow near the sides of the column), and the large pressure drop across the column. These factors contribute to separation efficiencies on the order of 25,000–50,000 plates m^{-1}.

To increase the resolution of HPLC, in 1984 Terabe developed micellar electrokinetic capillary chromatography (MEKC) as a special application of capillary zone electrophoresis (CZE) that, unlike CZE, was capable of separating neutral compounds. CZE separates charged species in capillaries with an i.d. of 25–100 μm and filled with a running buffer of known pH and ionic strength. The application of an electric field across the length of the capillary causes the individual solute ions to migrate toward the electrode of opposite charge. In addition, the charged nature of the interior wall of the fused-silica capillary induces charge in the mobile phase. This electrical double layer contributes to a pH-dependent electroosmotic flow (eof) of buffer toward the cathode as the buffer is attracted by the electrode. The eof is characterized by a flat flow profile, in contrast to the parabolic profile of laminar flow in HPLC. In CZE, charged solutes are separated on the basis of differences in their electrophoretic mobilities. Neutral solutes, however, co-migrate with a velocity dictated by the eof and cannot be separated by CZE. Ionic detergents with long alkyl chains (e.g., sodium dodecyl sulfate [SDS]) were therefore added to the CZE running buffer at a concentration above the critical micelle value to generate a micellar pseudophase into which neutral solutes can partition. These detergent micelles are characterized by a hydrophobic interior and a charged hydrophilic exte-

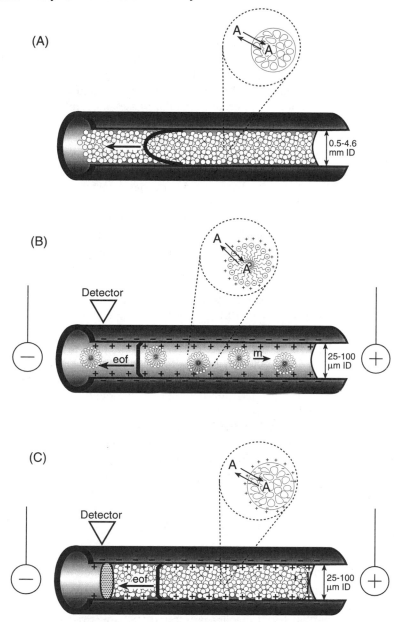

FIGURE 1. Comparison of three high-resolution techniques for the separation of neutral organic molecules. (A) In conventional HPLC, compound A partitions between the mobile phase and the bead. (B) In micellar electrokinetic capillary chromatography (MEKC), the partition is between the two moving phases: the solute (moving as a result of electroosmotic flow) and the detergent micelles (moving in the opposite direction). (C) In capillary electrochromatography (CEC), compound A partitions between the charged mobile phase and the bead.

rior. The choice of detergent and electrode polarity across the capillary in an MEKC separation are such that the direction of the eof and the direction of migration of the charged micelles are opposite (Figure 1B). In addition, the pH of the running buffer is maintained above pH 5–6 to ensure a strong eof. Neutral solutes are separated with efficiencies greater than 100,000 plates m^{-1} on the basis of differential partitioning between the running buffer and micellar phases. Several review articles describing MEKC have appeared in the literature (4, 5), and particular attention has been paid to using the technique in chiral separations (6, 7).

Capillary electrochromatography (CEC) is another method by which both neutral and charged molecules can be separated. It is a hybrid of conventional packed-column HPLC and CZE. The technique has developed rapidly over the past decade and has been described and reviewed in the chromatographic literature (8–11). Separations by CEC (Figure 1C) are carried out in fused-silica capillaries with 25–100 μm i.d. packed with 1–3 μm diameter silica-based, reversed-phase porous particles. The packing material is held in place within the capillary by two frits composed of fused silica, and placed at the capillary inlet and just before the detection cell of the capillary. The liquid eluant (aqueous or organic) that carries solutes through the packed capillary is driven electroosmotically, rather than hydraulically as in conventional HPLC. The surface charge along the inner wall of the capillary combines with the surface charge on the silica particles to create an electrical double layer at each solid–liquid interface. The application of an electric field of approximately 100–300 V cm^{-1} across the length of the capillary induces an electroosmotic flow in the direction of the cathode with a velocity of up to 3 mm s^{-1}. Neutral solute molecules partition between the stationary phase and mobile phase and are separated into well-defined bands along the length of the capillary on the basis of differences in their individual partition coefficients. Charged analytes are separated by a combination of partitioning and electrophoretic migration mechanisms. The lack of a pressure drop along the length of the capillary, the small diameter of the packed beads, and the flat profile of electroosmotic flow combine to yield separation efficiencies of 100,000–200,000 plates m^{-1}. An advantage of CEC over MEKC is that the hydrophobic phase is chemically bonded to the silica support and will not interfere with solute detection by UV, mass spectrometry, or NMR.

Qualitative Analysis by Infrared Spectroscopy

Once compounds have been separated, true analysis begins. The first method for qualitative analysis that I will discuss is infrared (IR) spectroscopy, a powerful detection method of interest to synthetic and analytical chemists alike. In the mid-IR region (4000–650 cm^{-1}), this technique can provide important functional group information (4000–1300 cm^{-1}) and structural information (1300–650 cm^{-1}) for a variety of organic molecules. The information contained in a Fourier transform IR (FTIR) spectrum may be used to confirm the identity of a compound, or in reaction monitoring to signal the completion of a synthetic step. When used as a detector in

HPLC, the FTIR spectra are normally collected and saved for later reconstruction using Gram-Schmidt orthogonalization (*12*) or suitable alternative methods (*13*). Real-time monitoring of component bands eluting from the chromatographic column may be achieved in-line by first detecting the elution of a compound with UV, and then taking an IR spectrum of that sample. IR spectroscopy is primarily a qualitative technique, because the narrow absorption bands that characterize IR spectra usually lead to deviations from Beer's law. If quantitation is desired, then the chosen IR absorption band should have a reasonable bandwidth and intensity, and be well resolved from nearby absorptions so that an accurate integration can be performed.

Coupling FTIR spectrometers to liquid chromatographic (LC) systems is not entirely straightforward. As a given functional group will absorb similarly whether it is a part of the solvent or solute, solvent subtraction in IR is problematic. In addition, the interfaces must be designed to have as small a volume as possible so as to minimize band broadening outside of the column. The two types of HPLC–FTIR interfaces commonly used are flow-through and solvent-elimination. Flow-through interfaces couple the chromatographic effluent directly with the inlet of the FTIR spectrometer; thus, they are simple and capable of real-time analysis, but have low sensitivity and the problem of subtracting solvent absorbances remains. Solvent-elimination systems eliminate the latter problem by removing the chromatographic eluant from the sample prior to infrared analysis. They are sensitive systems, but often complex, and in general incompatible with real-time analysis. The details of these two types of interface are described below.

Flow-Through Interfaces

The most straightforward approach to coupling HPLC with FTIR spectrometry would be simply to treat the FTIR spectrometer as an on-line detector and pass the effluent from the chromatographic column through a flow cell mounted within the sample beam of an FTIR spectrometer. Indeed, the earliest reported LC–FTIR spectrometry systems used flow-through cells in exactly this manner (*14–16*). These early systems were, however, of little practical value for routine compound analysis by HPLC because many common chromatographic solvents absorb across the mid-IR region. Background absorption by the mobile phase narrows the useful range of the mid-IR that can be used for analyte detection. In addition, solvents that strongly absorb infrared radiation severely limit the path lengths that can be used in simple flow cells, further reducing the sensitivity of this detection technique. Figure 2 shows the regions of the mid-IR in which a variety of common solvents used in HPLC show strong absorption. Of the more common normal-phase solvents, chloroform and dichloromethane provide the widest spectral window, and allow the use of liquid-flow cells with path lengths of up to 1 mm. Brown and Taylor have demonstrated that substitution of $CDCl_3$ for $CHCl_3$ opens the spectral window around 3000 cm^{-1}, where C–H stretching absorbances occur, and no loss of solute spectral information occurs when $CDCl_3$ is used in combination with flow cells with path lengths of 0.2 mm (*17*). Chloroform and dichloromethane may, however,

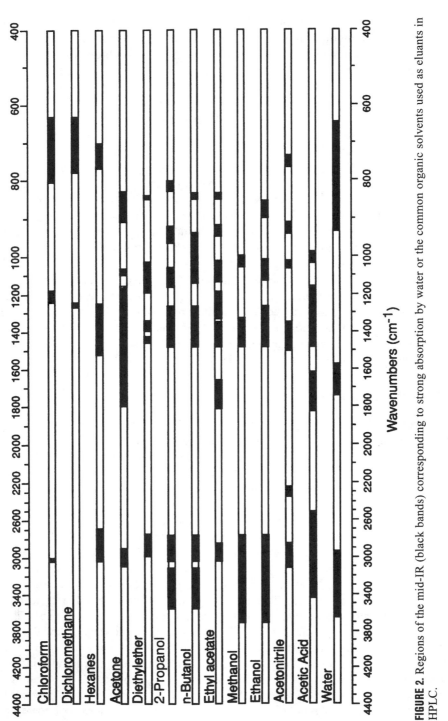

FIGURE 2. Regions of the mid-IR (black bands) corresponding to strong absorption by water or the common organic solvents used as eluants in HPLC.

be too polar to be used by themselves as eluants for chromatographic separations of certain compounds of pharmaceutical interest. Addition of a hydrocarbon solvent, such as hexane, to chloroform will increase the retention of polar solutes on silica, but will also compromise spectral information in the 1250–1530 cm^{-1} and 2700–3100 cm^{-1} regions. In reversed-phase HPLC, it is the strong absorption of water across much of the mid-IR range that makes solute detection by FTIR difficult. Indeed, flow cells designed for use with solvents containing high percentages of water require path lengths of 25 μm or less (*18*).

The use of microbore (1 mm i.d.) rather than analytical (4.6 mm i.d.) chromatographic columns reduces solute elution volumes and so can increase analyte concentration 20-fold, helping to compensate for the reduced sensitivity of narrow path-length flow cells in HPLC–FTIR systems (*19*). Additional advantages of microbore-scale chromatographic separations include reduced solvent consumption and smaller injection requirements. Taylor and colleagues developed a flow cell with zero dead volume for use with microbore chromatographic columns in normal-phase HPLC–FTIR. A schematic of the flow-cell design is shown in Figure 3 (*20*). The flow cell has a path length of 0.45 mm, an internal volume of 0.33 μL, and can detect less than 50 ng of 2,6-di-*tert*-butylphenol using measurements at 3641 cm^{-1}.

Optimization of flow-cell design for use in reversed-phase HPLC–FTIR applications has centered on the use of attenuated total reflectance (ATR) technology to reduce the effective cell path lengths to levels that keep solvent interference

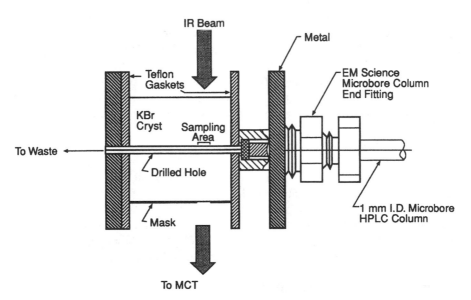

FIGURE 3. Schematic of a flow cell for HPLC–FTIR with zero dead volume. (Reproduced from reference 20. Copyright 1984 American Chemical Society.)

Abbreviation: MCT, mercury–cadmium–telluride.

in the mid-IR region at acceptable levels. In an ATR FTIR experiment, the sample is placed on the ATR cell's exterior surface, and IR radiation is reflected along the interior walls multiple times as it traverses the length of the cell. In contrast to typical FTIR flow cells whose path lengths correspond to the sample thickness, the IR radiation passing through an ATR cell penetrates only a few wavelengths into the exterior sample region per internal reflection. McKittrick et al. (21) compared the performance of micro and ultramicro cylindrical internal reflectance (CIRCLE, Spectra Tech Inc., Stamford, CT) cells for the detection of caffeine and theophylline after reversed-phase HPLC. The micro CIRCLE cell used had an internal volume of 24 μL and an effective path length of 4.65 μm (7.5 reflections), while the ultramicro cell had an internal volume of 1.75 μL and an effective path length of 3.18 μm (4.9 reflections). The extremely short path lengths of these cells do, however, compromise solute sensitivity. The detection limits for caffeine with the micro and ultramicro cells were 0.1 mg and 0.5 mg, respectively, when it was injected directly into the column. To overcome problems of column overloading over the solute range of 0.1 to 20 mg, the authors used a 7.8 mm i.d., C-18 reversed-phase column. Jinno and colleagues have demonstrated the use of deuterated solvents in microbore HPLC–FTIR as an approach to open up spectral windows in the mid-IR that are normally closed because of solvent opacity (22–24). By replacing H_2O with D_2O in reversed-phase eluants, they observed solute absorptions in the 3000–3800 cm^{-1} region. Further, the use of deuterated solvents in reversed-phase separations offers the additional advantage of enhanced capacity factors for some solutes (22).

In addition to the path-length requirements for flow-cell design in HPLC–FTIR discussed above, the materials of construction must also be considered. Flow cells for HPLC–FTIR should be made of materials compatible with the mobile phase eluants the cell will come in contact with, and the cell windows should be made of materials with high refractive indices and low wavenumber cutoffs. Flow cells designed for use in normal-phase HPLC applications may be constructed of KCl, KBr, CsBr, or CsI. For reversed-phase applications, flow cells may be constructed of ZnSe, as in the CIRCLE cells mentioned earlier, or silver chloride or BaF_2.

Advantages of Flow-Through FTIR Interfaces

The primary advantage of flow-through HPLC–FTIR interfaces is their simplicity. A flow-through interface provides a direct link between the HPLC column outlet and the FTIR spectrometer, enabling real-time analysis of the column effluent. Further, the dead-volumes of the typical FTIR cells range from 2–25 μL, so they do not contribute significantly to band broadening that can occur outside of the HPLC column. Finally, flow-through interfaces do not suffer from the potential for either losses of volatile components or thermal degradation of solutes.

Disadvantages of Flow-Through FTIR Interfaces

Flow-through interfaces suffer from low sensitivity under reversed-phase chromatographic conditions, which virtually eliminates the practicality of this mode of HPLC for the analysis of combinatorial libraries. When used with normal-phase HPLC methods, flow-through interfaces provide greater analyte sensitivity, but still suffer from potential losses of solute information due to regions of solvent opacity in the infrared. In addition, the mathematical complexity of performing spectral subtractions during gradient elution limits the practicality of using these interfaces with isocratic HPLC analysis.

Solvent-Elimination FTIR Interfaces

To overcome the problems associated with flow-through HPLC–FTIR interfaces (solvent opacity and loss of solute sensitivity) a number of HPLC–FTIR interfaces have been designed to remove the liquid eluant from the solutes of interest prior to FTIR spectrometry. The three interfaces for solvent elimination described to date are a solute-deposition interface (with deposition on a substrate compatible with IR followed by diffuse reflectance FTIR spectroscopy [DRIFT]), a monodisperse aerosol generation interface for combining liquid chromatography with FTIR (MAGIC-LC), and a continuous recording reflection–absorption interface. The characteristic features of each of these interfaces are discussed below.

In the HPLC–DRIFT approach, pioneered by Kuehl and Griffiths (25), the chromatographic eluant is deposited into a series of diffuse-reflectance cups filled with an appropriate substrate such as KBr. The diffuse-reflectance spectra of deposited solutes are recorded following evaporation of the chromatographic eluant. The DRIFT technique is used to measure the absorption of infrared radiation by solutes adsorbed onto a reflective substrate. While this system has demonstrated improved sensitivities over flow-through interfaces with microbore HPLC (26), the system suffers from several disadvantages. The HPLC–DRIFT interface does not allow a continuous analysis of the chromatographic eluant, and the quality of the diffuse-reflectance spectra is easily affected by changes in the particle size of the KBr substrate that may occur during solvent evaporation. In addition, aqueous-based eluants will dissolve KCl and KBr diffuse-reflectance substrates, so reversed-phase separations are not practical. However, Kalasinsky et al. (27) adapted this technique so that it is compatible with reversed-phase HPLC by adding a postcolumn treatment of the mobile phase with 2,2-dimethoxypropane. Water in reversed-phase eluants reacts with 2,2-dimethoxypropane on an equimolar basis to form two moles of methanol and one mole of acetone. The products of this reaction are more volatile than water and readily evaporate from the surface of the KCl substrate. Alternatives to chemical removal of water in reversed-phase HPLC–DRIFT systems have included the use of insoluble diffuse reflectance substrates such as industrial-grade diamond powder (28), and solute extraction in methylene chloride prior to deposition on KCl (29). Solvent can also be removed using a thermospray desolvation apparatus, with solvent then introduced into the DRIFT accessory of an FTIR spectrometer using a moving belt (30). The complexities of these HPLC–DRIFT systems makes them impractical for routine sample analysis, however.

The MAGIC-HPLC/FTIR interface was developed by Willoughby and Browner (31), and was adapted from a similar interface developed for coupling

HPLC with mass spectrometry. A schematic of a MAGIC-HPLC/FTIR system used for biopolymer conformation studies in reversed-phase HPLC (RP-HPLC) separations (*32*) is shown in Figure 4. The interface consists of three components: a monodisperse aerosol generator; a desolvation chamber at atmospheric pressure and ambient temperature; and a momentum-based particle-enrichment separator. During operation, the HPLC column effluent is converted into a highly uniform droplet stream by the monodisperse aerosol generator. Helium gas is used to disperse the droplet stream in the desolvation chamber, where the solvent molecules are removed by evaporation. Rapid desolvation at ambient temperatures is achieved within a few milliseconds, driven by the rapid increase in the surface area:volume ratio experienced by the aerosol droplets during evaporation. The momentum chamber removes the mobile-phase eluant vapor and helium under vacuum, and the solute particles are deposited on the surface of an infrared transparent window (e.g., KCl or KBr). The collection plate is automatically moved during analysis so that each solute is deposited on a clean portion of the window. The collection of solute FTIR spectra is performed off-line. The MAGIC interface has been shown to remove solvents that are as difficult to vaporize as 100% water at room temperature, and the interface is compatible with HPLC flow rates of up to 1.0 mL min^{-1} (*33, 34*). Although only ~30% of an eluting solute is deposited on the collection plate,

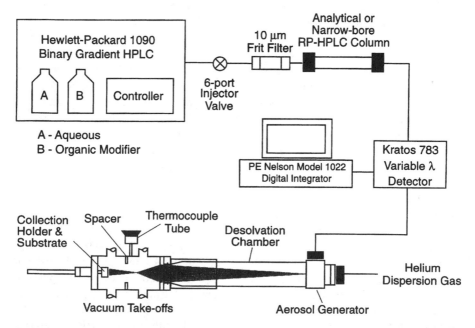

FIGURE 4. Diagram of a MAGIC-HPLC/FTIR system. After separation by the HPLC system (top), solvent is removed by aerosol generation and solute is deposited on the substrate for later analysis by FTIR. (Reproduced from reference 32. Copyright 1996 American Chemical Society.)

MAGIC-HPLC/FTIR can detect as little as 10–100 ng, so the instrument is suitable for trace analysis. In addition, as desolvation occurs at ambient temperature, the instrument can be used for the analysis of thermally labile compounds.

A reflection–absorption interface for HPLC–FTIR that records continuously was developed by Gagel and Biemann (*35, 36*) in the mid-1980s. The effluent from a narrow-bore HPLC column is continuously sprayed onto the surface of a rotating reflective disc (e.g., an aluminum mirror). Solvent evaporation is achieved by means of a stream of heated nitrogen blown across the surface of the deposited eluant aerosol. Continuous rotation of the reflective disc in concert with translation of the nebulizer along the disc's radius produce a spiral pattern of deposited solute bands. As the disc rotates, the FTIR spectra are recorded by reflectance–absorbance spectroscopy using a reading device housed in the sample compartment of the spectrometer. A diagram of the continuous LC–FTIR interface is shown in Figure 5. The continuous-recording LC–FTIR system is the only solvent-elimination interface that has been commercialized, with Lab Connections (Marlborough, MA) selling it as the LC Transform. The commercial instrument replaces the heated gas nebulizer of the original design with an ultrasonic nebulizer and vacuum desolvation chamber, and the original aluminum mirror is replaced with an aluminum-backed germanium disc. This interface is compatible with both normal- and reversed-phase separations. Representative separations in the original studies used 1 mm i.d. columns at mobile-phase flow rates of 30 μL min^{-1} (*35, 36*), and useful spectra were obtained following injection of 31 ng of phenanthrenequinone (*36*). A somewhat similar direct-deposition apparatus using a concentric flow nebulizer was reported by Lange et al. (*37*). The greater desolvation efficiency of the concentric flow nebulizer allowed the use of higher mobile-phase flow rates in reversed-phase HPLC. Neither of these interfaces, however, allowed the use of nonvolatile buffers in the chromatographic mobile phase. Somsen et al. incorporated a liquid–liquid extraction (LLE) device upstream of the solute-deposition assembly to improve the compatibility of the

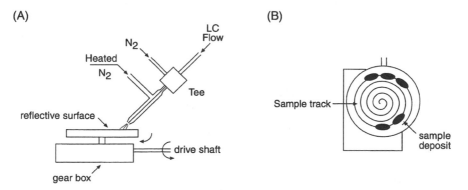

FIGURE 5. Diagram of an HPLC–FTIR interface capable of continuous recording. (A) Side view and (B) view of the deposited components. (Reproduced from references 35 and 36, respectively. Copyright 1986 and 1987 American Chemical Society.)

direct deposition HPLC–FTIR interface with reversed-phase eluants (*38*). Representative FTIR spectra of acenaphthenequinone, diuron, phenanthrenequinone, and linuron obtained with the LC–LLE–FTIR interface are shown in Figure 6. The LC–LLE–FTIR system allows the use of nonvolatile, buffered, aqueous eluants (e.g., 10 mM potassium phosphate, pH 7) at flow rates of up to 200 μL min^{-1}. However, the low extraction efficiencies of polar solutes into dichloromethane compromise the sensitivity of the LC–LLE–FTIR interface.

Advantages of Solvent-Elimination FTIR Interfaces

Solvent-elimination HPLC–FTIR interfaces provide nearly complete elimination of the mobile phase from the solute components. These interfaces are compatible with both normal-phase and reversed-phase eluants and can accommodate the use of volatile buffers in the HPLC mobile phase. Gradient elution may be performed without complicating the acquisition of the solute FTIR spectra. Many of the solvent-elimination interfaces described to date are sufficiently sensitive to identify not only the major component bands eluted from the column, but also the low-level impurities in library samples.

Disadvantages of Solvent-Elimination FTIR Interfaces

The major disadvantages of solvent-elimination HPLC–FTIR interfaces stem from the complexity of these systems and their general incompatibility with real-time analysis of the HPLC column effluent. In addition, volatile solutes can be lost during the solvent-elimination step, and the use of heat to aid solvent evaporation can lead to degradation of thermally labile compounds. Components of the chromatographic solvent may become entrapped in the solute band during the deposition process, and thus appear in the sample spectra, and irregularities in solute deposition can lead to spectral distortions. The limited capacity of these interfaces to rapidly remove aqueous-based eluants restricts the practicality of performing reversed-phase separations on analytical-scale HPLC columns at flow rates above 0.5 mL min^{-1}. Consequently, reversed-phase HPLC–FTIR analysis of combinatorial libraries employing a solvent-elimination interface should be carried out on narrowbore (2.0 mm i.d.) or microbore (1 mm i.d.) columns at mobile-phase flow rates of 0.5 mL min^{-1} or less.

Qualitative Analysis by NMR Spectroscopy

Nuclear magnetic resonance (NMR) spectroscopy is undoubtedly the most information-rich spectroscopic technique available for probing the structure of organic compounds in solution. NMR spectroscopy can be used to differentiate between most structural, conformational, and optical isomers. The chemical shift information and coupling constant parameters provided by ^{1}H NMR allow specific protons within an organic molecule to be assigned unambiguously. By performing heteronuclear ^{1}H/^{13}C shift and homonuclear shift correlated experiments, the exact structure of an

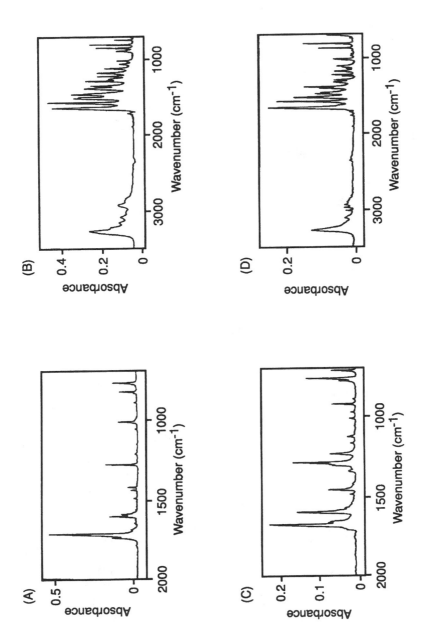

FIGURE 6. Representative FTIR spectra from an LC–LLE–FTIR analysis of (A) acenaphthenequinone, (B) diuron, (C) phenanthrenequinone, and (D) linuron. (Reproduced from reference 38. Copyright 1996 American Chemical Society.)

organic molecule may be determined. In contrast to mass spectrometry, NMR spectroscopy is a noninvasive and nondestructive technique. A sample may be recovered after an NMR analysis and used in subsequent analyses.

Despite the strengths of NMR as a powerful spectroscopic method, it is difficult when using most NMR techniques to distinguish between a compound of interest and any impurities or contaminants that may be present in the sample. Structure elucidation of mixtures of unknown organic compounds with overlapping ^{1}H NMR signals is difficult, especially in spectral regions that are crowded with many resonances. What is needed for the NMR analysis of combinatorial libraries is a means of distinguishing between the individual sample components during spectral acquisition. This section describes two techniques, both quite new but promising, that have been applied successfully to the analysis of complex sample mixtures by NMR. The first technique involves the use of flow-through interfaces to directly couple HPLC with NMR, and thus requires solvent interference to be eliminated. The second approach distinguishes between individual components of a sample solution based on differences in their diffusion rates, and does not require a separation step to be performed prior to NMR analysis.

Flow-Through Interfaces for NMR

The first interfaces for coupling liquid chromatography with ^{1}H NMR were described in the late 1970s (39, 40), but these systems used iron magnets with low field strengths, and detection sensitivity was poor (e.g., a 100-μg limit of detection). Research performed during the early 1980s by three independent research groups led to the first applications of on-line HPLC–NMR using superconducting magnets (41–43). These flow-through interfaces all included a hollow glass tube, through which the chromatographic eluant flows, and a radio-frequency (rf) coil attached to the outside. This apparatus is placed into the bore of the superconducting NMR magnet. These three early HPLC–NMR interfaces differed in the design of the rf coil and in the internal volume and geometry of the flow-through glass tube. The continuous-flow cell designed by Bayer and co-workers in Tübingen, Germany (42) was later improved and commercialized by Bruker (Karlsruhe, Germany).

A diagram of the commercial continuous-flow NMR probe and a typical arrangement for coupling HPLC with NMR are shown in Figures 7A and 7B, respectively. The flow cell is a U-shaped glass tube with a slight bulge at the measuring end. An rf coil is attached to the outside of the glass tube along the region defined by the bulge. The HPLC–NMR probe is preceded by a switching valve (see Figure 7B) that allows spectra to be collected in either continuous-flow or stopped-flow mode. Miniaturization of the basic flow-cell design shown in Figure 7A has been described (44, 45) for applications using nanoscale separation techniques (e.g., CZE, CEC, and microcapillary HPLC). In addition, so-called "inverse" continuous-flow probes are designed with an additional coaxial coil, matched to the ^{13}C resonance frequency, surrounding the primary ^{1}H detection coil. The latter design enables heteronuclear ^{1}H and ^{13}C shift-correlated experiments to be performed on

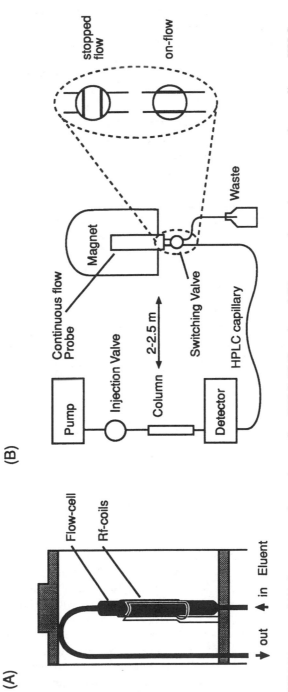

FIGURE 7. Diagrams of (A) the geometry of a continuous-flow HPLC–NMR probe and (B) an experimental arrangement for coupling an HPLC system with an NMR spectrometer. (Reproduced with permission from reference 47. Copyright 1995 Elsevier Science Publishers B.V.)

compounds eluting from the HPLC column. A continuous-flow, on-line interface for collecting ^{13}C NMR spectra of compounds separated by HPLC was developed by Stevenson and Dorn (*46*). This system uses dynamic nuclear polarization (DNP) enhancement of ^{13}C nuclei through a solid–liquid intermolecular transfer (SLIT) of polarization from a silica-phase immobilized nitroxide (SPIN) system to the ^{13}C nuclide of analyte molecules in the liquid stream. The DNP technique provides a 10–100-fold enhancement of the ^{13}C-NMR signal, making it compatible with continuous-flow semipreparative chromatographic separations. A schematic of the continuous-flow, on-line HPLC–^{13}C-NMR system is shown in Figure 8.

In contrast to conventional NMR probes, which typically spin the sample tube at 20 Hz to expose the sample to a more uniform magnetic field (B_0), the continuous-flow probes used in HPLC–NMR systems do not rotate the sample within the field of the superconducting magnet. As a consequence, the NMR line widths achieved with continuous-flow HPLC–NMR interfaces operated in stopped-flow mode are significantly greater than those obtained with conventional NMR probes. For example, the signal line width of chloroform at the height of the ^{13}C satellites in acetone is 3–4 Hz when measured using conventional ^{1}H NMR instrumentation, whereas continuous-flow probes give values of 12–16 Hz (*47*).

Additional line-broadening is observed when continuous-flow HPLC–NMR probes are used in on-flow rather than stopped-flow mode because the nuclei have limited residence times within the detection coils of the probe when compared to their spin–lattice and spin–spin relaxation times. The spin–lattice (T_1) and spin–spin (T_2) relaxation times of a nucleus in flowing mode are related to their static values and the residence time (τ) of the nucleus in the detector cell by equations 1 and 2, respectively (*48*).

$$\frac{1}{T_{1,\text{flow}}} = \frac{1}{T_{1,\text{static}}} + \frac{1}{\tau} \qquad (1)$$

$$\frac{1}{T_{2,\text{flow}}} = \frac{1}{T_{2,\text{static}}} + \frac{1}{\tau} \qquad (2)$$

As constant flow rates are generally used in HPLC, τ is dependent on the internal volume of the flow cell used in the HPLC–NMR probe. Using a 250-MHz NMR instrument, Bayer et al. found little degradation in the signal intensity for CHCl$_3$ in acetone-d_6 at flow rates of up to 5 mL min^{-1} with a flow cell having an internal volume of 126 μL ($1/\tau = 0.66$ Hz) (*42*). If, however, the flow cell had a volume of 44 μL, the CHCl$_3$ line width was 1.05 Hz, almost double the static line width (0.55 Hz) at a flow rate of 1 mL min^{-1}. Dorn reported that the sample should reside in the field of the superconducting magnet for approximately five times the spin–lattice relaxation time (T_1) before the total magnetization is measured, and that the sample should be resident in the flow cell for a total of 1–6 s (*41*). When used

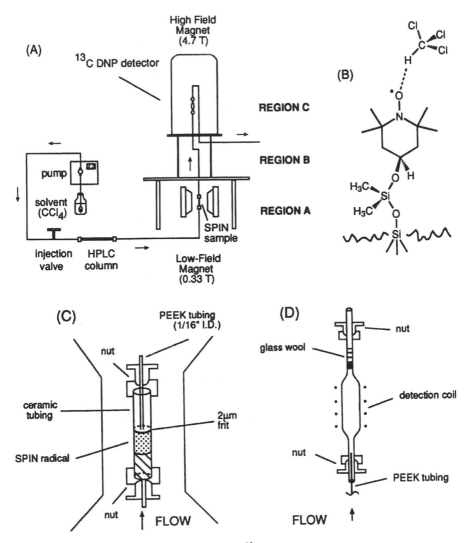

FIGURE 8. Diagram of a continuous-flow HPLC/^{13}C-NMR (DNP) instrument (A) illustrating the polarization buildup (region A), sample transfer (region B) and NMR detection (region C) portions of the system. (B) The SPIN system used, with CHCl$_3$ as the sample analyte. (C) The low-field electron paramagnetic resonance (EPR) flow cell in region A. (D) The high-field NMR flow cell in region C. (Reproduced from reference 46. Copyright 1994 American Chemical Society.)

Abbreviation: PEEK, poly ether ether ketone.

with typical chromatographic flow rates of 1–2 mL min^{-1}, these requirements constrain the internal volume of HPLC–NMR flow cells to 35–100 μL. These cell volumes are of the same order as typical chromatographic peak widths and can lead to band broadening outside of the HPLC column, and thus a significant loss of resolution between solute peaks that elute close to one another. The dispersion effects of flow cells with detection volumes in the range 1–400 μL have been determined from peak width measurements of mixtures of dansylated amino acids separated on 250 × 4.6 mm i.d. analytical HPLC columns (49). HPLC–NMR flow cells with detection volumes greater than 48 μL contribute to increased plate heights (an indication of the column's separation efficiency) for solutes with capacity factors less than 2.5. Band broadening that occurs outside of the separation column can be minimized in HPLC–NMR systems by placing the outlet of the HPLC column close to the inlet of the flow cell. In the system designed by Laude and Wilkins (50) the HPLC column was placed within the bore of the cryomagnet and connected to the inlet of a flow cell (with a volume of 28 μL) by a length of poly(tetrafluoroethylene) (PTFE) tubing having an internal volume of 2–3 μL. In addition to reducing chromatographic dispersion, this arrangement provided essential premagnetization of the solutes before detection within the flow cell. Although minor adjustments to the NMR shim currents were required, insertion of the stainless steel column within the magnet did not degrade the quality of the NMR spectra recorded (50). Commercial HPLC–NMR interfaces premagnetize samples by placing the chromatographic column under the magnet, and the column outlet is connected to the flow-cell inlet with a short length of narrow-bore tubing.

The volume of sample within the region of the rf coil of a continuous-flow HPLC–NMR probe is only slightly smaller than the total volume of the glass cylinder containing the detection area, and this high filling factor (which constitutes an advantage over conventional NMR sample probes) helps offset the negative effects of failing to spin the sample cell within the magnetic field. Nonetheless, the sensitivity of NMR detection in liquid chromatographic applications is low compared with that of other detection methods. The limits of detection that can be achieved with ^{1}H NMR continuous-flow probes are dependent on the number of equivalent nuclei within the detection volume, the strength of the magnetic field, the filling factor of the detection coil, and the spin–lattice relaxation (T_1) times of the detected protons. Albert (47) reported a detection limit of 500 ng for the benzyl CH_2 protons of n-butylbenzylphthalate obtained with a flow cell with a volume of 120 μL on a 600-MHz NMR instrument. An upper detection limit of 10–20 μg for compounds with molecular weights of 100–300 daltons was reported by Dorn (41) on a 200-MHz instrument, and lower limits of detection can be expected with higher frequency instruments. Detection under stopped-flow conditions is approximately an order of magnitude more sensitive than in on-flow mode, because of the improvements in the signal-to-noise ratio that can be achieved with long acquisition times. The signal-to-noise improvements offered by stopped-flow HPLC–NMR can make up for the longer experimental times required when performing trace component analyses.

A number of techniques are available for eliminating interference from proton-containing chromatographic eluants in HPLC–NMR systems. The most direct approach is to use deuterated or halogenated solvents in the HPLC mobile phase, but the high cost of many deuterated solvents makes this approach impractical for routine reversed-phase separations. The alternative is to use solvent-suppression techniques. Presaturation to suppress the interfering solvent signals arising from water, acetonitrile, and low-pH phosphate buffers requires a relatively lengthy saturation period during which sample spectra cannot be acquired, and consequently is limited to stopped-flow analysis (*42*). Binomial suppression methods, such as the 1-1 hard pulse and 1331 pulse sequences (*51, 52*), are compatible with on-flow analysis and can be used to suppress multiple solvent resonances simultaneously. A 1-1 hard-pulse solvent-suppression technique was used by Laude et al. (*53*) to eliminate solvent signals from acetonitrile, methanol, and water in reversed-phase chromatographic separations. Pulsed-field gradients (PFGs) also provide a useful means of suppressing background water signals in 1D and 2D NMR experiments (*54–56*). Presaturation solvent-suppression schemes for use with two-dimensional nuclear Overhauser enhancement spectroscopy (NOESY) have also been described (*57*). Developments in solvent suppression in systems coupling liquid chromatography and NMR detection are described in a recent review article (*58*).

In on-flow HPLC–NMR experiments, data are accumulated over predefined time intervals (e.g., 10–30 s), and the Fourier-transformed spectrum is represented as a row of a two-dimensional plot of chemical shift versus retention time. In a typical run, up to 256 or 512 rows may be generated. A 2D ^{1}H NMR plot from the reversed-phase HPLC separation of the positional isomers of *O*-(2-fluorobenzoyl)-D-glucopyranuronic acid is shown in Figure 9 (*59*). ^{1}H NMR signal intensities are represented by the contours in the two-dimensional plot, and the band at 4.83 ppm is due to HDO in the chromatographic eluant that has been suppressed by a NOESY presaturation technique. The data in Figure 9 were obtained from injection of a total of 250 μg of the glucuronide isomers. The ^{1}H NMR spectrum of each row of the two-dimensional HPLC–NMR plot may be plotted individually, as illustrated in Figure 10 for a series of vitamin A acetate isomers separated on a cyanopropyl column in *n*-heptane (*47*). In stopped-flow HPLC–NMR experiments, the switching valve connecting the column effluent to the inlet of the continuous-flow probe is actuated by the response of a UV detector placed between the column outlet and NMR probe inlet. Stopped-flow experiments effectively trap solutes of interest for extended periods of time and allow more lengthy ^{1}H homonuclear, or ^{1}H/^{13}C heteronuclear shift-correlated experiments to be performed.

Application of HPLC–NMR

On-line HPLC–NMR has been applied extensively in drug metabolism studies in the pharmaceutical industry (*59–63*). The relative ease with which information-rich spectra can be obtained using this technique has enabled the rapid identification of important drug metabolites in a variety of complex matrices, including blood, urine,

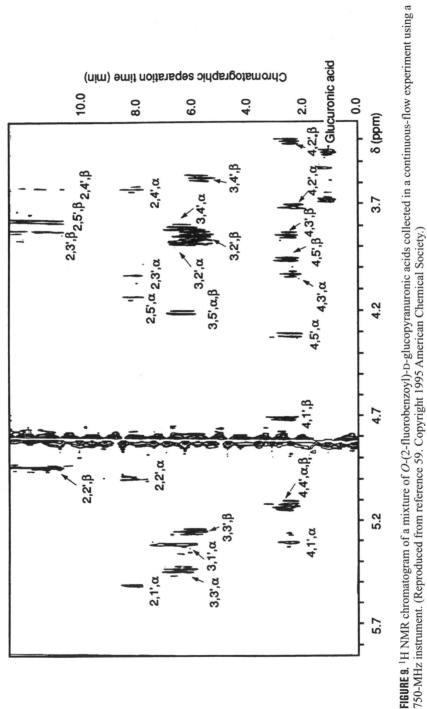

FIGURE 9. ^{1}H NMR chromatogram of a mixture of O-(2-fluorobenzoyl)-D-glucopyranuronic acids collected in a continuous-flow experiment using a 750-MHz instrument. (Reproduced from reference 59. Copyright 1995 American Chemical Society.)

tissue, and bile. The identification of single drug metabolites from biological samples is made easier if a chromatographic separation step is performed prior to ^{1}H NMR detection. The separation step does not, however, have to yield pure components for analysis, as chromatographic deficiencies can be compensated for by the structural selectivity of NMR. The merits of HPLC–NMR in drug metabolite analyses can certainly be applied to the characterization of combinatorial libraries of compound mixtures, or single components.

Advantages of Qualitative Analysis by HPLC–NMR

The combination of HPLC and NMR excels in the amount of structural information it provides in a single analysis. Virtually any NMR experiment that can be performed in a conventional sample tube can be duplicated by either on-flow or stopped-flow techniques. While not a mature technique, continued improvements in interface design and advances in NMR instrumentation will make HPLC–NMR an increasingly viable tool for the characterization of combinatorial libraries of organic compounds.

Disadvantages of Qualitative Analysis by HPLC–NMR

The main disadvantages of using NMR as a detector in HPLC are its low sensitivity and high system and operating expenses. In addition, the use of nondeuterated solvents in the chromatographic eluant requires solvent-suppression techniques, and this causes a loss of spectral information at the chemical shifts corresponding to solvent resonances.

Nonchromatographic 2D NMR Techniques

In 1992, Morris and Johnson (*64*) introduced a 2D NMR technique that allows ^{1}H NMR spectra of discrete chemical species present in a mixture to be displayed based on differences in their diffusion coefficients. Morris and Johnson applied the term diffusion-ordered 2D NMR spectroscopy (DOSY) to this technique, and expanded on their original work in a 1993 paper (*65*). The first step of the DOSY experiment involves generating the 2D NMR data using pulsed-field-gradient (PFG) NMR techniques. All signals from a given molecule will show the same diffusion coefficient, which allows ^{1}H NMR signals from the same molecule to be correlated. The second step in the DOSY experiment is to transform and display the acquired NMR data sets with chemical shift information plotted in one dimension and diffusion coefficients in the second dimension. Successful data reconstruction depends on imposing a set of constraints to limit the range of possible solutions to the data set. These include establishing upper and lower limits for possible solute diffusion rates, and imposing an assumption that the decay constants for the exponential components are positive (*65*). A DOSY spectrum for a mixture containing HOD, glucose, adenosine triphosphate (ATP), and sodium dodecylsulfate (SDS) micelles in D_2O (*65*) is shown in Figure 11A along with individual ^{1}H NMR spectra of components present in the mixture (Figure 11B).

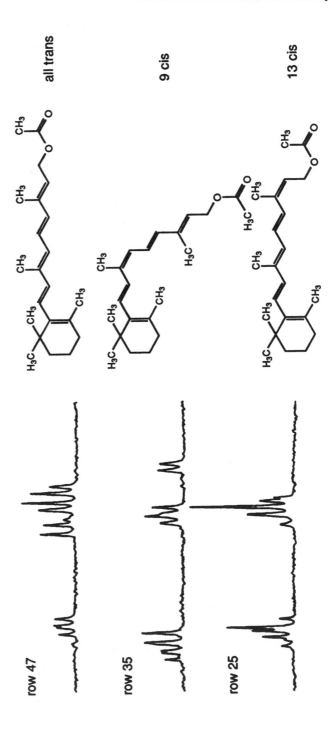

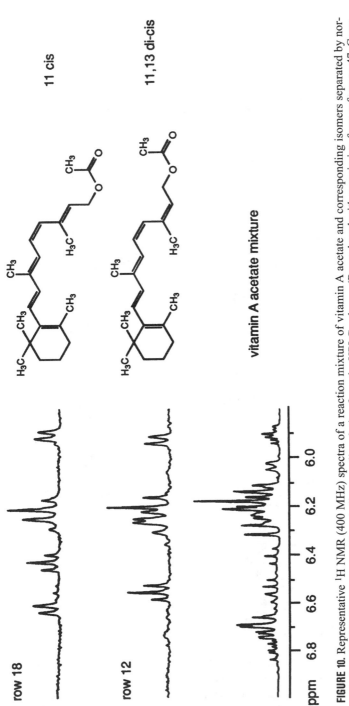

FIGURE 10. Representative ¹H NMR (400 MHz) spectra of a reaction mixture of vitamin A acetate and corresponding isomers separated by normal-phase HPLC. Lower row numbers indicate earlier elution from the HPLC column. (Reproduced with permission from reference 47. Copyright 1995 Elsevier Science Publishers B.V.)

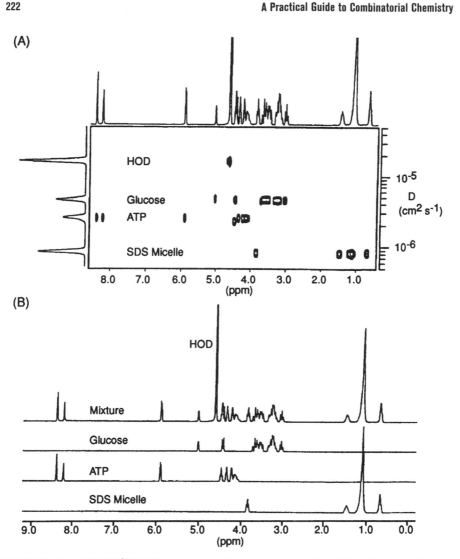

FIGURE 11. (A) A DOSY ^{1}H NMR spectrum of a mixture containing HOD, glucose, ATP and SDS in D_2O, with chemical shift plotted against diffusion coefficients. (B) Individual ^{1}H NMR spectra generated from the 2D data set. (Reproduced from reference 65. Copyright 1993 American Chemical Society.)

Application of DOSY NMR

The DOSY technique is a recent development in NMR and has not been applied as extensively to mixtures analyses as have HPLC–NMR methods, but the few reports that have appeared in the literature demonstrate the potential strengths of the technique. Lin et al. (*66*) applied DOSY to quantitative studies of the complex formation between the *cis* and *trans* isomers of the dipeptide phenylalanylproline by βcyclodextrin. Barjat et al. (*67*) were able to resolve and identify a number of the components in a perchloric acid extract of gerbil brain using a high-resolution DOSY technique.

Qualitative Analysis by Mass Spectrometry

Mass spectrometry (MS) is without question one of the most sensitive analytical techniques available for the routine characterization of organic compounds, with many instruments capable of detecting compounds in the low picomole to femto-mole range. In addition to its sensitivity, MS provides useful molecular weight and structural information that can be used to follow reaction progress, or to confirm the synthesis of a final product. MS is increasingly being embraced by combinatorial chemists as the premier technique for reaction monitoring and final product analysis. In fact, some proponents of combinatorial chemistry have gone so far as to speculate that MS may obviate the need for characterization of compound libraries by NMR and FTIR spectroscopies (*68*).

Many organic molecules, especially those of pharmaceutical and agricultural interest, are relatively nonvolatile, or thermally labile, or both, and many of these compounds are best separated by HPLC. Consequently, many of the automated sample-handling devices for MS instruments have been designed exclusively for use with liquids. Central to MS, however, is the requirement that the analytes exist as ions in the gas phase prior to mass analysis and detection. Hence, successful compound analysis by HPLC–MS depends primarily on the design of appropriate interfaces that will both accommodate the introduction of analytes in solution and generate appropriate gas-phase ions for mass analysis and detection.

Interfaces for HPLC–MS

All HPLC–MS systems are designed to perform four basic tasks. First, the chromatographic eluant is stripped from the solute molecules. The solute molecules are then converted into gas-phase ions. These solute ions are introduced into the mass analyzer, where they are separated on the basis of differences in their mass-to-charge (m/z) ratio. Finally, the separated ions are detected by an electron multiplier, and the signal is amplified and recorded. Mass analysis and detection are carried out under high vacuum, while the desolvation and sample ionization processes may be performed at atmospheric pressure. Four widely used interfaces for coupling HPLC instruments with mass spectrometers are shown schematically in Figure 12. These interfaces all remove the HPLC effluent from the solutes of interest, and then ionize the solutes for detection by the mass analyzer. The four interfaces differ, however, both in the ionization methods used, which impacts the types of solutes that are compatible with each technique, and in their capacity to efficiently desolvate the chromatographic eluant, thus imposing flow-rate restrictions on each device. A brief description of each interface follows.

A.

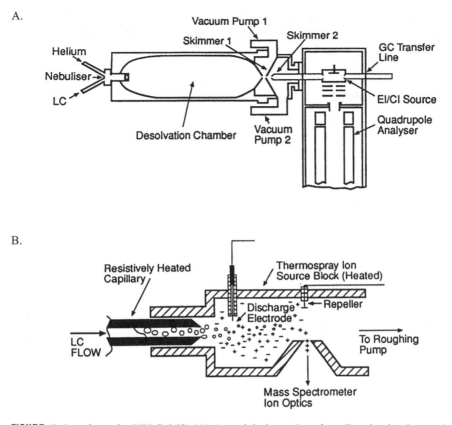

B.

FIGURE 12. Interfaces for HPLC–MS. (A) A particle-beam interface. Desolvation by a nebulizer is followed by skimmers and an EI or CI source. (B) A thermospray interface, in which desolvation and ionization occur simultaneously as solute ions are expelled from the charged mobile phase. (Reproduced with permission from reference 69. Copyright 1991 VG Instruments.)

Continued on next page

The particle-beam interface (Figure 12A) (*69*) has its roots in the monodisperse aerosol generator for interfacing chromatography (MAGIC) interface (*70, 71*) developed in the mid-1980s. This interface was designed for use with quadrupole mass spectrometers and is composed of a desolvation chamber held at ambient temperature and pressure, followed by a series of skimmers leading to the vacuum regions of the mass analyzer and detector. The interface connects directly to the inlet of an electron ionization (EI) or chemical ionization (CI) source. The HPLC column effluent to be analyzed is combined with a stream of helium gas, and the resulting aerosol enters the desolvation chamber where the mobile-phase eluant is vaporized. The decrease in droplet volume relative to droplet surface area that

C.

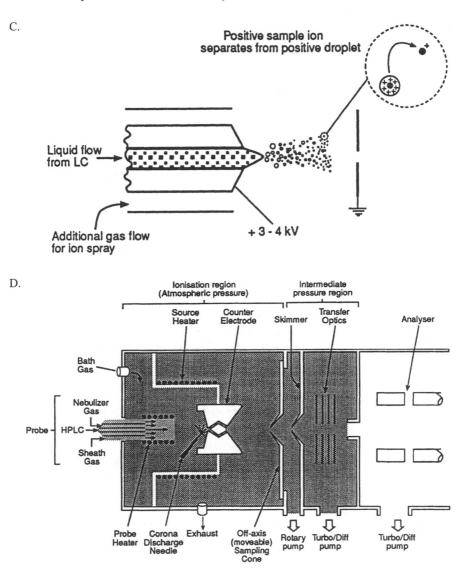

FIGURE 12—*Continued.* Interfaces for HPLC–MS. (C) An electrospray interface. A difference in voltage between the capillary outlet and an opposing counterelectrode charges the emerging droplets, resulting in desorption of the solute as ions. (D) An interface for atmospheric pressure chemical ionization (APCI). A nebulizer creates an aerosol, which is heated before the solute is chemically ionized by the corona discharge needle. (C is reproduced with permission from reference 74. Copyright 1993 Academic Press. D is reproduced with permission from reference 83. Copyright 1993 International Scientific Communications Inc.)

accompanies evaporation favors rapid vaporization of the aerosol. The mixture of solvent vapor, helium, and solute particles enters the first evacuated region at high velocity through a narrow nozzle. A series of two skimmers separate the jet of solute particles from solvent vapor and helium, and the gases are pumped out of the interface by vacuum pumps 1 and 2. The beam of solute particles enters the ion source of the mass spectrometer, where collision with a heated plate results in instantaneous vaporization of the solute. The solute vapor is then ionized by EI or CI and passed to the inlet of a quadrupole mass analyzer. The particle-beam interface is compatible with HPLC flow rates of 0.1–1.0 mL min^{-1}. Detection limits for particle-beam instruments lie in the low nanogram range, making it the least sensitive of the four HPLC–MS interfaces. In addition, the harsh nature of the EI ionization process results in extensive solute fragmentation. Such EI spectra can, however, be used to search libraries, and can greatly aid in the identification of unknown compounds.

The thermospray interface (Figure 12B) (69) accomplishes eluant desolvation and solute ionization in a single step. The modern thermospray interface consists of a resistively heated capillary tube housed in a heated ion-chamber, and its development dates to the pioneering work of Marvin Vestal and colleagues during the mid-1980s (72). The chromatographic eluant enters the thermospray interface through the capillary tube, which is maintained at a temperature near the boiling point of the HPLC mobile phase. Partial volatilization of the eluant within the capillary causes the eluant to expand, forming a high-speed jet of vapor and solvent droplets. Additional solvent evaporation occurs within the heated block of the thermospray interface, which is maintained under vacuum by a roughing pump. The thermospray technique is usually performed with chromatographic eluants that contain a volatile electrolyte (e.g., ammonium acetate), which is either added directly to the mobile phase or combined with the eluant stream after it exits the column. Although the solvent aerosol is electrically neutral overall, with statistical fluctuations individual droplets in the aerosol have an excess of positive or negative charge from the added electrolyte. As the size of the electrically charged solvent droplets decreases with solvent loss, the electrical field within each droplet increases. Eventually, the electrical field gradient across a droplet becomes great enough to cause the ejection of solute ions from the droplet surface. Solute ionization can be aided by passing an electrical discharge through the solvent vapor, effectively forming a CI plasma. Once formed, the solute ions are directed into the mass spectrometer through a small opening in a sampling cone with the aid of a repeller electrode placed opposite the mass spectrometer inlet. Thermospray sources can handle chromatographic flow rates of up to 1–2 mL min^{-1}, making the interface compatible with analytical-scale HPLC separations that use standard 4.6 mm i.d. columns. The thermospray process results in the formation of molecular ions, but the technique results in more fragmentation than is usually observed with electrospray ionization. Disadvantages of the technique include variation in the appearance of thermospray mass spectra from instrument to instrument, and, on a given instrument, as a function of ion-source pressure or temperature (*73*). In addition, the technique can be troublesome when applied to the analysis of thermally labile compounds.

In the electrospray ionization (ESI) process (Figure 12C) (*74*), the chromatographic eluant is pumped through a narrow-bore capillary whose outlet is held at a potential of a few kilovolts relative to an opposing counterelectrode (*75*). As with the thermospray ionization technique, a volatile electrolyte is usually present in the chromatographic eluant. The liquid becomes charged as it emerges from the electrospray capillary, and Coulombic repulsion causes the stream to spread out as a plume of charged droplets. The mechanism leading to ion formation in electrospray is not well understood, but for the sake of simplicity it may be regarded as a process wherein the charged droplets decrease in size through both evaporation and explosive fragmentation (the latter driven by Coulombic repulsion forces), eventually reaching a radius at which the electrical field gradient is large enough for ionized solute molecules to desorb from the droplet surface. More detailed accounts of the probable mechanisms involved in the electrospray process have been described by Meng and Fenn (*76, 77*) and Kebarle and Tang (*78*). The application of an annular sheath gas surrounding the spray needle (termed ion spray) enhances the desolvation process, maintaining cleanliness of the vacuum system, and allowing higher chromatographic flow rates to be used. The electrospray process takes place at atmospheric pressure, and the sample ions formed are directed through a small orifice and a series of skimmers into the vacuum region of the mass spectrometer. The electrospray process is a soft ionization technique, yielding predominately $[M + H]^+$ or $[M - H]^-$ ions even with thermally labile and nonvolatile molecules. In addition, ESI can generate multiply charged ions, so high-molecular-weight biomolecules can be analyzed with relatively inexpensive quadrupole mass filters. Electrospray interfaces are generally operated at flow rates of 1–200 μL min^{-1}, with well-designed interfaces capable of tolerating 1.0 mL min^{-1}. Consequently, chromatographic separations on electrospray HPLC–MS instruments are performed using narrow-bore (2 mm i.d.) or microbore (1 mm i.d.) columns. If separations on analytical-scale (4.6 mm i.d.) columns are desired, then the HPLC column effluent may be split to reduce the flow rate entering the electrospray interface. The principles of the electrospray technique and its use with liquid chromatography have been described in a number of review articles (*79, 80*).

The coupling of LC to MS through an atmospheric pressure chemical ionization interface (APcI; Figure 12D) provides a convenient means for characterizing complex sample mixtures (*81, 82*). The HPLC column effluent enters the APcI interface (*83*) through a narrow capillary around which a concentric flow of nitrogen gas is provided to pneumatically generate a fine aerosol. An additional flow of nitrogen makeup gas helps direct the solvent/analyte aerosol towards the heated region of the APcI probe. The temperature of the APcI probe is maintained such that effective desolvation and solute volatilization are achieved. The vaporized aerosol then passes by a corona discharge needle held at a potential of approximately 2 kV, where chemical ionization of the solute is achieved at atmospheric pressure. The mobile phase solvent vapor serves as the reagent gas for the formation of the reactant ions necessary in the chemical ionization process. The solute ions produced through ion–molecule reactions with these reactant ions then enter the intermediate pressure region of the mass spectrometer through a sampling cone. Typically, the sampling cone is maintained at 10–100 V to help dissociate analyte–solvent cluster ions before the analyte ions pass through a skimmer on their way to the mass analyzer. The transport regions used in APcI and electrospray interfaces are nearly identical in design, and most modern systems allow rapid conversion between the two types of interface (*84*). As with ESI, APcI generally produces little fragmentation, although sample fragmentation can be induced by increasing the voltage applied to the sampling cone. ESI and APcI differ in the greater abundance of ions with multiple charges that are observed in ESI mass spectra. APcI interfaces are compatible with chromatographic flow rates as high as 1–2 mL min^{-1}. The sensitivity and ruggedness of APcI interfaces for HPLC–MS are greater than those obtainable on thermospray HPLC–MS instruments.

Quadrupole Mass Spectrometers

The preceding section has considered the method of ion formation. I now turn to how the mass of the ions is actually determined. Mass spectrometers may be classified according to their method of separating charged particles, with magnetic field deflection, quadrupole, and time-of-flight instruments being the most prevalent in analytical laboratories. Typical bench-top mass spectrometers familiar to most combinatorial chemists are quadrupole instruments. Quadrupole mass spectrometers are distinguished by their fundamental design as either quadrupole mass filter, or quadrupole ion storage (ion-trap) instruments.

The quadrupole mass filter is designed as a series of four rods lying parallel to one another. When the setup is viewed in cross section, a line between two of the rods describes the *x* axis, and a line between the other two rods describes the *y* axis. Ions entering from the inlet end of the mass spectrometer travel with constant velocity in the direction parallel to the quadrupole rods. Application of both a direct current (dc) voltage and a radio frequency (rf) voltage to the rods causes the sample ions to oscillate in a complex pattern along the *x* and *y* directions of the quadrupole mass filter. At any given rf voltage, only sample ions with a particular *m/z* ratio are capable of traversing the mass filter without striking the rods. These ions are then detected at the exit of the quadrupole mass filter. Ions with *m/z* ratios above or below the value producing a stable oscillation strike the quadrupole rods on their way through the mass analyzer and are lost. The mass spectrum is acquired by varying each of the rf and dc frequencies while keeping their ratios constant, such that sample ions traverse the mass filter in increasing order of *m/z*.

The design for the second type of quadrupole mass spectrometer, the ion trap, consists of a donut-shaped electrode horizontally encircling the sample chamber with two hemispherical end-cap electrodes placed above and below the center ring electrode. Paul-type ion traps achieve three-dimensional trapping of sample ions through the use of time-varying potentials applied between the end caps and the ring electrode. Application of a constant homogeneous magnetic field along the symmetry axis of the ring electrode in conjunction with a steady potential difference applied between the end caps is characteristic of Penning ion traps. Mass spectra are generated on ion-trap instruments by sequentially scanning sample ions out of the trap through application of an rf voltage ramp. The ejected ions are detected as they exit the ion trap. Interfaces for ion-trap mass spectrometers using both ESI (*85*) and APcI (*86, 87*) have been described.

Miniaturization: Microcapillary LC–MS and CE–MS

The amount of material typically injected into conventional HPLC–MS systems far exceeds the amount that can be detected by mass spectrometry, so the size of the sample used for library characterization by HPLC–MS could be reduced without sacrificing data quality through miniaturization of the chromatographic and inter-face equipment. Sample conservation is becoming an increasingly important con-cern as the sample requirements for screening assays are pushed below the levels required for conventional analytical characterization, and combinatorial library gen-eration is directed toward the microgram scale.

Each of the four HPLC–MS interfaces described above has been miniaturized for compatibility with chromatographic separations performed using 10–320 μm i.d. packed capillaries, using the instrument requirements for nanoscale liquid chro-matography described elsewhere (*88*). This subject has been thoroughly reviewed, with pertinent examples (*89*). In addition, the development of microelectrospray mass spectrometry has been described by Emmett and Caprioli (*90*). The develop-ment of high-efficiency micellar electrokinetic capillary chromatography (MEKC) and the recent emergence of capillary electrokinetic chromatography have provided a wealth of nanoscale capillary separation techniques that may be applied to the analysis of libraries of small organic molecules. Garcia and Henion demonstrated the first successful coupling of on-line gel-filled capillary electrophoresis with tan-dem mass spectrometry in 1992 (*91*). That same year, Parker et al. described the characterization of a series of macrolide antibiotics by nanoscale packed capillary liquid chromatography and capillary electrophoresis (CE) coupled with electrospray ionization mass spectrometry (ESIMS) (*92*), and later extended these techniques to the analysis of a series of 17 penicillins and cephems (*93*). Capillary electrochro-matography has been coupled with ESIMS for the separation and characterization of mixtures of pharmaceuticals (*94*) and peptides (*95*). Takada et al. (*96*) reported an APcI interface for CE–MS that is compatible with the use of relatively high concen-trations of nonvolatile buffers, such as 20-mM sodium phosphate. ESI is, however, susceptible to interference from the buffers and detergents commonly used in CE and MEKC. For example, the peak intensities of the $[M + H]^+$ ions for a series of seven tamoxifen metabolites analyzed by CE–ESIMS were reported to decrease by a factor of 3 when 7-mM SDS was added to the running buffer, but no further decrease in peak intensity was observed with 10–30-mM SDS (*97*).

Application of MS to Library Characterization

Mass spectrometry, with input either from a loop injection or an HPLC apparatus, has been used more extensively for both reaction monitoring and final product characterization of combinatorial libraries than any other analytical technique, including NMR and FTIR spectroscopies. Of the MS ionization techniques, ESI has been most commonly applied to the qualitative analysis of compound libraries. Examples taken from the recent literature include the characterization of mixtures containing up to 55 components directly infused into an ESIMS instrument (98), qualitative analysis of urea-linked diamine libraries (99), analysis of 4-aminoproline analogs by ion-spray mass spectrometry (100), and characterization of synthetic peptide libraries by HPLC combined with ion-spray MS (101). The use of MS in the analysis of combinatorial libraries is not simple, however. It is very important that the choice of an ionization technique be made with consideration for the chemical properties of the solutes of interest.

Figure 13A (102) provides a qualitative means for selecting an ionization technique most appropriate for a given set of compounds. The plot identifies the most suitable HPLC–MS interface based on solute molecular weight and relative polarity. For example, HPLC with particle-beam MS is best suited to the analysis of relatively nonpolar organic molecules within the mass range of 100–1000 daltons, while MS characterization of moderately polar solutes within this mass range may be performed using thermospray or APcI ionization techniques. ESIMS is appropriate for the analysis of low-molecular-weight organic compounds containing ionizable functional groups, and a number of derivatization protocols have been developed for incorporating ionic and solution-ionizable substituents into the structures of nonpolar solutes to enhance their detectability by ESIMS (103–105). Iwabuchi et al. (106) examined the relative sensitivities of thermospray, APcI, and ESI techniques, in both positive- and negative-ion mode, in the mass spectrometric analysis of a series of 13 hydroxymethylglutaryl coenzyme A (HMG CoA) reductase inhibitors and their analogs. The compounds vary in polarity, with log P values from -2.49 to 4.40. The results of this work are summarized in the plot of Figure 13B. These workers noted that, on average, ESI was ~10 times more sensitive than thermospray ionization, and 100 times more sensitive than the APcI technique.

Figure 13B illustrates that both APcI in positive-ion mode and ESI in negative-ion mode will likely show significant variations in response when solutes with widely different polarities are analyzed. This is an important point to consider when dealing with samples that contain multiple components (e.g., libraries of mixtures). For example, consider the two mass spectra in Figure 14, which were generated in this author's laboratory. Figure 14A is the positive-ion scan APcI mass spectrum of a 265-dalton compound synthesized by solid-phase methods, while Figure 14B is the negative-ion ESI mass spectrum of the same compound. The expected quasi-molecular ions of the target compound are observed in each mass spectrum. The APcI mass spectrum suggests the presence of a polymeric contaminant in the sample, presumably originating from the resin used in the solid-phase synthesis, that was not detected by negative-mode ESIMS. I do not present Figure 14 as any sort of

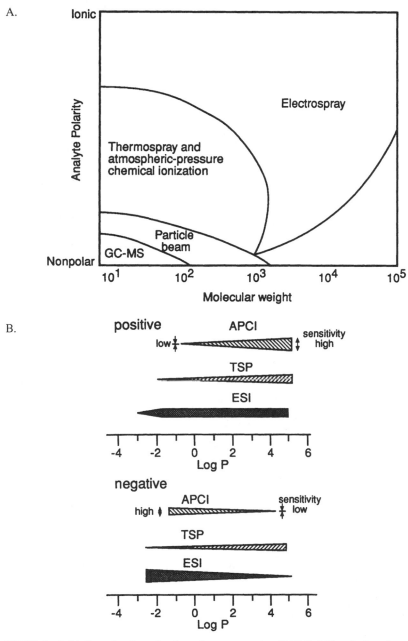

FIGURE 13. Guidelines for the selection of an appropriate HPLC–MS technique based on (A) compound molecular weight and polarity, and (B) comparison of the sensitivity of APCI, thermospray (TSP), and ESI (in both positive- and negative-ion modes) for a series of HMG CoA reductase inhibitors as a function of their Log P values. (A is reproduced with permission from reference 102. Copyright 1996 Advanstar Communications Inc. B is reproduced with permission from reference 106. Copyright 1994 John Wiley & Sons Ltd.)

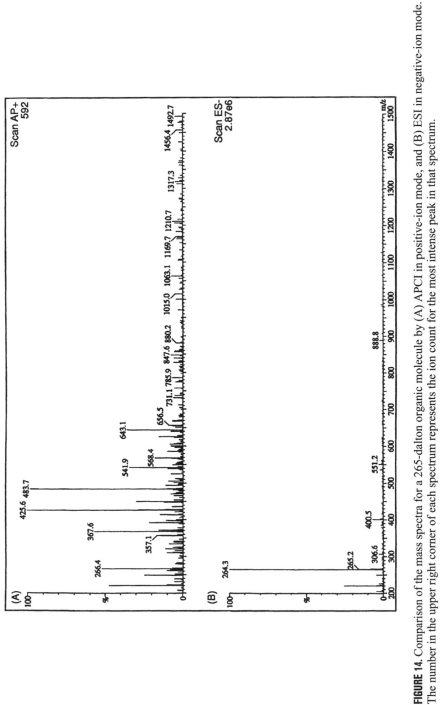

FIGURE 14. Comparison of the mass spectra for a 265-dalton organic molecule by (A) APCI in positive-ion mode, and (B) ESI in negative-ion mode. The number in the upper right corner of each spectrum represents the ion count for the most intense peak in that spectrum.

proof of the superiority of positive-ion APcI over negative-ion ESI for library characterization, but rather as an illustration of the potential for significant differences between mass spectra obtained from the same sample using the two techniques.

Tandem MS

HPLC is capable of resolving mixtures containing up to 50–100 components in a single gradient run, but routine analyses of compound libraries using standard gradient protocols may not guarantee baseline resolution of every component present in a given sample. Selective detection techniques, such as tandem MS (MS–MS), can provide the additional resolving power necessary for the analysis of complex sample libraries. The basic tandem MS experiment may be described by the dissociation of a parent ion (m_p^+) into a product ion (m_d^+) and a neutral fragment (m_n) as follows:

$$m_p^+ \rightarrow m_d^+ + m_n \tag{3}$$

Positive ions are used here for the purpose of illustration, and an analogous equation can be written for the dissociation of a negatively charged parent ion. Tandem MS techniques involve following the loss of the parent ion and formation of the product ion and neutral fragment, with the latter being inferred from the difference between the m/z of the parent and product ions.

The three most common MS–MS experiments performed on triple quadrupole mass analyzer instruments are illustrated in Figure 15. These tandem MS experiments are termed the product-ion scan (Figure 15A), the neutral-loss scan (Figure 15B), and selected reaction monitoring (Figure 15C). In each experiment, the first and third quadrupoles, Q1 and Q3, are used to scan parent and product-ion masses, respectively, while quadrupole Q2 serves as a collision chamber. A neutral target gas is admitted into Q2, where it induces fragmentation of the parent ions emitted by Q1 through a process known as collision-induced dissociation (CID). Argon, nitrogen, methane, or ammonia gas are often used individually or as mixtures in tandem MS experiments. During the CID process, the translational energy of the parent ion is converted into internal energy through collision with the target gas molecules. Collisional activation of the parent ion is followed by fragmentation.

The product-ion scan basically provides a fragmentation mass spectrum of a single parent ion. The m/z of the parent ion is fixed by mass analyzer Q1, and the masses of all of the product ions formed from the parent ion in mass analyzer Q2 are passed by mass analyzer Q3. A product-ion scan can be used to distinguish between multiple components introduced directly into the mass spectrometer as a mixture, or as a single eluting band containing multiple coeluting components from an HPLC column. By combining the resolving power of HPLC with this selective technique, the constituents of complex mixtures can be identified. Indeed, production scan MS–MS experiments have frequently been applied to the analysis of compounds originating from solid-phase synthesis (*107, 108*). Qualitative analysis of product mixtures by ESIMS–MS has been reported for samples cleaved from single

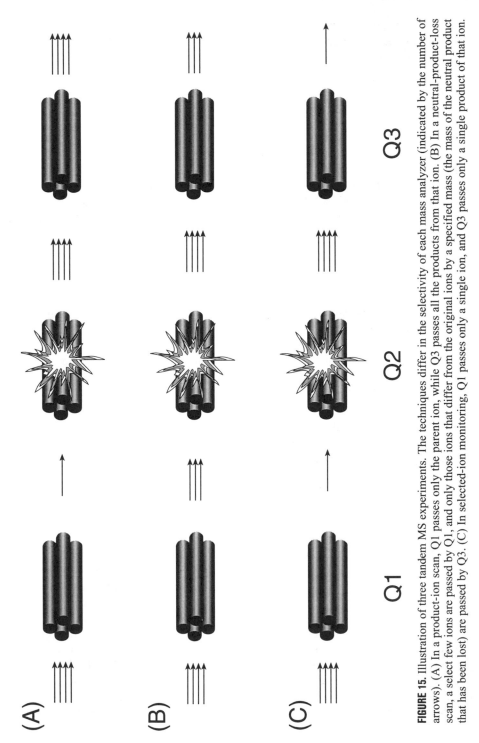

FIGURE 15. Illustration of three tandem MS experiments. The techniques differ in the selectivity of each mass analyzer (indicated by the number of arrows). (A) In a product-ion scan, Q1 passes only the parent ion, while Q3 passes all the products from that ion. (B) In a neutral-product-loss scan, a select few ions are passed by Q1, and only those ions that differ from the original ions by a specified mass (the mass of the neutral product that has been lost) are passed by Q3. (C) In selected-ion monitoring, Q1 passes only a single ion, and Q3 passes only a single product of that ion.

resin beads and introduced by direct infusion without prior chromatographic separation (*109, 110*). The electrospray mass spectrum and electrospray MS–MS spectrum of an angiotensin II antagonist cleaved from a single Sasrin bead (*110*) are shown in Figures 16A and 16B, respectively.

Many organic molecular ions dissociate by loss of a small neutral molecule associated with the parent structure (e.g., loss of CO_2 from carboxylic acids or loss of NO• from nitroaromatic compounds). A neutral-loss scan involves setting mass analyzers Q1 and Q3 such that they scan for a constant difference in mass between the parent ion and the product ion. Mass analyzer Q1 passes a select range of incoming sample ions to a collision cell, where product ions are generated through CID. Of the product ions generated in Q2, only those with a specific mass difference from their respective parent ions are passed by mass analyzer Q3. The mass difference between scans of Q1 and Q3 equals the mass corresponding to the loss of a specific neutral fragment. A neutral-loss scan is particularly well suited to the analysis of combinatorial libraries, particularly those made up of compounds that share a common structural feature amenable to neutral fragment loss by CID. A recent example

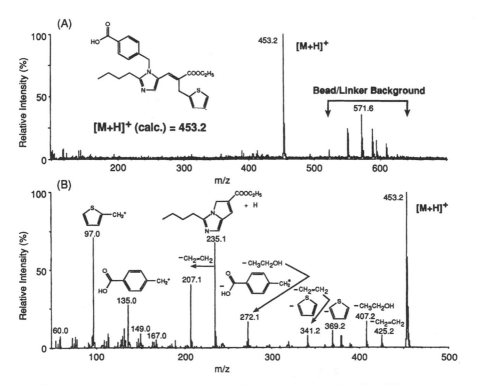

FIGURE 16. Analysis of angiotensin II cleaved from a single Sasrin bead by (A) electrospray MS, and (B) an MS–MS product-ion scan of the *m/z* 453.2 ion. (Reproduced from reference 110. Copyright 1996 American Chemical Society.)

of the application of this technique to the analysis of *N*-methylcarbamate pesticides was described by Volmer et al. (*111*). A series of *N*-methylcarbamate pesticides were analyzed by ESI LC–MS–MS, where the loss of methylisocyanate (mass 57) indicated that an *N*-methylcarbamate pesticide was in the sample extract.

A selected reaction monitoring experiment is performed by fixing mass analyzer Q1 to pass a single parent ion to collision chamber Q2, and setting mass analyzer Q3 to pass only one product ion fragment. While selected reaction monitoring provides less structural information than the typical mass spectrum, the technique delivers greatly enhanced sensitivity for the target analyte. This sensitivity increase is achieved by reducing background noise, rather than by increasing analyte signal intensity. The selected reaction monitoring technique is best suited for the analysis of trace amounts of compounds present in complex matrices and may have application for the qualitative analysis of compound mixtures cleaved from single beads.

Tandem MS may also be performed on ion-trap mass spectrometers (*112*). Such experiments involve tandem operations over time, in the same space, whereas tandem MS on triple quadrupole mass spectrometers involves tandem operations in different regions of space. For example, a product-ion scan carried out on an ion-trap mass spectrometer involves scanning the rf voltage to eject all ions in the trap up to but not including the ion of interest, reducing the rf voltage to provide a convenient low mass cutoff, inducing the trapped ion of interest to undergo collision with the background gas, and recording the product-ion spectrum by ramping the rf voltage a second time to eject the product ions from the trap. If desired, a specific product ion from the initial CID may be trapped and subjected to an additional CID cycle and this process repeated. The ability to perform such MS^n experiments on ion-trap mass spectrometers provides the combinatorial chemist with a powerful tool for structure elucidation.

Advantages of Qualitative Analysis by MS

The greatest advantage in using MS for the analysis of libraries of organic compounds is the universality of the technique. The only requirement is that the solute be converted to a gas-phase ion prior to mass analysis. Mass spectrometry is an extremely sensitive technique, with most ionization methods offering low picomole or femtomole limits of detection. Further, careful interpretation of the fragmentation patterns present in the mass spectrum, combined with tandem MS–MS experiments, can provide important structural conformation and aid in the identification of unknown compounds that may be present in the library sample. Finally, well-designed MS interfaces cause little loss of compound resolution, and so are well suited for coupling to HPLC instruments.

Disadvantages of Qualitative Analysis by MS

The main disadvantage of applying HPLC–MS as a qualitative technique for the characterization of combinatorial libraries lies in the dual requirement that the chromatographic eluant be removed from the analyte and the solute molecules be converted to a gas-phase ion prior to mass analysis and detection. The former places constraints on the eluant flow rates that can be used and excludes the use of common nonvolatile buffers and modifiers (e.g., sodium or potassium phosphates, perchlorates, and ion-pairing reagents) from the HPLC mobile phase. The need to volatilize and ionize solute molecules prior to mass analysis presents a formidable challenge to researchers wishing to analyze the large number of nonvolatile and labile compounds that can be separated by HPLC. For this reason, considerable use is made of condensed-phase ionization techniques such as thermospray ionization, ESI, and APcI. And while the mass spectrometer is a universal detector, no single interfacing or ionization technique is suitable for the analysis of mixtures that contain a variety of solutes and impurities with widely differing polarities.

Quantitation of Combinatorial Libraries

Quantitation by NMR Spectroscopy

Thus far I have discussed the detection of compounds, but the quantitation of these compounds presents different challenges. The appeal of NMR in quantitative measurements lies in its universal response to sample nuclei. The integral intensity of a given resonance band in an NMR spectrum is directly proportional to the number of nuclei contributing to that signal, a fact that obviates the need for compound-specific reference standards.

There are, however, many complexities and potential pitfalls in performing sample quantitation by NMR, and to appreciate these it is important to first understand the various steps involved in acquiring a spectrum. As an example, consider a basic single-pulse experiment. When a sample is placed within the NMR magnet, the nuclei in the molecule generate a bulk magnetization. This bulk magnetization is perturbed through the application of an rf pulse. This pulse is followed by a short delay to prevent overload of the receiver, after which the data is acquired. Data acquisition involves turning the receiver on and digitizing the free induction decay (FID) signal over the length of the acquisition time. The FID decays exponentially with a time constant of T_2^*, which is a function of the natural spin–spin (T_2) relaxation time and magnetic field (B_0) inhomogeneities. A consequence of the exponential decay of the FID is that the signal-to-noise ratio decreases with time, and long acquisition times can lead to an increase in the amount of noise present in the Fourier-transformed spectrum. Return of the perturbed spin system to equilibrium is governed by the spin–lattice (T_1) relaxation time, and an appropriate relaxation delay, during which no data is acquired, is required before application of the next rf pulse. In short, the acquisition time depends on T_2^*, while the pulse repetition time depends on T_1. For small molecules (i.e., with a molecular weight of less than 1000), T_1 is usually greater than T_2, but factors such as aggregation and chemical exchange can lead to an effective increase in T_2, and result in a more noisy spectrum. Sensitivity is improved by a factor of $\sqrt{N}$ through coherent addition of the FIDs obtained from the application of N rf pulses.

Quantitation by NMR spectroscopy involves addition of a suitable internal standard to the sample solution, followed by acquisition and integration of the NMR spectrum. With continuous-flow and stopped-flow HPLC–NMR systems, the internal standard may be added to the mobile phase, or mixed into the effluent after HPLC analysis. The major requirement placed on the choice of internal standard is that it not interfere with integration of any of the sample resonances in the spectrum, and it should be compatible with the chromatographic conditions employed. Hexamethyldisiloxane is a useful internal standard for ^{1}H NMR quantitation because of its small chemical shift (0.07 ppm) relative to tetramethylsilane. The compound 1,3,5-trioxane has also been reported as an internal standard for the ^{1}H NMR quantitation of aspirin, phenacetin, and caffeine in pharmaceutical preparations (*113*). A response factor is required to relate the number of moles of protons giving rise to a particular sample resonance signal in the ^{1}H NMR spectrum to the integrated signal of the internal standard. The internal standard response factor (K) is calculated as follows:

$$K = \frac{n_{is} M_{is}}{H_{is}} \tag{4}$$

where n_{is} represents the number of protons assigned to the internal standard signal used for quantitation, M_{is} is the molar concentration of internal standard in the sample solution, and H_{is} is the integral intensity of the internal standard resonance. Multiplication of the integrated intensity for the sample resonance by response factor K, followed by division by the number of protons assigned to the sample resonance, yields the molar sample concentration. The calculations for HPLC–NMR are complicated by the requirement that the sample integrals be summed over the elution volume of the chromatographic peak, but, when analyses are performed properly, it is possible to keep quantitation errors to less than a few percent.

A significant problem in obtaining accurate quantitation in NMR is related to variable spin–lattice relaxation times in the sample. If the spins in the sample molecule have different T_1 values, quantitation errors will arise if the rf pulse rate is faster than relaxation for any of the spins. Consequently, it is best to use a relaxation delay of at least 4 to 5 times the longest T_1 in the sample to ensure complete equilibration between rf pulses. If milligram quantities of sample are available for analysis, then accurate quantitation by ^{1}H NMR can be performed in a reasonable time. If, however, sample quantity is limited, the time required to collect multiple scans for signal averaging, coupled with the lengthy relaxation delays between scans, will result in routine work being intolerably long. The inefficiency of performing NMR quantitation with a limited amount of sample becomes more acute if the measurements are based on nuclei that are naturally of low abundance, such as ^{13}C. For example, for the quantitation of levels of *S*-carboxymethyl-L-cysteine and its metabolites in human urine by ^{13}C NMR, reported by Meese et al. (*114*), each sample required the accumulation of 100,000 FIDs over 24 h.

There are a number of additional experimental factors that can lead to spectral distortions in NMR, and consequently degrade the accuracy of sample quantitation. A thorough discussion of the pitfalls involved in performing quantitative analyses by NMR can be found in references 115 and 116. Low-molecular-weight organic compounds with short T_2 relaxation times (e.g., some organic salts) exhibit large spectral line widths, which can corrupt the precision of integration. Integration can also be compromised by poor adjustment (i.e., shimming) of the applied magnetic field. Large quantitation errors can also result from exceeding the dynamic range of the instrument's analog-to-digital converter. Problems associated with dynamic range usually result from attempts to quantitate small impurities in the presence of large amounts of sample.

The preceding discussion applies only to static measurements. When dealing with continuous-flow HPLC–^{1}H NMR systems, the effects of residence time within the NMR probe on sample quantitation must be considered. Equations 1 and 2, which were presented earlier in this chapter, describe the observed T_1 and T_2 relaxation times for flow systems in terms of their respective static values and the sample residence time (τ). Haw et al. (*117*) have modeled the effect of τ on effective spin–lattice relaxation times for samples with various T_1 values flowing through a cylindrical detection cell. An important finding of this work is that short τ values level the effective spin–lattice relaxation times of the spins (*see* Figure 17A), suggesting that quantitation errors may be minimized by using high flow rates or reducing the volume of the measurement cell. The use of low-volume flow cells in such continuous-flow systems is actually desirable from the viewpoint of solute resolution. Haw et al. (*117*) confirmed their findings by passing a solution containing known concentrations of dodecane and *m*-xylene through a flow probe at flow rates of 0.5 and 1.0 mL min^{-1}, while collecting ^{1}H NMR spectra using a variety of pulse intervals ranging from 0.5 to 6 s. Ratios of the total signal from dodecane (methyl plus methylene) to the α-methyl signal from *m*-xylene as a function of pulse interval were normalized with the known concentration ratio to prepare the graph in Figure 17B. A flow rate of 0.5 mL min^{-1} through the 40-μL flow cell required a pulse interval of 4.5–5.0 s to achieve quantitation to within ± 2% of theory, while the same level of accuracy could be achieved at the higher flow rate with a shorter (3.0–3.5 s) pulse interval.

Quantitation by HPLC with Evaporative Light Scattering Detection

The quantitative analysis of libraries often involves partial purification of compounds before they are quantified. HPLC, with its high separation efficiency, is an attractive method for the separation phase. The actual quantification is, however, more problematic. As I have noted, the coupling of HPLC and NMR is in its infancy, and the other known techniques that can interface with HPLC systems result in differential responses to equivalent quantities of different compounds.

A.

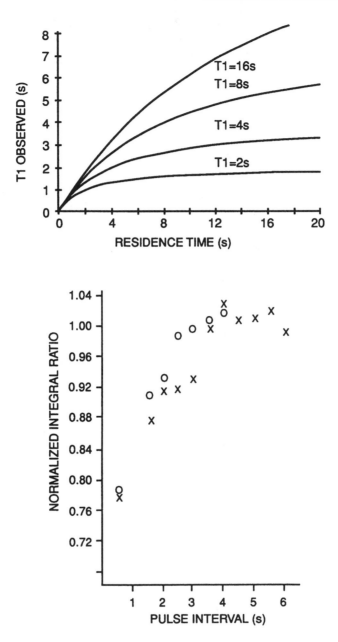

B.

FIGURE 17. Theoretical plots of (A) T_1(obs) as a function of residence time for solutes with various T_1 values and (B) normalized integral ratio (m-xylene α-CH$_3$/total dodecane protons) vs. pulse interval for flow rates of 0.5 ($\times$) and 1.0 ($\circ$) mL min^{-1}. The higher flow rate requires shorter pulse intervals for gathering accurate measurements. (Reproduced with permission from reference 117. Copyright 1982 Academic Press.)

Compounds generated through combinatorial synthesis are often quantitated using HPLC in combination with low-wavelength UV and/or MS detection (*99, 100, 118, 119*). While HPLC with low-wavelength UV detection is a convenient method for qualitatively assessing sample purity and yield, it cannot be used for sample quantitation without the appropriate use of well-characterized reference standards because of the potential for significant differences in molar absorptivity between compounds within the sample library. A similar argument can be made against quantitation of combinatorial libraries by MS based on the significant differences in ionization efficiency that may exist between library components. In the absence of a detection method capable of providing an equivalent response to all compounds in a combinatorial library, the individual array components would have to be quantitated based on the response of an appropriate series of external, or internal standards. The considerable effort that would be required to synthesize and characterize reference standards for each member of a compound library makes this approach an intolerable solution to the problem of library quantitation. Although no truly universal detection method for HPLC exists, there are several detection methods that have been demonstrated to provide similar responses to compounds within a given structural class (e.g., refractive index, flame ionization, and evaporative light scattering). The advantages of using HPLC with evaporative light scattering detection (ELSD) for the rapid quantitation of compound libraries are described below.

Ford and Kennard were the first to combine evaporative light scattering detection with HPLC for the analysis of a series of low-molecular-weight polymers (*120*). Subsequent commercialization and improvements in detector design have resulted in the successful application of this instrument to the analysis of a wide variety of compounds, including alkylsurfactants and alkylarylsurfactants (*121*), triglycerides (*122*), (phospho)lipids (*123*), sugars (*124*), pharmaceutical salts (*125*), and pharmaceutical excipients (*126*). The major components of an evaporative light scattering detector are shown in Figure 18. HPLC column effluent enters the detector through a narrow-bore tube, where it is mixed with a high-velocity stream of nitrogen gas and nebulized. The nebulized sample stream then enters a heated drift tube. The drift tube is maintained at a temperature sufficient to cause rapid evaporation of the mobile-phase solvent, leaving a narrow mist of nonvolatile solute to be swept toward the detection system by a stream of nitrogen carrier gas. The detection system consists of a laser (with a wavelength of 670 nm) and a silicon photodiode, oriented at an angle of 90° to each other and the central axis of the cylindrical drift tube. Passage of the narrow band of solute particles through the laser beam results in scattering of the source. The intensity of the scattered light, registered by a photodiode, is proportional to the amount of solute present in the chromatographic band eluting from the HPLC column.

Although routine operation of the detector is relatively straightforward, the underlying processes involved in the nebulization, desolvation, and subsequent light scattering by solute particles within the instrument are interdependent and quite complex. Mourey and Oppenheimer (*127*) modeled the response characteristics of an evaporative light scattering detector using angular light scattering theory in com-

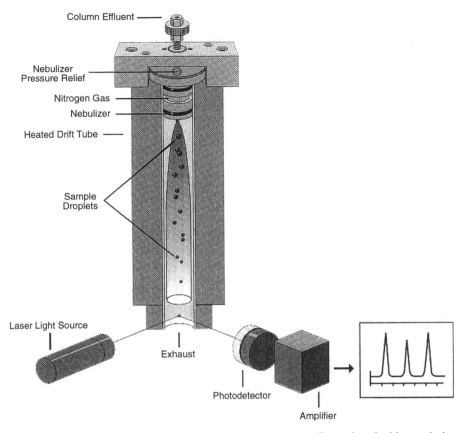

Column Effluent

Nebulizer
Pressure Relief

Nitrogen Gas

Nebulizer

Heated Drift Tube

Sample
Droplets

Laser Light Source

Exhaust

Photodetector

Amplifier

FIGURE 18. Schematic of an evaporative light scattering detector. (Reproduced with permission from Alltech Associates, Inc.)

bination with nebulization theory and particle size distribution statistics. Their treatise expanded on the earlier work of Charlesworth (*128*), who used geometric optics to model detector response. Of note is the detector's sigmoidal response over extended concentration ranges, which can be linearized by plotting the area under the peak-response curve against solute concentration on a log–log scale.

To compare ELSD with other techniques, known amounts of a series of seven steroids were quantitated using UV, ELSD (both with β-estradiol as an external standard and following HPLC), and ^{1}H NMR (using an internal standard) (Table I) (*129*). The quantitation errors were calculated on the basis of differences between the theoretical concentrations of the prepared samples and the values calculated on the basis of the response of an internal or external standard. The chromatographic separations were performed by normal-phase HPLC on a cyanopropyl-bonded column. Quantitation by ^{1}H NMR was performed on a 400-MHz instrument, and each steroid was prepared separately in a solution of $CDCl_3$ containing N,N,N',N'-tetramethyl-D-tartaramide as an internal standard. Quantitation with low-wavelength UV detection (210

TABLE I. Comparison of Quantitation Errors Produced Using UV, ELSD, and ^{1}H NMR

Steroid To Be Quantified	UV[a]	ELSD[b]	ELSD[c]	^{1}H NMR
Estrone	–7%	4%	14%	6%
Prednisolone	51%	7%	d	2%
Prednisone	52%	13%	13%	1%
Cortisone	58%	14%	11%	4%
Hydrocortisone	61%	8%	9%	4%
4-Androstene-3,17-dione	70%	22%	16%	3%
Pregnenolone	78%	1%	21%	1%

NOTE: Quantitation based on the use of β-estradiol as an external standard was performed for UV and ELSD. ^{1}H NMR quantitation was performed using N,N,N',N'-tetramethyl-D-tartaramide as an internal standard.

[a]UV was at 210 nm.

[b]After isocratic HPLC.

[c]After gradient HPLC.

[d]Prednisolone was not included in the sample mixture separated by normal-phase gradient HPLC.

nm) produced the greatest quantitation errors, which are attributed to the significant differences among the molar absorptivities of the steroids examined. Greater accuracy was obtained using ^{1}H NMR, with errors of 1–6%. The errors observed when using ELSD were significantly lower than those obtained by UV at 210 nm, and were only slightly greater than the errors found with ^{1}H NMR quantitation.

A unique advantage of ELSD over UV and MS detection methods is its compatibility with gradient HPLC. There is negligible baseline perturbation during gradient operation with ELSD under both reversed-phase and normal-phase conditions (*121, 130, 131*). Further, comparison of the errors determined from the evaporative light scattering detector's response to the seven steroids under isocratic and gradient HPLC conditions suggests that the response characteristics of this detector are not affected significantly by changes in eluant composition. Detection limits for evaporative light scattering detectors lie in the submicrogram range.

Data Handling Requirements for Library Analysis

If the traditional analytical mainstays of the synthetic chemist (NMR, FTIR, MS, and HPLC) are to be applied to the characterization of combinatorial libraries on a routine basis, then appropriate data-management tools are necessary to review, process, and archive the data in an organized manner. The logistics of managing the analytical data from even simple combinatorial libraries of single components can become overwhelming and increase in complexity when dealing with large libraries made up of mixtures. In addition, issues concerning the organization and accessibility of analytical data are complicated by the fact that many instrument vendors store data in proprietary file structures, although efforts are being made to standardize the specifications for raw and processed analytical data. Although laboratory information management systems (LIMS) have been used in analytical laboratories since 1982, they are not, by themselves, designed to handle large volumes of raw data originating from different instruments. Historically, the primary purpose of a LIMS has been to manage only a select portion of the information generated for each sample and handle the administrative tasks needed to run an analytical laboratory. Several approaches have been described for integrating chromatographic data-acquisition systems with LIMS (*132*), and it is inevitable that such systems will be expanded to include support for capturing data from mass spectrometers and NMR instruments.

In addition to the sample scheduling and administrative features common to many LIMS packages, there are a number of data-handling tasks that will be required of software packages designed to manage the analytical data generated from combinatorial libraries. An effective system for managing information electronically would allow spectra and chromatographic data from a variety of instrument platforms to be viewed in a common program environment. Furthermore, such an electronic system would permit the analytical data to be organized and searched on the basis of chemical structure in addition to the text and numeric search capabilities offered with conventional database packages. The use of hyphenated chromatographic and spectroscopic methods to characterize library samples requires that these dissimilar types of analytical data be logically linked within the database to facilitate data interpretation. Further, the process of data interpretation itself is an area deserving attention with regards to automation. Indeed, researchers at ArQule, Inc., have applied a neural network software package, Braincel, to the task of automating the interpretation of chromatographic and MS data obtained from the analysis of combinatorial libraries (*133*).

There are two commercial software packages that perform some of the functions necessary for managing the data generated from library analysis. Target and Envision are products of Thru-Put Systems, Inc. (Orlando, FL), and Grams/32 is an upgrade of the older Grams/386 product from Galactic Industries Corporation (Salem, NH). Target is designed to process raw data files from GC, LC, and MS instruments. Grams/32 is a more powerful tool for the combinatorial chemist, because it handles a wider range of the industry's standard file formats from a broader range of analytical instruments, including MS, HPLC, NMR, FTIR, Raman, near-IR, and UV–vis instruments.

References

1. Snyder, L. R.; Kirkland, J. J. *Introduction to Modern Liquid Chromatography*; John Wiley and Sons: New York, 1979.
2. Bidlingmeyer, B. A. *Practical HPLC Methodology and Applications*; John Wiley and Sons: New York, 1992.
3. Dorsey, J. G.; Cooper, W. T.; Wheeler, J. F.; Barth, H. G.; Foley, J. P. *Anal. Chem.* **1994**, *66*, 500R–546R.

4. Altria, K. D. *J. Chromatogr.* **1993**, *646*, 245–257.
5. Corstjens, H.; Billiet, H. A. H.; Frank, J.; Luyben, K. *J. Chromatogr. A* **1995**, *715*, 1–11.
6. Kuhn, R.; Hoffstetter-Kuhn, S. *Chromatographia* **1992**, *34*, 505–512.
7. Rogan, M. M.; Altria, K. D.; Goodall, D. M. *Chirality* **1994**, *6*, 25–40.
8. Smith, N. W.; Evans, M. B. *Chromatographia* **1994**, *38*, 649–657.
9. Elmer, T.; Unger, K. K.; Tsuda, T. *Fresenius J. Anal. Chem.* **1995**, *352*, 649–653.
10. Dittmann, M. M.; Wienand, K.; Bek, F.; Rozing, G. P. *LC-GC* **1995**, *13*, 800–814.
11. Lelievre, F.; Yan, C.; Zare, R. N.; Gareil, P. *J. Chromatogr. A* **1996**, *723*, 145–156.
12. de Haseth, J.; Isenhour, T. *Anal. Chem.* **1977**, *49*, 1977–1981.
13. Wang, C.; Spars, D.; Williams, S.; Isenhour, T. *Anal. Chem.* **1984**, *56*, 1268–1272.
14. Vidrine, D. W.; Mattson, D. R. *Appl. Spectrosc.* **1978**, *32*, 502–506.
15. Vidrine, D. W. *J. Chromatogr. Sci.* **1979**, *17*, 477–482.
16. Shafer, K. H.; Lucas, S. V.; Jakobsen, R. J. *J. Chromatogr. Sci.* **1979**, *17*, 464–470.
17. Brown, R. S.; Taylor, L. T. *Anal. Chem.* **1983**, *55*, 1492–1497.
18. Robertson, R.; de Haseth, J.; Browner, R. *Appl. Spectrosc.* **1990**, *44*, 8–13.
19. Johnson, C. C.; Taylor, L. T. *Anal. Chem.* **1983**, *55*, 436–441.
20. Johnson, C. C.; Taylor, L. T. *Anal. Chem.* **1984**, *56*, 2642–2647.
21. McKittrick, P. T.; Danielson, N. D.; Katon, J. E. *J. Liquid Chromatogr.* **1991**, *14*, 377–393.
22. Jinno, K.; Fujimoto, C. *J. Liquid Chromatogr.* **1984**, *7*, 2059–2071.
23. Fujimoto, C.; Uematsu, G.; Jinno, K. *Chromatographia* **1985**, *20*, 112–116.
24. Jinno, K. *J. High Resolut. Chromatogr. Chromatogr. Commun.* **1982**, *5*, 364–367.
25. Kuehl, D.; Griffiths, P. R. *J. Chromatogr. Sci.* **1979**, *17*, 471–476.
26. Conroy, C. M.; Griffiths, P. R.; Jinno, K. *Anal. Chem.* **1985**, *57*, 822–825.
27. Kalasinsky, V. F.; Whitehead, K. G.; Kenton, R. C.; Smith, J. A. S.; Kalasinsky, K. S. *J. Chromatogr. Sci.* **1987**, *25*, 273–280.
28. Castles, M. A.; Azarraga, L. V.; Carreira, L. A. *Appl. Spectrosc.* **1985**, *40*, 673–680.
29. Conroy, C. M.; Duff, P. J.; Griffiths, P. R.; Azarraga, L. V. *Anal. Chem.* **1984**, *56*, 2636–2642.
30. Robertson, A. M.; Litlejohn, D.; Brown, M.; Dowle, C. J. *J. Chromatogr.* **1991**, *588*, 15–24.
31. Willoughby, R.; Browner, R. *Anal. Chem.* **1984**, *56*, 2625–2631.
32. Turula, V. E.; de Haseth, J. A. *Anal. Chem.* **1996**, *68*, 629–638.
33. Robertson, R. M.; de Haseth, J. A.; Kirk, J. D.; Browner, R. F. *Appl. Spectrosc.* **1988**, *42*, 1365–1368.
34. Robertson, R. M.; de Haseth, J. A.; Browner, R. F. *Mikrochim. Acta (Wein)* **1988**, *2*, 199–202.
35. Gagel, J. J.; Biemann, K. *Anal. Chem.* **1986**, *58*, 2184–2189.
36. Gagel, J. J.; Biemann, K. *Anal. Chem.* **1987**, *59*, 1266–1272.
37. Lange, A. J.; Griffiths, P. R.; Fraser, D. J. J. *Anal. Chem.* **1991**, *63*, 782–787.
38. Somsen, G. W.; Hooijschuur, E. W. J.; Gooijer, C.; Brinkman, U. A. Th.; Velthorst, N. H.; Visser, T. *Anal. Chem.* **1996**, *68*, 746–752.
39. Watanabe, N.; Niki, E. *Proc. Jpn. Acad.* **1978**, *54*, 194–199.
40. Bayer, E.; Albert, K.; Nieder, M.; Grom, E.; Keller, T. *J. Chromatogr.* **1979**, *186*, 497–507.
41. Dorn, H. C. *Anal. Chem.* **1984**, *56*, 747A–758A.
42. Bayer, E.; Albert, K.; Niedler, M.; Gromm, E.; Wolff, G.; Rindlisbacher, M. *Anal. Chem.* **1982**, *54*, 1747–1750.
43. Laude, D. A.; Wilkins, C. L. *Trends Anal. Chem.* **1986**, *5*, 230–235.
44. Wu, N.; Peck, T. L.; Webb, A. G.; Magin, R. L.; Sweedler, J. V. *Anal. Chem.* **1994**, *66*, 3849–3857.
45. Behnke, B.; Schlotterbeck, G.; Tallarek, U.; Strohschein, S.; Tseng, L. H.; Keller, T.; Albert, K.; Bayer, E. *Anal. Chem.* **1996**, *68*, 1110–1115.
46. Stevenson, S.; Dorn, H. C. *Anal. Chem.*, **1994**, *66*, 2993–2999.
47. Albert, K. *J. Chromatogr. A* **1995**, *703*, 123–147.

48. Albert, K.; Bayer, E. *Trends Anal. Chem.* **1988**, *7*, 288–293.
49. Albert, K.; Nieder, M.; Bayer, E.; Spraul, M. *J. Chromatogr.* **1985**, *346*, 17–24.
50. Laude, D. A.; Wilkins, C. L. *Trends Anal. Chem.* **1986**, *5*, 230–235.
51. Clore, G. M.; Kimber, B. J.; Gronenborn, A. M. *J. Magn. Reson.* **1983**, *54*, 170–173.
52. Turner, D. L. *J. Magn. Reson.* **1983**, *54*, 146–149.
53. Laude, D. A., Jr.; Lee, R. W. K.; Wilkins, C. L. *Anal. Chem.* **1985**, *57*, 1464–1469.
54. Hurd, R. E. *J. Magn. Reson.* **1990**, *87*, 422–428.
55. Wider, G.; Wuethrich, K. *J. Magn. Reson., Ser. B* **1993**, *102*, 239–241.
56. Hwang, T. L.; Shaka, A. J. *J. Magn. Reson., Ser. A* **1995**, *112*, 275–279.
57. Nicholson, J. K.; Foxall, P. J. D.; Spraul, M.; Farrant, R. D.; Lindon, J. C. *Anal. Chem.* **1995**, *34*, 793–811.
58. Sinallcombe, S. H.; Patt, S. L.; Keifer, P. A. *J. Magn. Reson., Ser. A.* **1995**, *117*, 295–303.
59. Sidelmann, U. G.; Gavaghan, C.; Carless, H. A. J.; Spraul, M.; Hofmann, M.; Lindon, J. C.; Wilson, I. D.; Nicholson, J. K. *Anal. Chem.* **1995**, *67*, 4441–4445.
60. Albert, K.; Kunst, M.; Bayer, E.; DeJong, H. J.; Genissel, P.; Spraul, M.; Bermel, W. *Anal. Chem.* **1989**, *61*, 772–775.
61. Spraul, M.; Hofmann, M.; Dvortsak, P.; Wilson, I. D. *Anal. Chem.* **1993**, *65*, 327–330.
62. Spraul, M.; Hofmann, M.; Lindon, J. C.; Farrant, R. D.; Seddon, M. J.; Nicholson, J. K.; Wilson, I. D. *NMR Biomed.* **1994**, *7*, 295–303.
63. Spraul, M.; Hofmann, M.; Lindon, J. C.; Nicholson, J. K.; Wilson, I. D. *Methodol. Surv. Bioanal. Drugs* **1994**, *23*, 21–32.
64. Morris, K. F.; Johnson, C. S., Jr. *J. Am. Chem. Soc.* **1992**, *114*, 3139–3141.
65. Morris, K. F.; Johnson, C. S., Jr. *J. Am. Chem. Soc.*, **1993**, *115*, 4291–4299.
66. Lin, M.; Jayawickrama, D. A.; Rose, R. A.; DelViscio, J. A.; Larive, C. K. *Anal. Chim. Acta* **1995**, *307*, 449–457.
67. Barjat, H.; Morris, G. A.; Smart, S.; Swanson, A. G.; Williams, S. C. R. *J. Magn. Reson.* **1995**, *108*, 170–172.
68. Gordon, E. M.; Barrett, R. W.; Dower, W. J.; Fodor, S. P. A.; Gallop, M. A. *J. Med. Chem.* **1994**, *37*, 1385–1401.
69. Mellon, F. A. *VG Monographs* **1991**, *2*, 1–19.
70. Willoughby, R. C.; Browner, R. F. *Anal. Chem.* **1984**, *56*, 2626–2631.
71. Winkler, P. C.; Perkins, D. D.; Williams, W. K.; Browner, R. F. *Anal. Chem.* **1988**, *60*, 489–493.
72. Vestal, M. L.; Fergusson, G. J. *Anal. Chem.* **1985**, *57*, 2373–2378.
73. Honing, M.; Barcelo, D.; Van Baar, B. L. M.; Brinkman, U. A. Th. *Trends Anal. Chem.* **1995**, *14*, 496–504.
74. Chapman, J. R. *A Practical Guide to HPLC Detection*; Academic: New York, 1993; p. 186.
75. Whitehouse, C. M.; Dreyer, R. N.; Yamashita, M.; Fenn, J. B. *Anal. Chem.* **1985**, *57*, 675–679.
76. Meng, C. K.; Fenn, J. B. *Org. Mass Spectrom.* **1991**, *26*, 542–549.
77. Fenn, J. B. *J. Am. Soc. Mass Spectrom.* **1993**, *4*, 524–535.
78. Kebarle, P.; Tang, L. *Anal. Chem.* **1993**, *65*, 972A–986A.
79. Fenn, J. B.; Mann, M.; Chin, K.; Wong, S. F.; Whitehouse, C. M. *Mass Spectrom. Rev.* **1990**, *9*, 37–70.
80. Bruins, A. P. *J. Chim. Phys. Phys.-Chim. Biol.* **1993**, *90*, 1335–1344.
81. Wachs, T.; Conboy, J. C.; Garcia, F.; Henion, J. D. *J. Chromatogr. Sci.* **1991**, *29*, 357–366.
82. Duffin, K. L.; Wachs, T.; Henion, J. D. *Anal. Chem.* **1992**, *64*, 61–68.
83. Bajic, S.; Doerge, D. R.; Lowes, S.; Preece, S. *Am. Lab.* **1993**, *February*, 40B, 40D, 40F, 40H, 40O.
84. Andrien, B. A.; Boyle, J. G. *Spectroscopy* **1995**, *10*, 42–44.

85. Henion, J. D.; Wachs, T.; Mordehai, A. V. *J. Pharm. Biomed. Anal.* **1993**, *11*, 1049–1061.
86. Mordehai, A. V.; Hopfgartner, G.; Huggins, T. G.; Henion, J. D. *Rapid Commun. Mass Spectrom.* **1992**, *6*, 508–516.
87. Mordehai, A. V.; Henion, J. D. *Rapid Commun. Mass Spectrom.* **1993**, *7*, 205–209.
88. Chervet, J. P.; Ursem, M.; Salzmann, J. P. *Anal. Chem.* **1996**, *68*, 1507–1512.
89. Tomer, K. B.; Moseley, M. A.; Deterding, L. J.; Parker, C. E. *Mass Spectrom. Rev.* **1994**, *13*, 431–457.
90. Emmett, M. R.; Caprioli, R. M. *J. Am. Soc. Mass Spectrom.* **1994**, *5*, 605–613.
91. Garcia, F.; Henion, J. D. *Anal. Chem.* **1992**, *64*, 985–990.
92. Parker, C. E.; Perkins, J. R.; Tomer, K. B.; Shida, Y.; O'Hara, K.; Kono, M. *J. Am. Soc. Mass Spectrom.* **1992**, *3*, 563–574.
93. Parker, C. E.; Perkins, J. R.; Tomer, K. B. *J. Chromatogr.* **1993**, *616*, 45–57.
94. Lane, S. J.; Boughtflower, R.; Paterson, C.; Underwood, T. *Rapid Commun. Mass Spectrom.* **1995**, *9*, 1283–1287.
95. Schmeer, K.; Behnke, B.; Bayer, E. *Anal. Chem.* **1995**, *67*, 3656–3658.
96. Takada, Y.; Sakairi, M.; Koizumi, H. *Anal. Chem.* **1995**, *67*, 1474–1476
97. Lu, W.; Poon, G. K.; Carmichael, P. L.; Cole, R.B. *Anal. Chem.* **1996**, *68*, 668–674.
98. Dunayevskiy, Y.; Vouros, P.; Carell, T.; Winter, E. A.; Rebek, J., Jr. *Anal. Chem.* **1995**, *67*, 2906–2915.
99. Hutchins, S. M.; Chapman, K. T. *Tetrahedron Lett.* **1995**, *36*, 2583–2586.
100. Bray, A. M.; Chiefari, D. S.; Valerio, R. M.; Maeji, N. J. *Tetrahedron Lett.* **1995**, *36*, 5081–5084.
101. Metzger, J. W.; Wiesmuller, K. H.; Gnau, V.; Brunjes, J.; Jung, G. *Angew. Chem. Int. Ed. Eng.* **1993**, *32*, 894–896.
102. Volmer, D. A.; Vollmer, D. L. *LC-GC* **1996**, *14*, 236–242.
103. Martin, J.; Quirke, E.; Adams, C. L.; Van Berkel, G. J. *Anal. Chem.* **1994**, *66*, 1302–1315.
104. Wilson, S. R.; Tulchinsky, M. L.; Wu, Y. *Bioorg. Med. Chem. Lett.* **1993**, *3*, 1805–1808.
105. Wilson, S. R.; Wu, Y. *J. Am. Soc. Mass Spectrom.* **1993**, *4*, 596–603.
106. Iwabuchi, H.; Kitazawa, E.; Kobayashi, N.; Watanabe, H.; Kanai, M.; Nakamura, K. *Biol. Mass Spectrom.* **1994**, *23*, 540–546.
107. Gordon, D. W.; Steele, J. *Bioorg. Med. Chem. Lett.* **1995**, *5*, 47–50.
108. Brown, B. B.; Wagner, D. S.; Geysen, H. M. *Mol. Diversity* **1995**, *1*, 4–12.
109. Stankova, M.; Issakova, O.; Sepetov, N. F.; Krchnak, V.; Lam, K. S.; Lebl, M. *Drug Dev. Res.* **1994**, *33*, 146–156.
110. Brummel, C. L.; Vickerman, J. C.; Carr, S. A.; Hemling, M. E.; Roberts, G. D.; Johnson, W.; Weinstock, J.; Gaitanopoulos, D.; Benkovic, S. J.; Winograd, N. *Anal. Chem.* **1996**, *68*, 237–242.
111. Volmer, D. A.; Vollmer, D. L.; Wilkes, J. G. *LC-GC* **1996**, *14*, 216–224.
112. March, R. E.; Strife, R. J.; Creaser, C. S. In *Practical Aspects of Ion-Trap Mass Spectrometry*; March, R. E.; Todd, J. F., Eds.; CRC: Boca Raton, FL, 1995; Vol. 3, pp 27–88.
113. Eberhart, S. T.; Hatzis, A.; Rothchild, R. *J. Pharm. Biomed. Anal.* **1986**, *4*, 147–154.
114. Meese, C. O.; Specht, D.; Ratage, D.; Eichelbaum, M.; Wisser, H. *Fresenius J. Anal. Chem.* **1993**, *346*, 837–840.
115. Szantay, C., Jr. *Trends Anal. Chem.* **1992**, *11*, 332–344.
116. Rabenstein, D. L.; Keire, D. A. In *Modern NMR Techniques and Their Application in Chemistry*; Popov, A. I.; Hallenga, K., Eds.; M. Dekker: New York, 1991; Chapter 5, pp 323–369.
117. Haw, J. F.; Glass, T. E.; Dorn, H. C. *J. Magn. Reson.* **1982**, *49*, 22–31.
118. Patek, M.; Drake, B.; Lebl, M. *Tetrahedron Lett.* **1995**, *36*, 2583–2586.
119. Ostresh, J. M.; Husar, G. M.; Blondelle, S. E.; Dorner, B.; Weber, P. A.; Houghten, R. A. *Proc. Natl. Acad. Sci. U.S.A.* **1994**, *91*, 11138–11142.

120. Ford, D. L; Kennard, W. *J. Oil Colour, Chem. Assoc.* **1966**, *49*, 299–313.
121. Bear, G. R. *J. Chromatogr.* **1988**, *459*, 91–107.
122. Stolyhwo, A.; Colin, H.; Guiochon, G. *Anal. Chem.* **1985**, *57*, 1342–1354.
123. Sotirhos, N.; Throngren, C.; Herslof, B. *J. Chromatogr.* **1985**, *331*, 313–320.
124. Marrae, R.; Dick, J. *J. Chromatogr.* **1981**, *210*, 138–145.
125. Peterson, J. A.; Risley, D. S. *J. Liq. Chromatogr.* **1995**, *18*, 331–338.
126. Lafosse, M.; Elfakir, C.; Morin-Allory, L.; Dreux, M. *J. High Res. Chromatogr.* **1992**, *15*, 312–318.
127. Mourey, T. H.; Oppenheimer, L. E. *Anal. Chem.* **1984**, *56*, 2427–2434.
128. Charlesworth, J. M. *Anal. Chem.* **1978**, *50*, 1414–1420.
129. Kibbey, C.E. *Mol. Diversity* **1996**, *1*, 247–258.
130. Asmus, P. A.; Landis, J. B. *J. Chromatogr.* **1984**, *316*, 461–472.
131. Stolywho, A.; Colin, H.; Guiochon, G. *J. Chromatogr.* **1983**, *265*, 1–18.
132. McDowall, R. D. *LC-GC* **1996**, *14*, 28–31.
133. Zambias, R. A. Presented at the Cambridge Healthtech Institute Program on Rapid Compound Characterization, San Diego, CA, May 1996.

Deconvolution Tools in Solution-Phase Synthesis

Xavier Williard and André Tartar

When natural products are screened for new pharmacological functions, the detection of an activity must be followed by purification and structure determination. Combinatorial chemistry, however, offers the possibility of generating mixtures of structurally diverse compounds using a predetermined format, greatly simplifying the identification of a biologically active compound within the mixture. There are two basic approaches to identifying active compounds in combinatorial chemistry. Tagging involves linking an identifier (either directly or, more often, indirectly) to each type of molecule present in the library; unfortunately the chemistry of the tag may prevent the use of certain types of assays or modify the recognition by target biological macromolecules. Here, we concentrate on the second approach, deconvolution, in which the library is partitioned into a series of sublibraries that are organized in such a way as to allow the identification of an active compound without the necessity for a tag.

Combinatorial Libraries are *N*-Dimensional Matrices

The basis for deconvolution strategies and the reason they work is that, as outlined by Pirrung (*1*), any combinatorial library can be conceptually represented as an *N*-dimensional matrix. Each axis of the matrix has as many elements as the number (n) of building blocks present in each set used in the synthesis, and N is the number of steps of combinatorial synthesis involved in the preparation of the library. Two examples are given in Figure 1. The first matrix (Figure 1a) corresponds to two steps of combinatorial synthesis using four different building blocks in position A ($n_A = 4$) and four different building blocks in position B ($n_B = 4$). This library con-

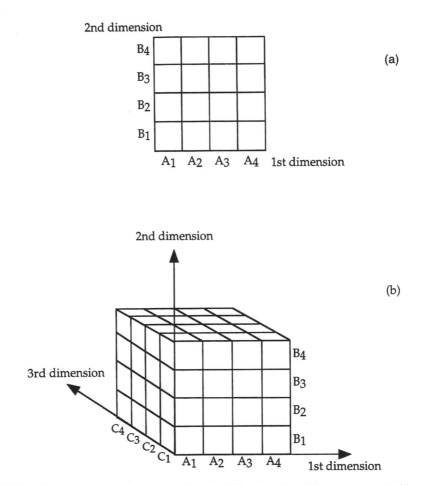

FIGURE 1. Matrix representation of combinatorial libraries. (a) A library constructed in two steps ($N = 2$) with four building blocks used at each step ($n_A = n_B = 4$) has 16 compounds ($A_n - B_n$). (b) A library constructed in three steps ($N = 3$) with four building blocks used at each step ($n_A = n_B = n_C = 4$) has 64 compounds ($A_n - B_n - C_n$).

tains $n_A \times n_B = 16$ compounds, and each compound can be localized in a different cell of the matrix (Ax, By). The second matrix (Figure 1b) illustrates a library constructed with three steps ($N = 3$) of combinatorial synthesis, each involving four building blocks ($n = 4$). This library contains 64 compounds ($n^N = 4^3$), and each compound can be localized in a different cell of the matrix defined by the combination Ax, By, Cz. Although graphical representation is more difficult for $N > 3$, any combinatorial library can be conceptually represented as an N-dimensional matrix in which each compound of the library is localized in a different cell.

　　To obtain all the information available from a library, in our case the biological activity of each compound, this value would have to be assessed for each of the cells by determining the activity of each pure compound (and this is the procedure that is followed for spatially addressable combinatorial libraries (*2*) prepared by parallel synthesis). In general, however, the purpose of random screening of compound libraries is not to obtain complete information on all compounds present in these libraries. For example, the simplest question that can be asked is whether there is at least one active compound present in the library. Provided that the activity of the active compound can be detected above the background of the mixture, answering this question requires that only one activity determination be performed on a pool containing all the compounds. The next step is to determine the structure of the active compound present in this library. This determination requires additional information from additional activity-determination experiments, but this information still represents a small subset of the complete information that can theoretically be obtained from the library (the activity of each individual compound). Thus the principle of all deconvolution methods is to limit the number of biological experiments, so that only those that will allow the determination of the structure of the most active compound are performed.

Several different deconvolution methods have been proposed. All of them are based on the partition of the complete library into a series of sublibraries, which are then tested individually. The three major methods, which are described in the following sections, are iterative deconvolution, positional scanning, and orthogonal libraries.

Iterative Deconvolution

Iterative deconvolution involves the detection of an activity in the full library, followed by the synthesis and screening of a series of sublibraries containing a decreasing number of compounds until a unique structure is identified. The basic concept is illustrated in Figure 2, which shows as an example a small library constructed using three coupling steps with a mixture of four building blocks used at each step. This yields $4 \times 4 \times 4 = 64$ trimers. According to the general convention used by Houghten (*3*), in which X represents the mixture of building blocks in a position (four building blocks in this case) and O defines one individual building block (among the four), the complete initial library can thus be represented as XXX.

If a biological activity is detected in this mixture, three steps of deconvolution will be sufficient to identify the structure of the active compound. The first step is to determine the building block that is required in the first position. Four sublibraries (represented as OXX) are prepared, with the first position occupied by a defined building block and the last two positions still containing a mixture of the four building blocks. These four different sublibraries, each containing 16 ($1 \times 4 \times 4$) trimers, are then screened to determine which of the building blocks at the first position confers the biological activity. If the most active sublibrary contains A_3 at this position, four new sublibraries (represented by A_3OX) will be prepared in which the first position is defined as A_3, the second position is one of four defined building

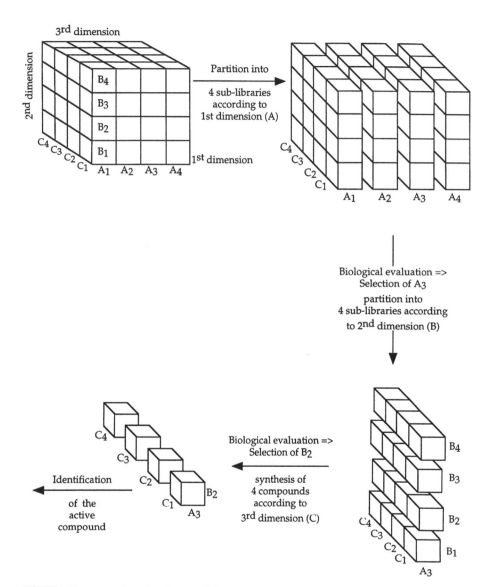

FIGURE 2. Three-step iterative deconvolution of a three-position ($N = 3$), 64-compound combinatorial library $A_nB_nC_n$ with $n_A = n_B = n_C = 4$.

blocks, and the third position still consists of the mixture of four building blocks. The determination of the biological activity of these four new sublibraries will indicate which of the building blocks in the second position confers the activity. If the most active sublibrary contains B_2 in the second position, the third deconvolution step will involve the synthesis of four additional sublibraries (of one compound each, and represented as A_3B_2O), and the evaluation of their biological activity will identify the structure of the active compound. The matrix representation of these three steps of deconvolution is shown in Figure 2. This method generates less complex mixtures at each deconvolution step, so the active sublibraries are enriched relative to their predecessors.

In this example, with three deconvolution steps each involving the preparation of four sublibraries, 12 sublibraries are required to identify an active compound from the 64 possibilities. The efficiency of the method increases sharply as the size of the library increases. The iterative deconvolution of a library that was synthesized in N steps with n building blocks used at each step (and thus containing n^N compounds) requires the preparation and biological evaluation of $N \times n$ sublibraries (over N iterations). Thus a hexapeptide library built from 20 natural amino acids contains 20^6 or 64,000,000 peptides, but its deconvolution requires the synthesis and evaluation of only $6 \times 20 = 120$ sublibraries.

The main limitation in cases such as this is the difficulty in detecting a biologically active compound in a mixture containing so many compounds. In general, however, such libraries contain compounds of limited diversity, and any active compound is generally associated with several suboptimal binders. It is crucial for the chemist to realize that the combined effect of the active and partially active compounds will generate a detectable biological activity, although it may make selecting the best binder difficult. The impact of this effect can be monitored by following the evolution of biological activities during the deconvolution. At each step, an additional position is defined, thus increasing the ratio of the concentration of the active compound to the concentration of inactive compounds. In an ideal library composed of totally unrelated compounds and with n building blocks in one position, the concentration of the active component, and so the biological activity of the mixture, should increase by a factor of n after the deconvolution of this position. In most cases reported to date, biological activities did not improve this much.

Freier et al. (4) have analyzed the evolution of activity when deconvoluting oligomer libraries for 15 cases reported in the literature (Table I). Using these values, they define a suboptimal binding factor (SBF) value, which represents the ratio of the theoretical activity for the initial sublibrary (based on the assumption that the winner is the only active compound among the pool of compounds present in the initial sublibrary) to the observed activity for the initial sublibrary. A high SBF indicates a high incidence of suboptimal binders in the initial sublibrary, whereas a value of 1 would indicate that the winner was the only active compound in the initial sublibrary.

The presence of large numbers of suboptimal binders must be taken into account during deconvolution as each step involves an irreversible choice, yielding more and more restricted arrays of compounds. The simplest approach in the decon-

TABLE 1. Suboptimal Binding Factor (SBF) is a Measurement of the Apparent Number of Active Compounds in a Sublibrary

Target for Assay[a]	Library Composition	Number of Monomers	Oligomer Length	Number of Fixed Positions[b]	Subset Complexity[c]	Subset Activity[d]	Winner Activity[d]	SBF[d]
Cellular Screens								
HIV	Phosphorothioate DNA	4	8	2	4096	20 μM	0.3 μM	61
HSV	Phosphorothioate DNA	4	8	1	16,384	70 μM	0.4 μM	94
S. aureus	Peptide	18	6	2	104,976	450 μg mL^{-1}	3.4 μg mL^{-1}	793
S. aureus	Peptide	19	6	2	130,321	1730 μg mL^{-1}	11 μg mL^{-1}	829
Antibody Binding								
pAB-FMRF	Peptide	15	4	0	50,625	1400 μg mL^{-1}	0.5 μg mL^{-1}	18
pAB-pep3	Peptide	16	6	0	16,777,216	6500 μg mL^{-1}	0.08 μg mL^{-1}	206
mAB-19B10	Peptide	18	6	2	104,976	250 μM	0.03 μM	13
mAB-125-10F3	Peptide	19	6	2	130,321	20 μM	0.004 μM	26
Nucleic Acid Binding (Experimental)								
Ha-ras RNA	2'-O-methyl RNA	4	9	1	65,536	10 μM	0.01 μM	66
Protein Binding or Activity								
HIV-protease	Peptide	22	4	1	11,132	4400 μM	1.4 μM	3.5
Opioid receptor	Peptide	19	6	2	130,321	3.452 μM	0.025 μM	1057
Opioid receptor	Peptide	19	6	2	130,321	2.1 μM	0.005 μM	310
Nucleic Acid Binding								
18-mer target	RNA	4	9	0	262,144	0.020 μM	5.5×10^{-7} μM	7.1
9-mer target	RNA	4	9	0	262,144	0.034 μM	1.1×10^{-6} μM	8.0
6-mer target	RNA	4	9	0	262,144	7.9 μM	0.047 μM	1561

[a]Abbreviations: HIV, human immunodeficiency virus; HSV, herpes simplex virus.

[b]Number of fixed positions in the initial round of deconvolution.

[c]The number of unique molecules in each initial subset (the complexity) is $M^{(L-F)}$, where N is the number of monomers, L is the oligomer length, and F is the number of fixed positions.

[d]The concentration of the most active subset in the initial round necessary for half-maximal activity (subset activity), the activity of the final compound (winner activity), and the complexity were used to calculate the SBF of the most active sublibrary during the first step of deconvolution. (SBF = winner activity × subset complexity/subset activity.)

SOURCE: Adapted from Freier et al. (4).

volution strategy assumes that, when comparing several mixtures, the mixture with the greatest activity contains the most active compound. If, however, the building block selected in one position is not the one in the corresponding position of the most active compound, but the one that is shared by the largest number of suboptimal binders, the next steps of deconvolution will lead to a suboptimal binder. To avoid this problem it is important, at each step, to examine which building blocks are found in the three or four most active sublibraries. If two of these building blocks are structurally very different, then it is likely that two different paths exist for iterative deconvolution, and that these paths will eventually lead to totally different compounds. In this case, two independent deconvolutions must be performed using a different building block in each case (see the first of the examples in the section "Selected Examples", later in the chapter). Iterative deconvolution has proven its efficiency in a large number of cases, such as the isolation of nanomolar inhibitors of seven-transmembrane receptors (see example 2).

A major drawback of iterative deconvolution is the requirement for the preparation and biological evaluation of a new set of sublibraries at each step. (If the library is used again with a different assay, it is unlikely that the same sublibraries will be needed.) The number of steps can only be reduced by synthesizing more sublibraries. For example, the step in which the last randomized position of a library is deconvoluted can be eliminated by directly preparing an individual sublibrary for each different building block (avoiding the last recombine step of the divide–couple–recombine (DCR) method), and so generating sublibraries with formats such as XXO or XXXO. The dual positional format described by Houghten (*3*) is an extension of this strategy to the last two randomized positions. For example, a hexapeptide library can be prepared using four combinatorial steps in which mixtures of 20 amino acids are used, followed by two steps in which individual amino acids are used. This yields $20 \times 20 = 400$ different sublibraries with the general structure OOXXXX, which are then screened individually. Two advantages of this method are that there are only 4 instead of 6 steps at which sublibraries have to be resynthesized, and there are also 400-fold fewer compounds ($20^4 = 160,000$ rather than $20^6 = 64,000,000$) present in the pools that are submitted to the initial screening.

Erb et al. (*5*) have suggested that recursive deconvolution could be used to accelerate the iterative deconvolution process. Although this method was developed for use with solid-phase chemistry, it could be applied to solution-phase problems equally well. In this strategy, a portion of each resin sample is set aside after each coupling step and catalogued. The resins are then recombined before the next coupling. When an activity is detected in a pool of compounds, the partial libraries that have been set aside are used to speed up the deconvolution procedure.

Positional Scanning

This method was pioneered by Houghten with peptide libraries and has been the subject of several reviews (*6, 7*). It involves the initial synthesis of series of sublibraries in which one position is defined with a single building block while the remaining positions are composed of mixtures of building blocks. Thus, the number of sublibraries synthesized for each position i is n_i, the number of building blocks used in that position. The determination of the biological activity of these sublibraries yields information about the most important building block for each position. The same number of sublibraries is produced as in iterative deconvolution, but no resynthesis of sublibraries is required and no increase in biological activity is expected during positional scanning. According to the rules proposed by Houghten, a hexapeptide library containing the 20 natural amino acids at each position will be formatted for positional scanning as 120 (6 × 20) sublibraries corresponding to:

OXXXXX
XOXXXX
XXOXXX
XXXOXX
XXXXOX
XXXXXO

A simpler positional scan is shown as a matrix representation in Figure 3. In this case the library is constructed in two combinatorial steps ($N = 2$) from two sets of four building blocks (A at the first position and B at the second position; $n_A = n_B = 4$). The resulting library is represented as a two-dimensional (2D) matrix with 16 cells. Positional scanning of this library involves the division of the matrix into rows and columns. In each of the first four sublibraries, a single A building block is reacted with a mixture of the four B building blocks. In each of the second series of four sublibraries, a single B building block is reacted with a mixture of the four A building blocks. This method has been used by Smith et al. (8) for the synthesis and biological evaluation of amides and esters from a single coupling step. A similar approach, developed by Pirrung (1) to evaluate a library of 54 dimers (carbamates), is described in detail in the examples at the end of this chapter (example 3).

This representation can be extended to multidimensional matrices, as shown in Figure 4 for a three-position ($N = 3$) combinatorial library containing four different building blocks at each position ($n_A = n_B = n_C = 4$). The main advantage of positional scanning over iterative deconvolution is that the different positions of the library are evaluated simultaneously, and these results, found for each position, can be combined to directly identify active compounds. There is, however, no enrichment of the active species during the selection procedure with this method. This can be a problem if there are several active compounds in the library, as the screening does not provide any information on their connectivity. Thus, if several building blocks are found to be active at each position, the number of compounds that have to be individually prepared for biological testing corresponds to all possible combinations and so grows exponentially (e.g., three active amino acids at each position of a hexapeptide library would require the preparation of $3^6 = 729$ peptides).

Dooley and Houghten (7) proposed a number of alternative solutions to the synthesis of all possible peptides, including the synthesis of positional scanning

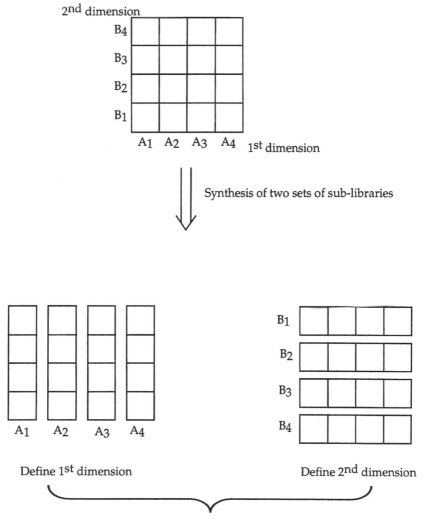

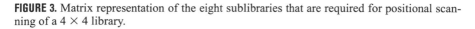

FIGURE 3. Matrix representation of the eight sublibraries that are required for positional scanning of a 4 × 4 library.

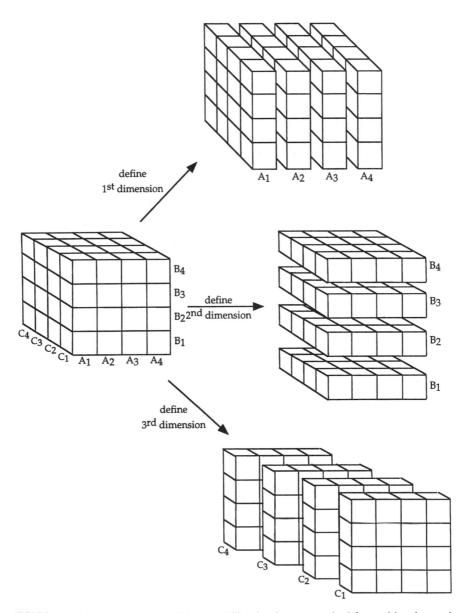

FIGURE 4. Matrix representation of the 12 sublibraries that are required for positional scanning of a 4 × 4 × 4 library. The complete set of 12 sublibraries is submitted to evaluation.

sublibraries such as $X_1X_2X_3X_4O$, in which X_n is limited to the subset of amino acids which proved to be effective for position n in the initial positional scan. This will increase the concentration of active compounds but requires additional synthesis. Another possibility is to pursue a classical iterative deconvolution using only the most active amino acids for each position. In this case, the best choice will be to start the iterative deconvolution from the position that gave the most variation in the biological assay when it was occupied with the different building blocks.

Orthogonal Libraries

Unlike iterative deconvolution but as with positional scanning, orthogonal libraries are screened simultaneously and allow the direct identification of active compounds without iterative resynthesis and selection steps.

This strategy involves the synthesis of two libraries, I and II, which contain the same compounds. The libraries are divided into sublibraries in such a way that the number of compounds in each sublibrary is equal to (or less than) the number of sublibraries in each library. For example, in Figure 5 a library containing 16 compounds is divided into four sublibraries of four compounds each. The principle of orthogonal partition is that all compounds that are present in the same sublibrary in library I are partitioned into different sublibraries in library II. Thus in Figure 5, each of the four compounds that are pooled in a sublibrary of library I are found in the four different sublibraries of library II. Initially, the biological activity of the four sublibraries of library I are tested. If an activity is detected in one of these sublibraries, then the four sublibraries of library II are screened. The fact that each sublibrary of II contains only one compound from a given sublibrary of I allows a straightforward identification of the active compound. This approach can be extended to matrices with more than two dimensions [see example 4 below for a three-dimensional matrix of 15,625 compounds (9)] provided that the number of sublibraries in each of libraries I and II is at least equal to the number of compounds present in each sublibrary. (For example, libraries containing 10,000 compounds can be organized into 2×100 sublibraries containing 100 compounds each.)

Although orthogonal libraries can be prepared by mixing individual compounds, they can more easily be assembled using classical methods of combinatorial chemistry. To obtain an orthogonal partition of compounds within the sublibraries, the mixtures of building blocks are partitioned orthogonally, so that any group of building blocks used to prepare library I shares only one building block with any group of building blocks used to prepare library II. Thus in Figure 5, library I was synthesized by pooling building blocks $A_1 + A_2$, $A_3 + A_4$, $B_1 + B_2$, and $B_3 + B_4$ while library II was synthesized by pooling building blocks orthogonally $(A_1 + A_3, A_2 + A_4, B_1 + B_3,$ and $B_2 + B_4)$. The orthogonal partition of building blocks leads directly to an orthogonal partition of compounds within the libraries. For larger libraries there are generally many possible ways to orthogonally partition groups of building blocks. However, to maximize the diversity within each subli-

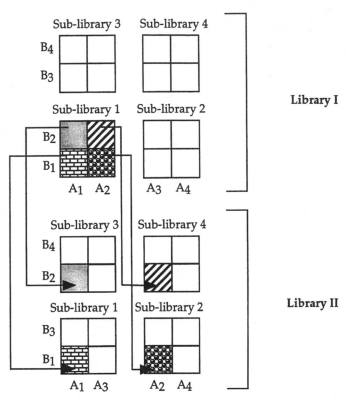

FIGURE 5. Composition of two orthogonal libraries corresponding to the general structure A_n-B_n with $n = 4$. Each of the four compounds pooled in library A1 (A_1B_1, A_1B_2, A_2B_1, A_2B_2) is partitioned into one of four different sublibraries of library B. As an example, detection of an activity in sublibrary 1 of library I and in sublibrary 3 of library II directly identifies the compound represented by the gray square as the active compound.

brary, it is important to select those partitions that maximize structural diversity within each group of building blocks.

The use of orthogonal libraries is not limited to polypeptides or other polymers that can be sequenced, and these libraries do not restrict the chemical strategies used to generate diversity. This method can even be applied to the rapid screening of pools of existing compound libraries of diverse origins.

Moreover, orthogonal libraries offer an internal validation during the screening, as any positive result in one library must be confirmed by a positive result in the orthogonal library, thus providing an internal control for the detection of false positive results. Another advantage is that an active compound will be pooled with different compounds in library I and library II, thus allowing detection of possible "suboptimal binding" effects (*4*). Unlike the two previous strategies, which can be applied to large libraries of low diversity (such as peptide or oligonucleotide libraries) in which the presence of several compounds of similar activity does not preclude the identification of an active species, orthogonal libraries are better suited to groups of compounds with high structural diversity. If several compounds with similar activities are present in the library, the number of ways to decode the read-out will rapidly increase. For example, if two compounds with the same activity exist in the library, they will generally be responsible for two hits in both library I and library II. These can be decoded as any of four possible structures, of which only two will be active. The number of possibilities increases with the square of the number of equally active compounds in each library. In such cases, iterative deconvolution is required.

Synthetic Strategies

Although solution screening offers a clear advantage in terms of the broad variety of assays that can be used, this strategy requires that compounds be available in a soluble form. After polymer-supported synthesis, some compounds are tested while still linked to their support, but for solution screening, a suitable linker between the compounds and the polymer must be designed. This linker, which has to be stable enough to withstand the reaction conditions used during synthesis, must be cleaved to release the products in a soluble form. This may restrict the molecular diversity of the products, as only a limited number of functional groups may be compatible with this last chemical step. On the contrary, solution-phase combinatorial chemistry yields compounds that can be directly screened. Moreover, with solution chemistry there is no requirement that at least one of the building blocks carry a functional group that can be linked to the support. A typical example is for a library of carbamates (example 3) prepared in solution using monofunctional isocyanates and alcohols. This synthesis yields several relatively unstable compounds, which would probably not have resisted the chemical reagents generally used to cleave linkers.

Purification from reagents used for the synthesis (e.g., coupling agents) is a problem, however. Given the structural heterogeneity of the compounds present in the library, any purification runs the risk of losing at least some compounds. Moreover, after solution synthesis each combinatorial compound is more or less diluted (depending on the number of members in the pool), but the reagents do not undergo such a dilution. Thus, a library of 100 amides synthesized using a carbodiimide as coupling agent between a set of 10 amines and 10 acids will contain 0.01 equiv of each amide but 1 equiv of the urea generated by the carbodiimide (if the reaction has reached completion).

To minimize these problems, solution synthesis should favor combinatorial chemistries using reagents that can be easily eliminated. The classical strategies involve reagents that generate volatile products, or polymer-supported reagents. In all cases, the compatibility of the different byproducts with the biological assays should be tested in a blank assay as a basic precaution before building any new combinatorial library. Such a test allows the chemist to make an informed choice between the different chemical possibilities.

Ensuring the Equality of Final Compound Concentrations

The deconvolution strategies described above rely on the determination of the biological activity (A) of mixtures of compounds. The result of such a determination depends not only on the biological activity of each individual compound but also on its molar representation in the mixture. As an example, a higher than expected concentration of a suboptimal binder can lead to an overestimation of the biological activity of the corresponding pool of compounds and introduce bias in the deconvolution. Two approaches to this problem are described below.

The Divide-Couple-Recombine (DCR) Method

In this method (10), polymer beads are divided equally into several batches. Each batch is then coupled with one of the building blocks before the different batches are pooled. If the division of the resin has been performed accurately, this method results in an equimolar distribution of the building blocks. This method will not work, however, unless the number of beads at the beginning of the synthesis is large enough. The magnitude of this problem increases dramatically with the size of the library.

Deviation from the ideal equimolar distribution of final products has been rigorously examined by Zhao et al. (11, 12). They analyzed the deviation from the ideal equimolar distribution of products in the final mixture of a DCR synthesis of a tripeptide library ($N = 3$), synthesized using 20 amino acids at each position ($n_A = n_B = n_C = 20$) and so containing $20 \times 20 \times 20 = 8000$ compounds. The synthesis used 90-μm beads, and the resin contained 2.86×10^6 beads g^{-1}. The deviation assessment takes into account the difference in abundance of every possible pair of compounds. The required number of beads is determined so that with a particular level [$(1 - \alpha) \times 100\%$] of confidence, each individual relative error is controlled to be less than a predetermined tolerable limit, L_2. The maximum ratio of the concentrations of any compound in the final mixture is less than $(1 + L_2)/(1 - L_2)$, where $0 < L_2 < 1$. Table II lists the required number of beads for different confidence levels and several different tolerable limits of $(1 - \alpha) \times 100\%$. For example, if we want the concentration ratio between the most and the least prevalent compounds to be <1.5, then $(1 + L_2)/(1 - L_2) = 1.5$ and $L_2 = 0.2$. Depending on the required confidence level, this would require 1.417 or 1.631 g of resin. If we want the ratio of all concentrations to be <1.1, this corresponds to an L_2 of 0.05, and the synthesis would require 22–26 g of resin. In contrast, if L_2 values are close to 1, only small quantities of resin are required in the synthesis, but this choice of L_2 will only ensure that all compounds are present (with the corresponding confidence level). Such ratios are acceptable when compounds are screened still linked to beads, as the requirement in this case is that each compound should at least be pres-

TABLE II. Required Beads Based on Individual Relative Errors

	Required beads (g)[a]	
Tolerable Limit, L_2	When $(1 - \alpha) = 0.95$	When $(1 - \alpha) = 0.99$
0.01	566.734	652.502
0.05	22.669	26.100
0.1	5.667	6.525
0.2	1.417	1.631
0.3	0.630	0.725
0.5	0.227	0.261
0.8	0.089	0.102
0.99	0.058	0.067

NOTE: The split–recombine synthesis involves $20 \times 20 \times 20$ reactions (i.e., three coupling steps generating a total of 8000 compounds).

[a]The beads have a diameter of 90 μm, and there are 2.86×10^6 beads in 1 g.

SOURCE: Reproduced with permission from reference 11.

ent on one bead, in order to be selected. For the solution phase, however, too high a ratio of the concentrations of different compounds (an L_2 value of 0.8 leads to a maximum ratio of 9 between the most and the least prevalent compound) precludes any accurate evaluation of the contribution of each individual compound to the biological activity of a pool, and so prevents the use of deconvolution strategies with such libraries.

Thus, considering the need to generate approximately equimolar mixtures and the amounts of resin that can easily be manipulated (10 to 100 g), the use of the DCR method in strategies involving deconvolutions will be limited to libraries containing between 10,000 and 100,000 compounds.

Coupling Mixtures of Building Blocks

The alternative to individual coupling reactions is to have mixtures of compounds in coupling reactions, yielding beads that carry mixtures of compounds. With this method the number of different compounds that can be prepared is limited only by the number of functional groups available on the beads, but, once again, achieving equimolar distribution of the compounds obtained after cleavage from the polymer is a problem. In this case, the factors controlling the abundance of different compounds are quite different. In the DCR method, excesses of reagents are used to force reactions to completion, but when coupling mixtures of building blocks the most reactive building blocks will be preferentially incorporated, leading to an unsatisfactory distribution of compounds in the final mixtures.

Solutions to this problem have been proposed for use with building blocks of similar reactivities, such as amino acids or nucleotides. One possibility is to perform a first coupling using only one equivalent of the mixture of building blocks per equivalent of reactive groups on the resin. Using a prolonged coupling time, even

TABLE III. Ratios of Amino Acids Necessary for Equimolar Coupling

Ala	Asp	Glu	Phe	Gly	His	Ile	Lys	Leu	Met	Asn	Pro	Gln	Arg	Ser	Thr	Val	Tyr
3.58	3.70	3.84	2.66	3.04	3.76	18.3	6.56	5.23	2.42	5.64	4.56	5.61	6.87	2.93	5.04	11.9	4.36

NOTE: Each number represents the mole percentage of amino acid derivative necessary for equimolar coupling when using a 10-fold excess of mixtures of Boc-protected amino acids and 1-hydroxybenzotria-zole ester derivatives in dimethylformamide (prepared in situ using diisopropylcarbodiimide).

SOURCE: Reproduced with permission from reference 13.

the less reactive building blocks will be able to react with reactive groups available on the resin. A second coupling step is then performed using an excess of the same building blocks (or a capping reagent) to force the reaction to completion. If a kinetically favored reaction occurs at this stage, it will only occur at the reactive groups that were left free after the first step. An improvement of this method, used when one of the building blocks displays too large a difference in reactivity, is to isolate part of the resin and to couple this building block separately from the mixture (partial DCR). This method has been used successfully with amino acid mixtures (amino acid analysis of the resulting peptide mixtures showed reasonable agreement with the expected values corresponding to an equimolecular representation of each amino acid).

A more accurate method for peptides proposed by Houghten (*13*) involves coupling mixtures of side-chain-protected Boc amino acids in a predetermined molecular ratio that compensates for the different coupling rates of the various amino acid derivatives. Table III gives an example of the molecular ratio for a mixture of 18 Boc-protected amino acids coupled in a 10-fold excess.

Although these methods offer the advantage over DCR of ease of synthesis and unlimited library size, they are less efficient as soon as building blocks of larger structural diversity are involved, as these building blocks generally display a wider range of reactivity than amino acids. Moreover, none of these methods can prevent the preferential "pairing" of two building blocks during couplings, corresponding to the preferential coupling of one of the building blocks of the mixture to one of the building blocks present on the polymer.

Selected Examples

To illustrate the different approaches outlined in this chapter, five examples have been selected from the literature.

Example 1

In this example (*14*), iterative deconvolution is used to select peptides selective for the μ opioid receptor from a hexapeptide library.

Composition of the Library

Six-residue peptides with free N-termini and amidated C-termini were prepared according to the dual positional format, so that the first two residues (at the N-termini) of each sublibrary were individually defined as one of the 20 natural amino acids and the last four positions consisted of equimolar mixtures of 19 of the 20 natural L-amino acids (with cysteine omitted). This generated 400 sublibraries, each represented by the general formula $O_1O_2XXXX-NH_2$, and each containing a total of 130,321 (19^4) different peptides. Thus the complete library contained $20 \times 20 \times 19^4 = 52,128,400$ hexapeptides.

Synthesis of the Library

The library was prepared by classical solid-phase peptide synthesis methods on a methylbenzhydrylamine polystyrene resin using t-Boc-protected amino acids. The XXXX–resin mixture was prepared as a unique batch using the DCR method, with each amino acid coupled individually.

The XXXX–resin was then divided into 400 equal portions, and two defined amino acids (O_1 and O_2) were coupled to each batch. Cleaved peptide mixtures were extracted with water, lyophilized, and reconstituted in water to yield a final concentration of 1–5 mg mL^{-1}, and used in standard in vitro receptor assays.

Biological Assay

Sublibraries were tested in a radioreceptor assay using an analog of methionine enkephalin ([^{3}H]-[D-Ala^2, MePhe^4, Gly-ol^5]enkephalin; DAMGO), which binds specifically and with high affinity to μ binding sites from rat brain membranes. IC$_{50}$ and ID$_{50}$ (effective concentration of peptide mixture required to achieve 50% inhibition of binding of the radioligand) values were determined in each case using serial dilutions of peptide mixtures starting from an initial concentration of 0.08 mg mL^{-1} of peptide mixture.

Design of the Sublibraries

The complete library was designed according to the dual positional format (O_1O_2XXXX). According to this strategy, the initial library consists of 400 sublibraries in which the first two amino acids are individually and specifically defined.

Identification of Active Compounds

The initial step of deconvolution involved the screening of the 400 O_1O_2XXXX sublibraries to determine their ability to inhibit the binding of 3-nM [^{3}H]-DAMGO to rat brain membranes. IC$_{50}$ values were then determined for each of the most effective peptide mixtures. The three most effective inhibiting peptide mixtures were YGXXXX-NH$_2$ (IC$_{50}$ = 3450 nM), WWXXXX-NH$_2$ (IC$_{50}$ = 3500 nM), and YPXXXX-NH$_2$ (IC$_{50}$ = 4050 nM), which were separately deconvoluted. Figure 6 illustrates the different deconvolution steps in the case of the YPXXXX-NH$_2$ sublibrary in which the four X positions were sequentially defined. This clearly illustrates that the most active compound (YPFGFR-NH$_2$, IC$_{50}$ = 13nM) would not have

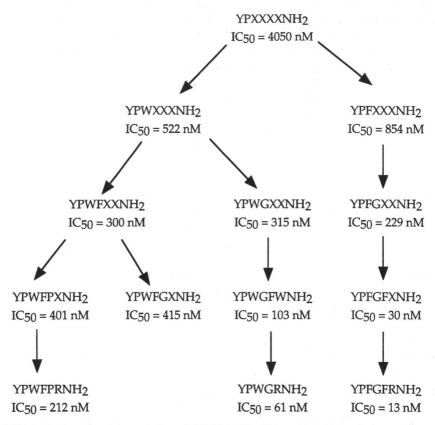

FIGURE 6. The iterative deconvolution of YPXXX-NH$_2$. IC$_{50}$ values (shown below each compound or compound mixture) were determined at each step, and the most potent mixtures were deconvoluted further. Note that the most potent of the final compounds comes from the sublibrary that was not the most potent after the first round of deconvolution.

been found if, at the first step of deconvolution, only the most active sublibrary (YPWXXX-NH$_2$) had been considered.

Example 2

This example (*15*) involves the discovery of nanomolar ligands for the α1-adrenergic receptor from a ~5000-component N-substituted glycine peptoid library using iterative deconvolution.

Composition of the Library

The general structure of the library is shown in Figure 7. It consists of trimers assembled from three sets of N-substituted glycine monomers and capped with one of three different N-terminal caps. For the synthesis of the N-substituted glycines, the total set of amines was partitioned into three groups that had either 3 hydroxylic building blocks, 4 aromatic building blocks, or 17 diverse building blocks. These sets of building blocks were designated O, A, and D, respectively. An additional source of diversity was to leave a free amine at the terminus or cap it as an acetamide or cyclohexylurea. Building blocks in the aromatic and hydroxylic sets were selected on the basis of their occurrence in known inhibitors of seven trans-membrane receptors, and building blocks in the diverse set were chosen for their structural diversity. Mixtures generated from the 6 possible permutations of the 3 sets of building blocks further diversified by 3 different *N*-terminal cappings generated the 18 sublibraries shown in Table IV. Each combinatorial mixture included 3O × 4A × 17D = 204 peptoid trimers and was capped in one of three possible ways. Each trimer in the library therefore contained at least one aromatic and one hydroxylic side chain (in addition, all possible dimers were included, adding 12 (3O × 4A), 51 (3O × 17D) or 68 (4A × 17D) additional compounds to pools with D, A, and O groups respectively). Thus, each of the 18 sublibraries contained either 216, 255, or 272 compounds.

Synthesis of the Library

The peptoid backbone was assembled in two steps from two readily available reagents, as shown in Scheme I. The first step consists of an acylation with bromo-acetic acid followed by nucleophilic displacement of bromide by a wide variety of primary amines. The repetition of these two steps allows the extension of the peptoid backbone. Only aliphatic hydroxyl, amine, and acid side chains required protection. The synthesis was performed according to the DCR strategy on an automated synthesizer using the isopycnic slurry method. Dichloroethane:DMF at 3:2 was used to split the resin. Each building block was coupled individually, after which all portions were combined. The complete library was generated through six independent syntheses, corresponding to different permutations of O, A, and D sets. After completion of the trimeric peptoids, resins were split into three equal portions, which had one of three different *N*-terminal caps added. These portions were not recombined.

TABLE IV. Composition of the 18 Sublibraries Representing the Complete Peptoid Library

	Identity of Cap	
H- = free amine	*Ac- = acetamido*	*Chu- = cyclohexylureido*
H-AOD-NH$_2$	Ac-AOD-NH$_2$	Chu-AOD-NH$_2$
H-ADO-NH$_2$	Ac-ADO-NH$_2$	Chu-ADO-NH$_2$
H-DAO-NH$_2$	Ac-DAO-NH$_2$	Chu-DAO-NH$_2$
H-DOA-NH$_2$	Ac-DOA-NH$_2$	Chu-DOA-NH$_2$
H-ODA-NH$_2$	Ac-ODA-NH$_2$	Chu-ODA-NH$_2$
H-OAD-NH$_2$	Ac-OAD-NH$_2$	Chu-OAD-NH$_2$

NOTE: Each sublibrary was made from all combinations drawn from the six permutations of three sets of building blocks (amines) and three *N*-terminal endings. Thus each compound contains at least one aromatic and one hydroxylic group.

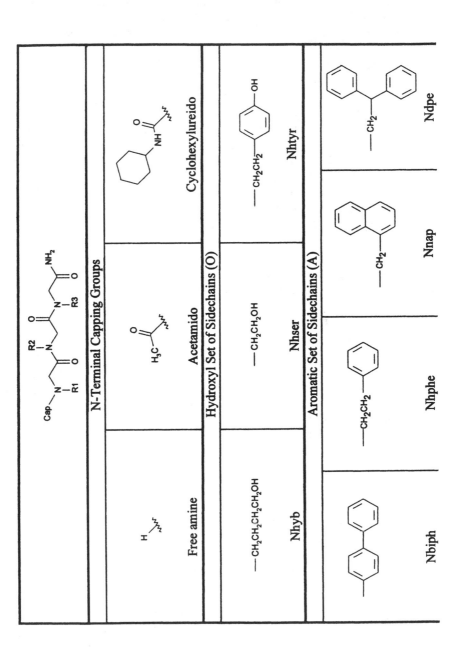

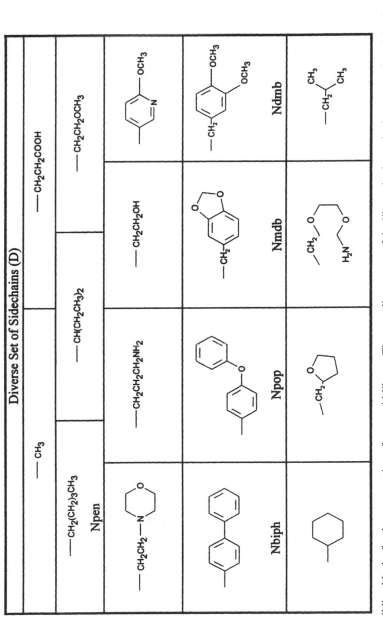

FIGURE 7. Building blocks for the construction of a peptoid library. The overall structure of the library is shown in the top panel, and the three caps are shown in the next set of panels. Below this are the side chains of the amino building blocks used, which were divided into three sets. O is a set of 3 hydroxylic building blocks, A is a set of 4 aromatic building blocks, and D is a set of 17 diverse monomers.

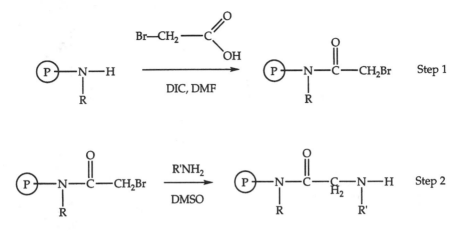

SCHEME I. Solid-phase synthesis of *N*-substituted glycine peptoids using a two-step addition cycle for each building block.

Biological Assay

Binding to the α1-adrenergic receptor was measured by competitive displacement of [^{3}H]-prazosin from a rat brain membrane preparation. The original eighteen sublibraries were tested at 100 nM per peptoid (21.6–27.2 μM total peptoids), and the smaller sublibraries were tested at 1 μM per peptoid.

Results

Individual active compounds were identified by iterative deconvolution (Figure 8). The organization of the library allowed the direct identification of the cap and the order in which the different types of side chains appear (although not the exact identity of any one side chain) in the most active peptoid by screening the 18 initial sublibraries. The H-ODA-NH$_2$ sublibrary was identified as the most inhibitory, indicating that the preferred peptoid had H as an N-Cap and a member of the hydroxyl set in the first peptoid position. This sublibrary was then resynthesized as four smaller sublibraries, each containing 68 compounds of the form H-XDA-NH$_2$, where X represents one of the three hydroxylic side chains of the O set or a solvent blank. A single sublibrary with a homotyrosine (NhTyr) side chain was responsible for most of the activity of the initial sublibrary (Figure 8). The next deconvolution step involved the preparation of 17 sublibraries of 4 compounds each. The compounds were of the format (H-NhTyr)-X-A-NH$_2$, with X now representing each of the 17 side chains from the diverse set. Two mixtures showed significant activities, and their side chains were fixed for the last round, which involved the synthesis of 8 individual peptoids each with a different side chain from the aromatic set. The most active compound was CHIR 2279, with the inhibition constant (K_i) of 5 ± 3 nM (Structure I).

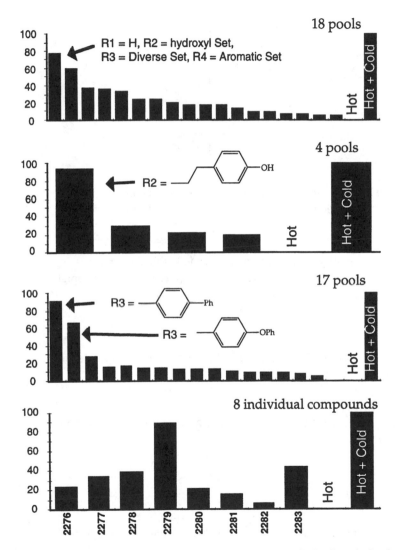

FIGURE 8. Iterative steps of deconvolution for the identification of high-affinity ligands for the α1-adrenergic receptor. At the top is the generic structure of the library members. In the first step, the identity of the cap and the preferred order of the types of side chains were identified. The subsequent steps deciphered the identity of the side chains one by one.

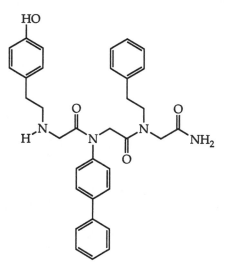

STRUCTURE I. Structure of CHIR 2279, the most potent inhibitor of the α1-adrenergic receptor, discovered from the library described in Figure 7. The compound has a K_i of 5 ± 3 nM.

A similar deconvolution performed with this library using an opiate receptor assay resulted in the identification of a potent μ-specific opiate receptor inhibitor (K_i = 6 nM).

Example 3

This example (*1*) covers the solution synthesis and screening of a nonpeptide "indexed" combinatorial library.

Composition of the Library

The complete library consisted of 54 carbamates corresponding to the general structure:

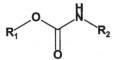

and synthesized using nine alcohols (R_1OH) and six isocyanates (R_2–N=C=O).

Synthesis of the Library

Compounds were prepared by solution synthesis by directly combining isocyanates and alcohols in the following reaction.

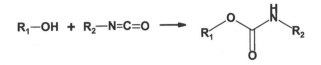

To eliminate kinetic effects, reactions were conducted with stoichiometric quantities of reagents or mixtures in tetrahydrofuran (THF) solution at 60–80 °C for 4 h in sealed pressure tubes. The solvent was then removed by a rotary evaporator.

Biological Assay

Mixtures were tested for their ability to inhibit acetylcholinesterase activity. An IC_{50} was determined for each mixture.

Design of the Sublibraries

Sublibraries were constructed according to the concept of an indexed library, which is similar to a two-dimensional positional scanning library (Figure 2). Two sets of sublibraries were assembled. In the first dimension, six sublibraries were prepared by reacting each individual isocyanate with the mixture of nine alcohols, while in the second dimension, nine sublibraries were prepared by reacting each individual alcohol with the mixture of six isocyanates.

Identification of Active Compounds

The 15 sublibraries described in the preceding paragraph were tested and their IC_{50} values determined. In the isocyanate dimension, potency decreases in the order Me > iPr > Et > tBu, while in the alcohol dimension, potency decreases in the order succinimide > benzotriazole > 4-hydroxybenzaldehyde. The most active compound was identified as O-succinimidyl N-methylcarbamate with an individual IC_{50} of 0.7 mM.

Example 4

This example (*9*) describes the discovery of a potent vasopressin V2 receptor antagonist from two orthogonal libraries of 15,625 trimers.

Composition of the Library

According to the general principle of orthogonal libraries described above, two libraries of the same soluble 15,625 trimers ($N = 3$) resulting from the combination of 25 building blocks at each position ($n_A = n_B = n_C = 25$) were independently prepared.

The building blocks were 23 D-amino acids and two nonchiral amino acids, glycine and isonipecotic acid. For the first library (library I), the 25 building blocks were partitioned into five sets, I_1 to I_5 (Figure 9). Three combinatorial reactions of these five sets yields $5^3 = 125$ sublibraries, each resulting from the incorporation of 1 set of 5 building blocks at each of the 3 variable positions and so containing $5^3 =$

	I_1	I_2	I_3	I_4	I_5
II_1	D-Leucine	D-Proline	D-Serine	(4-nitro)-D-phenylalanine	D-Isoglutamic acid
II_2	D-Arginine	D-Isoleucine	Glycine	D-Threonine	D-Isoglutamine
II_3	D-Glutamine	D-Tyrosine	D-Valine	D-e-Nicotinoyl-lysine	D-Histidine
II_4	D-Tryptophane	D-Asparagine	D-Glutamic acid	D-Methionine sulfoxyde	D-Alanine
II_5	Isonipecotic acid	D-Lysine	D-Tetrahydro-isoquinoleic acid	D-Aspartic acid	D-Phenylalanine

FIGURE 9. Orthogonal partition of the 25 building blocks used to construct a tripeptide library into five sets of five compounds (I_1 to I_5 for library I and II_1 to II_5 for library II).

125 different compounds. For the second library (library II), the same 25 building blocks were used, also yielding 125 sublibraries and the same 15,625 trimers, but the building blocks were partitioned orthogonally into 5 different groups, II_1 to II_5, as shown in Figure 9. As any set of 5 building blocks used for library I has only 1 building block in common with the sets of building blocks used for library II, any sublibrary from library I and any sublibrary from library II share only one compound.

Synthesis of the Library

The two libraries were synthesized by solid-phase synthesis according to the Boc/ benzyl strategy. At each step, each of the resulting peptidyl-modified resins was split into five portions, each of which was coupled to a different set of five building blocks. To avoid preferential incorporation of amino acids having the most favorable coupling rates, a first coupling was performed with only 1.1 equiv of acylating species for 1 equiv of amino groups available on the resin. A second coupling step was then performed to force the reaction to completion. Amino acid analysis on 10 sublibraries showed that the levels of most of the amino acids were within 10% of the expected value, the most divergent being within 25% of this value. After completion of the synthesis, peptides were cleaved from the resin by treatment with HF and lyophilized.

Biological Assay

Sublibraries were tested at a concentration of 25 μg ml^{-1}, which corresponds to ~0.2 μg ml^{-1} for each individual compound. The assay measured inhibition of the binding of 3 nM ^{3}H-AVP (tritiated arginine vasopressin) to porcine kidney epithelial cells.

Identification of Active Compounds

In the initial screening, three sublibraries with high inhibiting potency were identified in each library. These sublibraries were further tested at a concentration of 2.5 μg ml^{-1} and, on the basis of the potency of inhibition, the selection was narrowed to one sublibrary in I (I_3-I_2-I_4) and two sublibraries in II (II_5-II_3-II_1 and II_4-II_3-II_1). The trimer common to I_3-I_2-I_4 and II_5-II_3-II_1 was easily identified, as shown in Figure 10, as H-D-Tic-D-Tyr-(4-nitro)-D-Phe-OH, and turned out to be a potent inhibitor (IC_{50} = 63 nM). No inhibition activity could be detected with the candidate shared by I_3-I_2-I_4 and II_4-II_3-II_1, however. It was thus suspected that this activity arose as a by-product only when the reactions were performed in pools.

References

1. Pirrung, M. C.; Chen, J. *J. Am. Chem. Soc.* **1995**, *117*, 1240–1245.
2. Fodor, S. P. A.; Read, J. L.; Pirrung, M. C.; Stryer, L.; Lu, A. T.; Solas, D. *Science (Washington, D.C.)* **1991**, *251*, 767–773.

a)

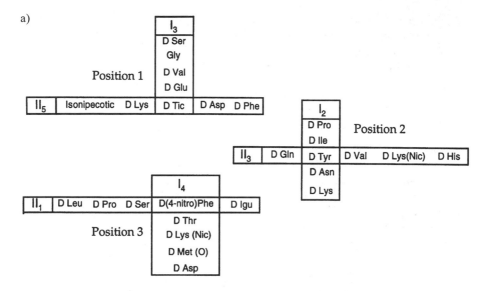

b)

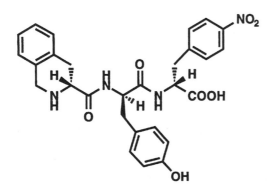

FIGURE 10. Screening of orthogonal libraries. (a) Identification of the trimer shared by I_3-I_2-I_4 and II_5-II_3-II_1. (b) The structure of this compound, H-D-Tic-D-Tyr-(4-nitro)-D-Phe-OH.

3. Houghten, R. A.; Pinilla, C.; Blondelle, S.; Appel, R. J.; Dooley, C. T.; Cuervo, J. H. *Nature (London)* **1991**, *354*, 84–86.
4. Freier, S. M.; Konings, D. A. M.; Wyatt, J. R.; Ecker, D. J. *J. Med. Chem.* **1995**, *38*, 344–352.
5. Erb, E.; Janda, K. D.; Brenner, S. *Proc. Natl. Acad. Sci. U.S.A.* **1994**, *91*, 11422–11426.
6. Pinilla, C.; Appel, R. J.; Blondelle, S.; Dooley, C. T.; Eichler, J.; Ostresh, J. M.; Houghten, R. A. *Drug Dev. Res.* **1994**, *33*, 133–145.
7. Dooley, C. T.; Houghten, R. A. *Life Sci.* **1993**, *52*, 1509–1517.
8. Smith, P. W.; Lai, J. Y. Q.; Whittington, A. R.; Cox, B.; Houston, J. G.; Stylli, C. H.; Banks, M. N.; Tiller, P. R. *Bioorg. Med. Chem. Lett.* **1994**, *4*, 2821–2824.

9. Déprez, B.; Williard, X.; Bourel, L.; Coste, H.; Hyafil, F.; Tartar, A. *J. Am. Chem. Soc.* **1995**, *117*, 5405–5406.

10. Lam, K. S.; Salmon, S. E.; Hersh, E. M.; Hruby, V. J.; Kazmierski, W. M.; Knapp, R. J. *Nature (London)* **1991**, *354*, 82–84.

11. Zhao, P. L.; Zambias, R.; Bolognese, J. A.; Boulton, D.; Chapman, K. *Proc. Natl. Acad. Sci. U.S.A.* **1995**, *92*, 10212–10216.

12. Zhao, P. L.; Nachbar, R. B.; Bolognese, J. A.; Chapman, K. *J. Med. Chem.* **1996**, *39*, 350–352.

13. Ostresh, J. M.; Winkle, J. H.; Hamashin, V. T.; Houghten, R. A. *Biopolymers* **1994**, *34*, 1681–1689.

14. Dooley, C. T.; Kaplan, R. A.; Chung, N. N.; Schiller, P. W.; Bidlack, J. M.; Houghten, R. A. *Peptide Res.* **1995**, *8*, 124–137.

15. Zuckermann, R. N.; Martin, E. J.; Spellmeyer, D. C.; Stauber, G. B.; Shoemaker, K. R.; Kerr, J. M.; Figliozzi, G. M.; Goff, D. A.; Siani, M. A.; Simon, R. J.; Banville, S. C.; Brown, E. G.; Wang, L.; Richter, L. S.; Moos, W. H. *J. Med. Chem.* **1994**, *37*, 2678–2685.

Equipment and Automation

Equipment for the High-Throughput Organic Synthesis of Chemical Libraries

Ralph A. Rivero, Michael N. Greco, and Bruce E. Maryanoff

As chemists have become aware of the potential of combinatorial approaches, there has been a significant increase in the sophistication of the synthetic chemistry involved. The sheer number of compounds involved means that new ways of handling high-throughput synthesis are required. This chapter focuses on the equipment currently available. Although standard synthetic equipment can be adapted to high-throughput applications, the use of specialized equipment, whether manual, semiautomatic, or fully automatic, is more efficient. Any apparatus must be able to handle the diversity of chemical reaction conditions, as well as keep track of the multitude of reactions.

The integration of molecular diversity generation and high-throughput screening is revolutionizing the discovery process in the areas of pharmaceuticals, agrichemicals, and materials science. This has in turn led to increased interest in the field of "supported" organic synthesis. Chemical platforms based on both insoluble (1–4) and soluble (4–6) polymers, with various linker technologies, have been quite effective for the synthesis of chemical libraries comprised of peptides, oligonucleotides, and small organic molecules, and the speed of synthesis of many of these libraries has been increased with "high-throughput synthesis" using both combinatorial and simultaneous-parallel synthetic methods.

The organic chemist has been very creative over the past several years in devising manual and semiautomated equipment for the high-throughput organic synthesis of target molecules, either as "single-pure" entities or as mixtures. Chemistry on solid supports, such as resin beads, has been especially useful. The use of automation will increase, and this should further improve the discovery process.

The challenge to manufacturers of automation equipment will be in developing instruments to serve as laboratory tools, rather than as research projects in themselves. Any equipment should be sufficiently flexible to perform a wide range of chemical conversions while remaining sufficiently simple for the average chemist to program and apply. The increased productivity already brought about by currently available equipment has elevated the general level of expectations for fully automated, small-molecule organic synthesizers.

Recent advances in solid-phase organic synthesis have afforded a set of powerful synthetic tools for finding novel therapeutic agents (7–9). This "brute-force" approach has significantly boosted the productivity of drug discovery in the pharmaceutical industry, although its coupling to structure-based and computational design approaches will probably deliver even more impressive gains. Because of the exciting potential of these new methods, the industry is making a major investment in the implementation of high-throughput organic synthesis and high-throughput screening. There is good reason to believe that this will streamline the process of delivering valuable therapeutic agents to patients, and reduce the investment required to do so.

Manual Methods

Chiron Mimotopes' Multipin SPOC (Solid-Phase Organic Chemistry)

Chiron's Multipin system (Chiron Mimotopes Pty. Ltd., Melbourne, Australia; 61-3-9565-1111), which evolved from the original pin apparatus of Geysen et al. (10, 11), uses polypropylene–poly(N,N-dimethylacrylamide/methacrylic acid) graft copolymer pins (12). The pins consist of a crown segment (13), which contains the graft copolymer prederivatized with an appropriate "handle", attached to a polypropylene stem (Figure 1). For a new synthesis the crowns, available with 1–1.5 μmol or 5–7 μmol loadings, are attached to the stem, and the resulting pin assembly is attached to a support block in an 8 × 12 format. Each of the 96 pins fits into a well of a 96-deep-well microtiter plate containing solutions of reagents (Figure 2). In this manner, simultaneous parallel synthesis of up to 96 individual compounds can be carried out, with the generation of both pin-bound and solution-phase products. The solutions are most conveniently concentrated using a concentrator equipped with a microplate rotor, yielding the final products in a 96-well microtiter-plate format.

The Multipin approach has been used in a number of interesting applications that require the synthesis of multiple peptides (14, 15). Recently, the equipment has been used for the synthesis of nonpeptide targets (16–18). For example, a library of 192 structurally diverse benzodiazepine derivatives was readily prepared (Scheme I) (16). Acid-cleavable compounds **1** and **2** were coupled to amine-derivatized pins (using dicyclohexylcarbodiimide and 1-hydroxybenzotriazole) yielding pin-bound benzophenones **3**. A sequence of peptide couplings and alkylation steps resulted in

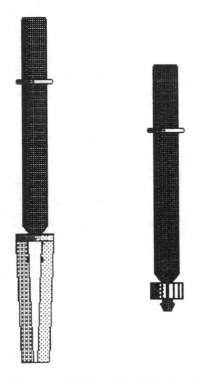

FIGURE 1. Individual pins with crowns (lightly shaded) that have 1–1.5 μmol (left) and 5–7 μmol (right) loading capacities. The crown segment, which is fully functionalized, is totally immersed in solvent. (Courtesy of Chiron Mimotopes Pty. Ltd.)

the synthesis of benzodiazepines **4** with an average yield per pin of 86 nmol. The Multipin method has also been used for the synthesis of β-turn mimetics (*17*) and for the optimization of solid-phase reactions (*18*). With this type of approach, ~10,000 spatially separate compounds have been prepared in parallel fashion using inexpensive laboratory equipment and readily available automation that was designed for the processing of microtiter plates in high-throughput screening efforts (*4*).

The "Tea-Bag" Method

Houghten (*19, 20*) first described "tea bags" and their use for sequential, multistep, solid-phase peptide synthesis in a parallel fashion (*21*). Polypropylene nets, or tea bags, are used to contain the resin, typically 50–100 mg. The tea bags are labeled and sorted according to the amino acid to be coupled to the resin, and couplings are conducted in parallel, with each amino acid, together with some coupling agents, in separate reaction vessels. The vessels are typically polypropylene screw-capped bottles of various sizes, depending on the number of bags used and the washing volume per bag (typically 3–5 ml). Tea bags can be washed separately or simultane-

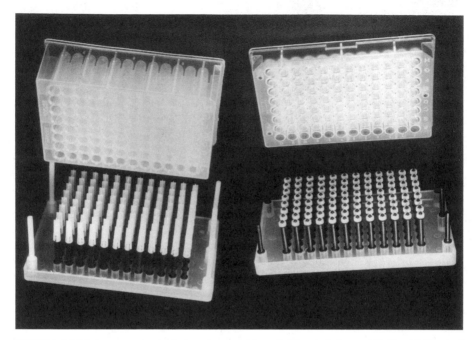

FIGURE 2. Multipin apparatus with per-pin capacities of either 1–1.5 μmol (left) or 5–7 μmol (right) shown with their corresponding deep-well microtiter plates. (Courtesy of Chiron Mimotopes Pty. Ltd.)

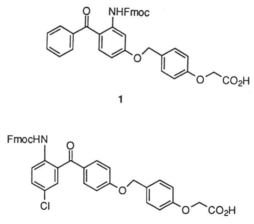

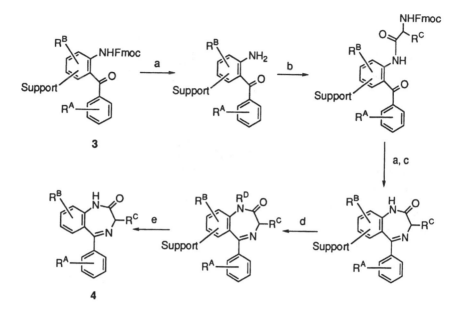

SCHEME I. Synthesis of benzodiazepinones. (a) 20% piperidine in dimethylformamide (DMF); (b) *N*-Fmoc-amino acid fluoride in CH_2Cl_2; (c) 5% acetic acid in DMF; (d) lithiated 5-phenylmethyl-2-oxazolidinone in DMF/trifluoroacetic acid (THF), 1:10 (v/v), followed by alkylating agent in DMF; and (e) TFA/H_2O/Me_2S, 19:1:2 (v/v).

Abbreviations: Fmoc, 9-fluorenylmethoxycarbonyl; TFA, trifluoroacetic acid.

ously, then combined in a common vessel for deprotection. Cleavage reactions are performed in the bags, allowing the product to escape the tea bag, and the final peptide products are isolated using common techniques such as centrifugation or lyophilization.

The tea-bag method has several advantages for the parallel synthesis of peptides. It is inexpensive and suitable for preparing relatively large amounts (30–50 mg) of peptides with ten or more residues. It is also flexible; solvents, reagents, coupling schemes, and reaction conditions can be easily changed. A disadvantage of this method is that it is manual, although automation has been introduced for the washing steps (*21*).

The tea-bag method has been used for structure–activity studies where large numbers of peptide analogues were required, for the elucidation of antigen–antibody interactions, and for the conformational mapping of proteins (*21*). For example, the method was used to identify the key residues of neuropeptide Y (NPY) necessary for binding to its receptor (*22*). A total of 67 analogues of the 17-residue peptide containing residues 1–4 and 25–36 of NPY separated by ε-aminocaproic acid (Aca) were prepared in which each residue was replaced by the corresponding D-enantiomer, glycine, or L-alanine. The bags, made of polypropylene mesh with

64-μm-diameter pores, were filled with 50–100 mg of resin and sealed with an electronic sealer.

For the synthesis of C-terminal amides, polystyrene resin loaded with 4-Fmoc-aminomethyl-3,5-dimethoxyphenoxyvaleric acid and an alanine spacer was used, while Wang resin was used to prepare C-terminal acids. The former linker provided the carboxamide, while the Wang resin provided the free acid. Both peptides were prepared from C- to N-terminal. Peptides were synthesized using an Fmoc strategy with a 10-fold excess of [(benzotriazol-1-yl)oxy]tris(dimethylamino)phosphonium hexafluorophosphate (BOP) and 1-hydroxybenzotriazole in a 0.2 M solution of dimethylformamide (DMF). Deprotections were carried out simultaneously in a 250-mL polyethylene bottle, while couplings were carried out in separate bottles according to the required amino acid. Some washing steps were effected simultaneously, while others were handled in separate vessels. Following final deprotection and washings, cleavages were performed by exposing the bags to the appropriate cleavage cocktails in separate vessels. Peptide amides were cleaved from the resin with a mixture of trifluoroacetic acid (TFA), thiocresol, and thioanisole (95:2:3, 5 mL), whereas C-terminal carboxylic acids were released by treatment with a mixture of TFA, CH_2Cl_2, thiocresol, and thioanisole (50:45:2:3). Following evaporation of solvent, the peptide products were dissolved in acetic acid, precipitated with ether, and collected by centrifugation and lyophilization. Purification by gel chromatography and/or HPLC afforded 12–71 mg of > 90% pure peptides. Evaluation of the peptides in an NPY receptor binding assay indicated that the C-terminal tetrapeptide of NPY (residues 33–36) is essential for receptor recognition.

Recently, the power of the tea-bag method was demonstrated through the construction of a combinatorial library consisting of 52,128,400 hexapeptides made up of D-amino acids (23). Using iterative selection, a novel, high-affinity peptide ligand for the mu opioid receptor was identified. Significantly, the discovered hexapeptide, Ac–ArgPheTrpIleAsnLys–NH$_2$, bears no sequence or structural resemblance to any known opioid peptide.

96-Deep-Well Filtration Array

A 96-well filtration apparatus for conducting multiple, simultaneous parallel syntheses was developed by researchers at Sphinx Pharmaceuticals (24). The apparatus consists of a polypropylene deep-well plate with a small hole at the bottom of each well (Figure 3). A polyethylene frit with 20-μm-diameter pores is positioned near the bottom of each well, and the entire plate is mounted on a gasket that seals the bottom of the individual wells. The gasket and plate are held together with a two-piece aluminum clamp, which is tightened at each corner with wing or knurl nuts. To speed filtration of resin washings and isolation of product solutions after cleavage, the deep-well plates can be fit into a vacuum plenum, which was designed to hold a rack of microdilution tubes, and a vacuum applied using an attachment at the bottom of the plenum. Alternatively, solvents can be gravity-filtered into a rack of microdilution tubes placed below the reaction plate.

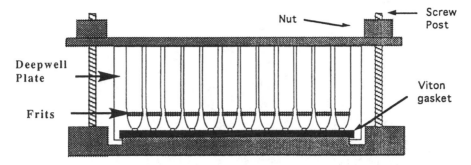

FIGURE 3. Deep-well reaction plate and clamp assembly (side view). (Reproduced with permission from reference 24. Copyright 1995 ESCOM Science Publishers B.V.)

A plate is loaded with the appropriate resin (~50 μmol/well) and clamped before the solvents are added using an eight-channel pipetter. More reactive reagents, such as acid chlorides, can be added individually by syringe. After a plate is loaded, the wells are capped with eight-well strip caps and the entire assembly is agitated by wrist-action, an oscillator, or a platform shaker. When reactions are complete, the clamp assembly is removed and the plate is placed on a receiving vessel. Strip caps are removed, the resins are washed with solvents delivered by an eight-tip manifold, and the solvent is filtered by either gravity or vacuum. Following completion of a synthetic sequence, the products are cleaved from the resin and solutions are collected into a rack of 96 polypropylene microdilution tubes placed below the plate (for gravity filtration) or in the vacuum plenum (for vacuum filtration). Concentration under reduced pressure is effected with a concentrator equipped with a rotor to accommodate deep-well plates.

This approach was demonstrated by the construction of a library of more than 600 phenolic compounds (Scheme II) (*24*). Resin-bound phenols **5** and **6** were prepared in separate batch syntheses; subsequent steps were carried out in the deep-well plates. Each of the eight resins, **5**, was reacted with one of 12 acid chlorides, 12 sulfonyl chlorides, and 12 isocyanates to give 96 (8 × 12) unique products from each class, represented by **7a–7c**, for a total of 288 compounds. Reaction wells containing silyl-protected phenol derivatives of **7a–7c** were treated with tetrabutylammonium fluoride, then with THF-saturated NaOMe in methanol (4:1) to release products **8–10** as solutions of phenoxide salts. Similarly, functionalized resin **6** was processed to yield the 96 products encompassed by general formula **11**. Acidification, salt removal, and concentration yielded 3–9 mg of the components of the phenolic libraries (~15–45% yield for plate synthesis). An application of this apparatus to the parallel synthesis of substituted biphenyl libraries has also been reported (*25*).

Diversomer Eight-Pin Synthesizer

Diversomer Technology, developed at Parke-Davis Pharmaceutical Research Division (Warner-Lambert Company), supports multiple, simultaneous parallel synthe-

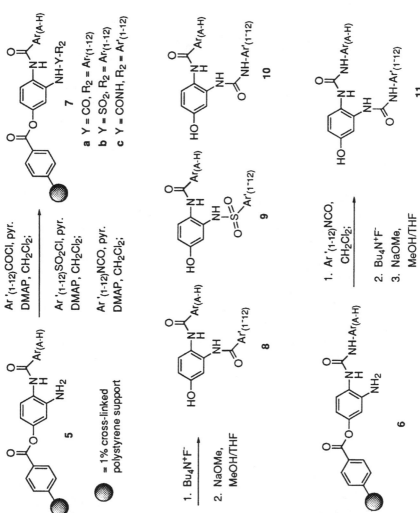

SCHEME II. Synthesis of phenolic libraries **8–11** in deep-well plates.

ses on a solid support (*26–28*). A diagram of the published apparatus, a version of which is commercially available (Chemglass, Inc., 3861 North Mill Road, Vineland, NJ 08360; 800-843-1794), is depicted in Figure 4. Up to 100 mg of resin can be placed inside the frit portion of the modified gas-dispersion tubes that serve as reaction "pins". These pins serve a similar function as the tea bags, keeping the resin constrained but allowing free entry and exit of reactant solutions. The eight pins fit into a corresponding reservoir vial containing solutions of reactants, with both pins and reservoirs being held together as a 4 × 2 unit. A manifold that encloses this assembly enables reactions to be carried out under an inert atmosphere. Introduction of a chilled gas into the manifold cools the extended portion of the pins to permit refluxing of solvent. Reagents can be injected at the top of each pin through a sealed gasket, and samples can be agitated by magnetic stirring, rotary shaking or sonication. Fully automated Diversomer equipment for the simultaneous synthesis of 40 samples has also been described (*29*).

Diversomer technology has been used for the simultaneous synthesis of 40-compound libraries of benzodiazepines and hydantoins (*26*). The synthetic route to

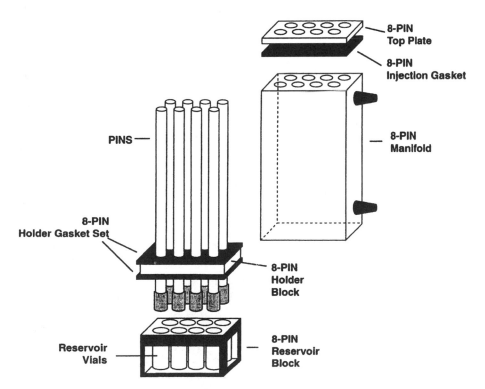

FIGURE 4. A Diversomer eight-pin synthesizer. (Reproduced with permission from reference 26. Copyright 1993 National Academy of Sciences.)

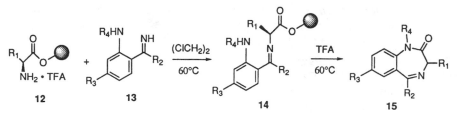

SCHEME III. Synthesis of benzodiazepines.

Abbreviation: DMAP, 4-(dimethyl)aminopyridine.

the benzodiazepines is shown in Scheme III. Five batches of Merrifield resins, each with a different amino acid attached (**12**), were each divided into eight aliquots of 99–107 mg and a single aliquot placed into each of the 40 pins. The pins were submerged into 40 reaction wells, with groups of five wells containing one of eight different benzophenone imines (**13**, 3–6-fold excess) such that each resin-bound amino acid reacts with a different benzophenone imine. On treatment of **14** with TFA at 60 °C, benzodiazepine targets **15** were formed and released from the resin. In this manner, 40 discrete benzodiazepines were prepared with purities of >90% (estimated by ^{1}H NMR) and in yields ranging from 9 to 63% (corresponding to 2–14 mg of each product).

Light-Directed Spatially Addressed Parallel Chemical Synthesis (LDS)

The LDS technology, developed at Affymax Research Institute, combines solid-phase chemistry using photolabile protecting groups with photolithography to generate high-density arrays of peptides and oligonucleotides on a silica surface (*30, 31*). The acronym VLSIPS (very large-scale immobilized polymer synthesis) is also used to refer to this technology. Each chemical entity on the surface-bound arrays can be assayed for useful properties such as binding to receptors. This novel technology is well suited to the synthesis of very large numbers of closely related peptides or oligonucleotides with a minimal amount of synthetic manipulations.

The LDS technology, as applied to peptide synthesis, is illustrated in Figure 5. Amino groups at the termini of linkers bound to a silica surface or "chip" (such as a glass microscope slide) are derivatized with the photolabile protecting group nitroveratryloxycarbonyl (NVOC) (*32*). Deprotection of predetermined sections of the surface is accomplished by directing light through a mask that has transparent and opaque regions. The entire surface is then exposed to a 1-hydroxybenzotriazole-activated ester of an NVOC-protected amino acid. Only the sections that were addressed by light in the previous masking step undergo coupling. The substrate is washed and a different region of the substrate is deprotected using a different mask. The process is repeated until the desired sequences are prepared. The pattern of masks used, along with the sequence of addition of amino acids, determines the identity and location of the final products. The maximum number of compounds prepared is a func-

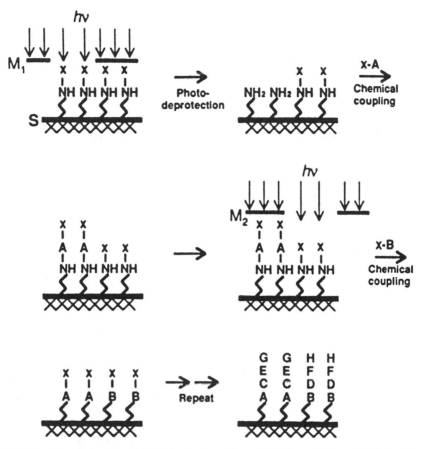

FIGURE 5. Peptide synthesis with LDS technology. A mask (M_1) results in spatially localized deprotection by light (top). This is followed by chemical coupling (middle, left) before the process is repeated with a different mask (M_2; middle right). (Reproduced with permission from reference 30. Copyright 1991 American Association for the Advancement of Science.)

tion of the physical dimensions of the array and of the photolithographic resolution. Arrays of up to 65,536 compounds in an area of ~1 cm^2 have been prepared.

Typically, fluorescence microscopy is used to detect binding interactions of a receptor of interest to a particular array. The array is incubated with a fluorescently tagged receptor, and fluorescence of the bound receptor at synthesis sites of the array is then measured. This approach has been used for epitope mapping of monoclonal antibodies (*30, 33*). Innovative extensions of LDS to the discovery of solid-state materials (*34*) and the synthesis of oligocarbamates (*35*) have been described. As this technology involves libraries of surface-bound compounds, it is open to the distortions that happen when ligands interact differently with a surface-bound ligand than with a ligand in a standard solution-phase binding protocol (*35*).

Automated Methods

The development of automated techniques has followed the use of the corresponding manual techniques in combinatorial chemistry. Thus, the automation of peptide synthesis is mature; the formation of an amide bond between a resin-bound amine and an appropriately protected amino acid has been optimized and can be carried out on a number of commercial instruments quite effectively. The automated delivery of subunits and reagents for amide coupling, and the processes of deprotection and cleavage from the solid support are quite simple. In consideration of the importance of peptides and peptide libraries we will, however, describe several of these commercially available instruments. A common feature of the instruments is a fritted reaction vessel. Resin washes are effected either by vacuum applied to the bottom of the reaction vessels or by nitrogen pressure applied to the top of the vessels.

Formation of an amide bond represents only a very small subset of the types of chemistries that the organic chemist would like to carry out in an automated fashion. The reagents, solvents, and conditions necessary to execute some of the nonpeptide-based syntheses recently described in the literature are complex, and the development of automation for the synthesis of small organic molecules is correspondingly immature.

The Gilson AMS 422 Peptide Synthesizer

This peptide synthesizer, currently only available in Europe, is manufactured by Gilson Medical Electronics (3000 W. Beltline Highway, Box 27, Middleton, WI 53562; 608-836-1551). The platform, which runs with Macintosh-compatible software, can incorporate up to 24 different amino acids (building blocks) in the simultaneous preparation of up to 48 different 15-mers in 40 hours. The reactions are run in a column reactor vessel equipped with a frit. Agitation is accomplished by bubbling nitrogen from the bottom of the reaction vessel. The instrument is specifically designed to carry out Fmoc-chemistry with in situ activation of acids by phosphorus-based reagents (e.g., PyBOP (benzotriazol-1-yl-oxytrispyrrolinophosphonium hexafluorophosphate), BOP), uronium reagents (e.g., HBTU (2-(1*H*-benzotriazol-1-yl)-1,1,3,3-tetramethyluronium hexafluorophosphate), TBTU (2-(1*H*-benzotriazol-1-yl)-1,1,3,3,-tetramethyluronium tetrafluoroborate)), or carbodiimides. The reactors produce 5–50 μmol of the final products. There is no rack provided to collect cleavage cocktails, so the user must collect the solutions manually.

Shimadzu PSSM-8 Automated Solid-Phase Multiple Peptide Synthesizer

The PSSM-8, manufactured by Shimadzu Corporation (Columbia, MD; 410-381-1227), can produce 5–50 μmol of eight different polypeptides. The instrument can incorporate up to 20 different amino acids (building blocks) and is also specifically designed to carry out Fmoc chemistry. The reaction vessels are disposable syringe tubes with filters, and the software is PC-driven. As with the AMS 422, the task of collecting cleavage solutions is left to the user.

Custom Multiple Peptide Synthesizer

Not being satisfied with the commercial instruments available in the early 1990s, Zuckermann and coworkers at Chiron Corporation did an excellent job of designing a fully automated, multiple-peptide synthesizer (*36, 37*) using commercially available components (Figure 6). It can prepare 36 individual peptides or equimolar peptide mixtures in a 6 × 6 array.

The key components of this instrument are a Zymark robot, and an Apple Macintosh II computer to control over 40 solenoid valves and to monitor sensors. Solvents and reagents are delivered through pressurized spigot lines to reaction vessels by the robot hands. A custom-designed reaction vessel rack constructed from high-density polyethylene sheets and rods contains one fritted glass resin mixing

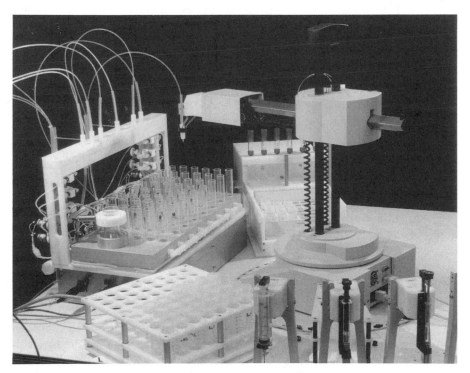

FIGURE 6. Chiron's custom multiple molecule synthesizer. See text for details. (Courtesy of Chiron Corp.)

chamber (with a capacity of 150 mL), and 36 1.5 × 10 cm and 12 2.5 × 10 cm glass reaction vessels, each fitted with a polyethylene frit with 20-μm-diameter pores. The smaller reaction vessels contain a bulge in the wall, 3 cm from the bottom, preventing resin from rising up the reactor vessel wall during mixing. Mixing is carried out by gentle bubbling of argon through the reactor vessel. The valves, plumbing, reaction vessels, and vacuum/pressure manifolds are all made of solvent-inert materials (Teflon, polypropylene, polyethylene, or glass). The reaction vessels are grouped in rows of six; there are two rows of large reaction vessels and six rows of small reaction vessels. Individual peptides are prepared by using standard Fmoc chemistry.

What separates this customized automated synthesizer from commercial models is its ability to generate equimolar mixtures of peptides. The generation of an equimolar peptide mixture requires two additional steps that involve the transfer of resin. The distribution and recombination of the resin, as required by the resin-splitting method, is performed with a modified 30-mL syringe robot hand. The resin is first distributed to a variable number of reaction vessels as an isopycnic slurry in 65% 1,2-dichloroethane/DMF. After the first coupling and washing protocol, the resin is transferred to the mixing chamber. The resin is mixed with argon bubbling, and the resin slurry is divided into equal portions by volume and added to each new reaction vessel. The exact slurry volume is detected by use of a liquid-level sensor on the mixing chamber. The small reaction vessels can hold up to 125 μmol of resin and the large up to 900 μmol. After subsequent deprotection, coupling, and (optional) capping steps, the resin is transferred back to the mixing chamber for recombination followed by distribution back to the reaction vessels. The number of monomers is limited only by the available work space.

To demonstrate the utility of the synthesizer, eight individual decapeptides were prepared, and had excellent purity as determined by reversed-phase HPLC (37). Mass spectral analysis was used to confirm the structure of the peptides. More recently, a library of peptoids (N-substituted glycine oligomers), members of which bound several seven-transmembrane G-protein-coupled receptors, was generated using this customized synthesizer (38).

A modified version of this instrument, with the capacity to control the reaction temperature between −10 and 100 °C, has been developed by Zuckermann and co-workers. This instrument (Figure 6) has been used to synthesize small, drug-like heterocyclic molecules (Zuckermann, R. N., Chiron Corporation, personal communication, 1996). It is laid out with the reaction block, containing a large mixing chamber, two rows of large reaction vessels (20 mL), and six rows of small reaction vessels (5 mL), on the left. To the right of the reaction block is the cleavage station where the the resin from six reaction vessels is transferred as a slurry to individual cleavage tubes for cleavage with a TFA cocktail. Cleavage solutions are then collected in a rack of 36 tubes. The front contains the monomer rack for holding up to 64 monomers and, to the right, the robot gripping hands with different syringes. The needle farthest to the left is used for resin dispensing.

Custom Automated Library Synthesizer

Workers at Selectide Corporation (1580 E. Hanley Boulevard, Tucson, AZ 85737) have also developed a library synthesizer that effectively performs split-synthesis (*39*) to generate libraries (Figure 7). The main components of the instrument are 20 individual fritted reaction vessels in which couplings are carried out, connected to a central mixing chamber. The resins are mixed together by filling the 20 individual reactors and the mixing chamber with "randomization solvent" and then blowing nitrogen through the frits at the bottom of the reactors and stirring the mixing chamber. The solvent allows for the previously segregated resins to mix as a slurry in the main

FIGURE 7. Selectide's custom automated library synthesizer (reactor assembly). The reaction vessels surround a central mixing chamber. (Courtesy of Selectide Corp.)

chamber. After complete mixing, stirring is halted, and sedimentation then redistributes the resin into the 20 reaction vessels ready for the next coupling step (*40*).

Custom Automated Multiple Oligonucleotide Synthesizer (AMOS)

Researchers at the Stanford DNA Sequence and Technology Center have developed an automated multiple oligonucleotide synthesizer for high-throughput, low-cost DNA synthesis (*41*). Their apparatus can simultaneously generate 96 different oligonucleotides in a 96-well microtiter-plate format using phosphoramidate chemistry. The 96 growing oligonucleotide chains, each mounted on a solid support, are in the individual wells below reagent valve banks that are used to deliver appropriate reagents into the wells. Each well has a filter frit, and a seal design is used to control the synthesis environment. Impressively, a plate of 96 20-mers can be prepared in <5 hours. The quality of the oligonucleotide synthesis was found to be comparable to that of commercial machines, with average coupling efficiencies routinely >98% across the entire plate. There has been no significant well-to-well variability observed with the more than 6000 oligonucleotides prepared.

Symphony/Multiplex System

The Symphony/Multiplex apparatus (Figure 8), manufactured by Protein Technologies, Inc. (Rainin Instrument Co., Mack Road, Box 4026, Woburn, MA 01888; 617-935-3050), is a new synthesizer capable of producing 12 independent biomolecules in reaction vessels holding up to 400 mg of resin (5 μM–0.1 mM). The glass or disposable reaction vessels are kept in an inert nitrogen atmosphere to minimize side reactions. Nitrogen bubbling through the bottom of fritted round reaction vessels is used for agitation. The instrument is compatible with Fmoc, Boc, or other bioorganic solid-phase synthesis chemistry. The instrument has a new matrix valve system that permits different solutions to be transferred simultaneously to different reaction vessels without the possibility of cross-contamination. This allows for the synthesis of 12 different biomolecules under conditions optimal for each. The control software runs on an IBM-compatible PC and is menu-driven. The automated TFA-cleavage protocol delivers the product biomolecules through individual tubing from each synthesis reactor directly to a centrifuge tube in a vented compartment. The material can be precipitated out with ether or concentrated to dryness. The instrument appears flexible enough to use tubes that can be taken directly to a speed vacuum. Although they are not currently available, resin-splitter and heating options are projected for the near future. This instrument should be flexible enough to be used for the synthesis of small, nonpeptide organic molecules.

Advanced ChemTech Biomolecular Synthesizer

The Model 357 Flexible Biomolecular Synthesizer (FBS), manufactured by Advanced ChemTech (5609 Fern Valley Road, Louisville, KY 40228; 800-456-

FIGURE 8. The Symphony/Multiplex synthesizer, with the reaction vessels on the left. (Courtesy of Protein Technologies, Inc.)

1403), is the only commercially available combinatorial and large-scale biomolecular synthesizer. Like the similar Chiron–Zymark custom automation equipment, the Model 357 mixes resin beads in an oversized well (600 mL) and distributes the resin as an isopycnic slurry by volume in order to generate combinatorial libraries. The polypropylene reaction block incorporates 36 9-mL fritted reaction wells and one 500-mL reaction well, which permits pooling and dividing of the resin as a slurry. There are standard protocols provided with the menu-driven, IBM-compatible software, and the user can also program custom protocols for different chemistries within the same run. The Model 357 uses variable speed vortex mixing and nitrogen bubbling, a controlled reaction environment, and an optional temperature-control system. The instrument possesses two robot arms: one to dispense solvents and reagents and the other to dispense resin slurries. Thirty-two 42-mL containers for amino acids or other subunits and five 400-mL containers for bulk reagents (such as diisopropylcarbodiimide and 20% piperidine/DMF) are located on the platform.

High-speed filtration is accomplished with positive nitrogen pressure over the top of the reactors.

This equipment was recently used to synthesize unnatural peptides (*42*). The six peptides were prepared using Wang resin and Fmoc-based solid-phase peptide chemistry by the procedure illustrated in Scheme IV. Because of the acid sensitivity of the Wang resin linkage, methods were developed for the introduction and removal of the Schiff base group under essentially neutral conditions. The *N*-terminal Fmoc of Gly-Leu-Wang resin was removed under standard conditions, and the free amine was reacted with benzophenone imine (1.5 equiv) in the presence of acetic acid (1.3 equiv). The resultant Schiff base activated amino acid was alkylated with various benzyl bromides using "Schwesinger bases" (*43*), and the imine was hydrolyzed to the free amine under mild conditions with $NH_2OH \cdot HCl$ in aqueous THF (pH 6).

The Advanced ChemTech Model 396 MBS (Multiple Biomolecular Synthesizer) and Model 496 MOS (Multiple Organic Synthesizer; Figure 9) have greater flexibility as their polypropylene blocks can accommodate 8, 16, 48, or 96 glass-fritted reaction vessels. This allows for the preparation of anything from 5 μmol to 1 mmol of a biomolecule. Multiple independent protocols can be used for all synthesis steps, allowing different syntheses to be carried out independently in each reaction vessel. Reagents and solvents are dispensed by two needles that pierce a septum covering the reaction vessels, and reactions are mixed with a vortex under an inert atmosphere. High-speed filtration for resin washes is accomplished with positive nitrogen or argon pressure over the top of the reaction vessels. Both instruments are controlled by versatile Microsoft Windows based software. The Model 396 MBS contains a rack that holds 32 42-mL polypropylene monomer containers, while the Model 496 MOS has a monomer rack flushed with inert gas that holds 36 50-mL polypropylene bottles. Both instruments contain five 400-mL bulk reagent

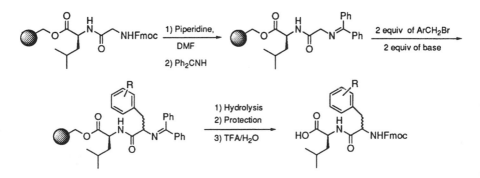

R = H, 2-Me, 3-Me, 4-Me, 4-CF$_3$, 4-NO$_2$

SCHEME IV. Solid-phase synthesis of unnatural amino acids and their incorporation into peptides.

bottles. The Model 496's bulk reagent bottles are also flushed with inert gas. The major difference between these models, in addition to the blanketing of monomers and reagents under an inert atmosphere in the Model 496, is the temperature control that is available with the Model 496 MOS. Depending on the choice of heat-transfer fluid, the Model 496 can control the temperature from –70 to 150 °C with equalization of up to 10 °C min⁻¹. Cleavage from the solid support requires the placement of a collection block under the reaction block. After the appropriate cleavage time, the reaction vessels are emptied into individual beakers.

Terrett et al. demonstrated the utility of the Model 396 by preparing a 30,752-compound combinatorial library related to peptide-like endothelin antagonists (*44*). Advanced ChemTech has nicely demonstrated the applicability of their instruments for the synthesis of small organic molecules. At two conferences in early 1996, Advanced ChemTech showed that the Model 496 could support a variety of organic reactions including palladium-catalyzed Suzuki coupling, oxidations, enolate alkylation chemistry, organometallic chemistry, and urea and urethane formation (*45, 46*).

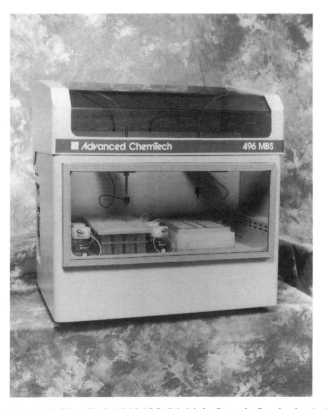

FIGURE 9. The Advanced ChemTech 496 MOS (Multiple Organic Synthesizer). (Courtesy of Advanced ChemTech.)

CombiTec Automated Synthesizer

The CombiTec automated solid-phase synthesizer, introduced early in 1996 by TECAN (TECAN US, P.O. Box 13953, Research Triangle Park, NC 27709; 800-338-3226), is the first commercial instrument to deviate from the fritted reaction vessel design. The CombiTec, advertised as a versatile organic molecule synthesizer, utilizes threaded round-bottom flasks that are screwed into Teflon reaction chamber modules and fitted with a fritted glass tube. Unlike the Diversomer apparatus, which uses the fritted tube to suspend the resin, this tube simply acts as a filtering straw to allow for the easy removal of solvents from the reaction vessel while keeping the resin in the reaction vessel. The Teflon reaction block is equipped with a cooling manifold and six reaction chamber modules.

Each module contains eight reaction chambers in a linear strip, giving the capacity for a total of 48 reaction vessels per reaction block. Reaction vessels with capacities of 5, 10, 15, and 25 mL can be accommodated in the block, so reactions can be scaled up or down. Chambers can be washed as a group of four, or individually using individual aspiration and dispensing tips. The reaction block can easily be removed from the instrument platform for off-line agitation, heating or cooling. The unit has a temperature range of −30 to 150 °C. The first-generation instrument will not have agitation while it is on the main platform. TECAN has found that good washing can be obtained with their vacuum–frit combination, which actually pulls the wash solvents through the resin while the resin is retained by the frit on the glass tube. Reagents and small volumes of solvents that are not handled by the bulk solvent delivery system are picked up by a robot from septum-capped tubes in specially designed racks on the main platform. There are a variety of racks available for monomers or reagents. The TECAN CombiTec synthesizer is built on the platform of the TECAN Genesis robotic sample processor, and has a combination of six syringe pumps and two six-way valves per dispensing tip, allowing a total of up to 11 different solvents or reagents to be addressed to each reaction chamber. The instrument is driven by Microsoft Excel software.

The entire instrument, without the off-line agitation or the heating–cooling module, is shown in Figure 10. The unique, two-part, flow-through design is illustrated in Figure 11.

After compound synthesis and washing, cleavage reagents can be added to the chamber. The cleavage solutions containing the products are removed from the reaction chamber by programming tip #1, the dispensing tips, to move to the bottom of the fritted tube. The syringe pump is drawn to pull the solution into the tubing. The tip is then removed from the reaction chamber and inserted into a clean vial, and the solution containing the product is dispensed.

Bohdan Automated Synthesizer

Bohdan Automation, Inc. (1500 McCormick Boulevard, Mundelein, IL 60060; 847-680-3939) has introduced an automated synthesizer that is capable of carrying out

FIGURE 10. The TECAN CombiTec automated synthesizer. (Courtesy of TECAN.)

both solution-phase and solid-phase chemistry. The Pathogenesis RAM synthesizer, resulting from a collaboration between Bohdan and Pathogenesis Corp., is a complete high-throughput synthesis station. The multifunctional workstation can prepare reagent solutions, add reagents to reactor vessels, work up reactions or cleave compounds if they are on a solid support, carry out liquid–liquid extraction, capture the weight of products in a Microsoft Excel table, and dissolve final compounds for assay and archiving. A separate reaction station agitates at the desired temperature under an inert atmosphere.

The reagent-preparation workstation (Figure 12) tares reagent vials (scintillation vials) and, after the scientist manually adds reagents to the vials and downloads the appropriate information (molecular weight and desired concentration), the proper amount of solvent is added and two vortex mixers are used to assure dissolution. A total of 48 reagent vials can be prepared. The reaction preparation mode (Figure 13) simply requires the manual addition of a reaction block to the platform. The modular reaction block holds 48 reaction vessels that are 16 mm × 60 mm or 16 mm × 100 mm. The desired reagents or monomers are then mapped into the reaction vessels. After all reagents are dispensed via tubing and canula by a robot, the reaction block is manually taken off-line to heat or cool and agitate under an argon atmosphere. While the instrument is primarily set up for solution-phase work-up via liquid–liquid extraction, cleavage solutions from solid-phase chemistry can be pulled from the reaction vessels with a fitted tube and then placed into the tubes in the dry-down rack. The dry-down racks are then placed in a speed vacuum for concentration.

Amide-bond formation, alkylation of alcohols, and the addition of isocyanates, acid chlorides and chloroformates to amines are among the solution-phase

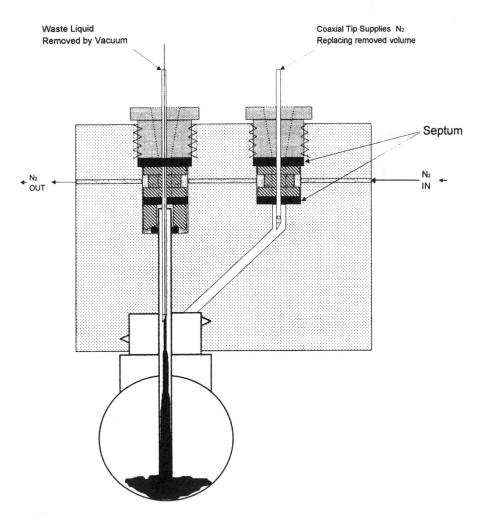

Waste Liquid
Removed by Vacuum

Coaxial Tip Supplies N₂
Replacing removed volume

Septum

N₂
OUT

N₂
IN

FIGURE 11. The CombiTec two-part flow-through design. The tip on the right dispenses reagents directly into the reaction vessel. The tip on the left withdraws solutions under vacuum through the fritted tube. The beads are blocked by the fritted tube, and so remain in the vessel. (Courtesy of TECAN.)

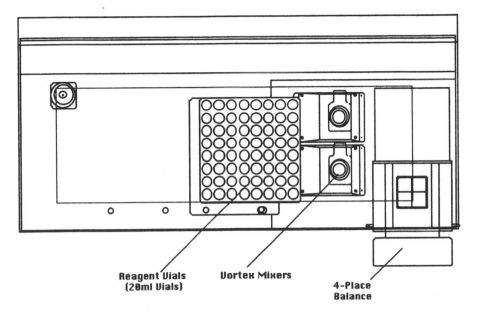

Reagent Vials
(20ml Vials)   **Vortex Mixers**   **4-Place**
                                            Balance

FIGURE 12. Top view of the Bohdan Synthesizer in reagent-preparation mode. The instrument automatically prepares standard solutions of reactants under an inert atmosphere. Reagents are weighed, appropriate amounts of solvent are added, and solubilization is ensured with a vortex mixer. (Courtesy of Bohdan Automation, Inc.)

reactions that have been demonstrated on this instrument. Acylation of a resin-bound amino acid was also cleanly accomplished. Although the instrument is capable of both solid- and solution-phase chemistry, it is very slow at conducting chemistry in the solid phase as solutions must be pulled through the fritted tube one reaction at a time.

The Nautilus 2400 Synthesizer

Argonaut Technologies, Inc. (887 Industrial Road, Suite G, San Carlos, CA 94070; 415-598-1350), introduced their 24-reaction-vessel solid-phase automated synthesizer in mid-1996. The glass and Teflon reaction vessels, available in 8- or 15-mL sizes, are contained or held in three separate modules under an inert atmosphere. The three modules can be independently agitated by rocking through an arc of 200°. Like the TECAN CombiTec apparatus, the Nautilus 2400 utilizes a fritted straw design for easy removal of solvents and reagents while the resin is maintained in the reaction vessel. The end of the Teflon straw is equipped with a 30-μm frit that is disposable. The Nautilus delivers precisely metered aliquots of reagents and solvents through a closed-fluid delivery system. Up to 192 different reagents, in volumes of between 50 μL and 5.8 mL, can be accurately delivered from a reagent autosampler. After exposure to a cleavage cocktail, the product-containing solutions

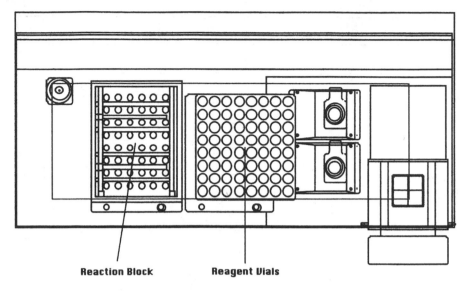

Reaction Block   **Reagent Uials**

FIGURE 13. Top view of the Bohdan Synthesizer in reaction-preparation mode. The instrument loads the RAM Synthesizer with starting materials for reactions. The reagents or building blocks are solubilized with the appropriate solvent and mixed with the vortex mixer. The robot then places the building blocks onto the deck. (Courtesy of Bohdan Automation, Inc.)

are collected in an integrated fraction collector. What separates the Nautilus 2400 from other synthesizers is the truly independent temperature control from one reaction to another. Independent digital controls regulate the temperature of each glass reaction vessel anywhere in the range of –40 to 150 °C. The entire system is controlled by a computer containing a Pentium 120 processor, and the software operates in a Microsoft Windows 95 environment.

The AccuTag-100 Combinatorial Chemistry System

Irori Quantum Microchemistry (11025 North Torrey Pines Road, La Jolla, CA, 92037; 619-546-1300) has introduced a system that takes advantage of the power of combinatorial chemistry, using the split-and-pool technique, but provides milligram quantities of discrete compounds. The AccuTag-100 system utilizes either rigid containers that hold resin (MicroKans) or MicroTube reactors made of polypropylene or a fluoropolymer. The former is reminiscent of the tea-bag method and the latter of the pin method. In both cases, a glass-encased radiofrequency tag is incorporated. The MicroKans hold ~30 mg of resin each and the Microtubes each provide ~30 μmol of product. The radio-frequency tag provides a unique identifier for each reactor and therefore for each compound being generated. By pooling and splitting microreactors in a process known as "directed sorting", one discrete compound is synthesized in each reactor. Unlike many of the known tagging methodologies, this technique is noninvasive and independent of the chemistry. While many methods for tagging have been described in the literature (*47–52*), this is the first commercialization of such technology. The entire system consists of a scanning station, synthesis manager software, reactors, radio-frequency tags and a Pentium PC.

The OntoBLOCK System

For a description of this system, refer to Chapter 12.

Safety Considerations

While the goal of this chapter is to familiarize the reader with various equipment that is intended to improve the efficiency of the synthetic organic chemist, the safety implications of using new techniques and methods should always be kept in mind. Traditional automated synthesizers used in the preparation of peptides and oligonucleotides do not pose many of the major safety issues associated with multiple organic synthesizers. The former equipment typically uses nonvolatile solvents, whereas multiple organic synthesizers typically use a wide variety of organic solvents, as well as a diversity of reactive reagents, some of which may have high vapor pressures, low flash-point temperatures, and/or be sensitive to exposure to air. The scientist must be conscious of the potential hazards of solvent exposure and reaction flammability while using manual or automated equipment. Sensible precautions should be taken, such as good ventilation, careful reaction setup with avoidance of closed systems, protection of reactants from air and/or moisture when necessary, avoidance of skin contact with chemicals and solvents, and adequate eye protection.

Conclusions

In this chapter, we have presented many of the recent innovations in equipment for the execution of high-throughput organic synthesis. This chemical technology has adapted and transformed traditional laboratory synthetic protocols so they can be used in new environments, principally those involving resin-supported organic chemistry. This new technology, whether manual or automated, is revolutionizing the way we conduct organic synthesis in the laboratory. The manual methods can be easily used by many scientists and have made a major impact on productivity. These methods have been used to produce a variety of structurally diverse molecules. Automation of organic synthesis, while still at an early stage of development, has already provided useful equipment for scientists to improve their ability to generate diverse structures rapidly. The Chiron custom automation system illustrates the creativity and innovation that can be contributed by the chemist. It is clear that instrument manufacturers need to concentrate on building versatile, easy-to-use instruments for the rapid synthesis of organic molecules. While these teams are busy, individual research chemists will continue to devise creative ways to improve the efficiency of organic synthesis in the laboratory.

References

1. Hermkens, P. H. H.; Ottenheijm, H. C. J.; Rees, D. *Tetrahedron* **1996**, *52*, 4527–4554.
2. Früchtel, J. S.; Jung, G. *Angew. Chem. Int. Ed. Engl.* **1996**, *35*, 17–42.
3. Terrett, N. K.; Gardner, M.; Gordon, D. W.; Kobylecki, R. J.; Steele, J. *Tetrahedron* **1995**, *51*, 8135–8173.
4. Thompson, L. A.; Ellman, J. A. *Chem. Rev.* **1996**, *96*, 555–600.
5. Han, H.; Wolfe, M. M.; Brenner, S.; Janda, K. D. *Proc. Natl. Acad. Sci. U.S.A.* **1995**, *92*, 6419–6423.
6. Han, H.; Janda, K. D. *J. Am. Chem. Soc.* **1996**, *118*, 2539–2544.
7. Thayer, A. M. *Chem. Eng. News* **1996**, February 12, 57–64.
8. Baum, R. *Chem. Eng. News* **1996**, February 12, 28.
9. Borman, S. *Chem. Eng. News* **1996**, February 12, 29–54.
10. Geysen, H. M.; Meloen, R. H.; Barteling, S. J. *Proc. Natl. Acad. Sci. U.S.A.* **1984**, *81*, 3998–4002.
11. Geysen, H. M.; Rodda, S. J.; Mason, T. J.; Tribbick, G.; Schoofs, P. G. *J. Immunol. Methods* **1987**, *102*, 259–274.
12. Maeji, N. J.; Valerio, R. M.; Bray, A. M.; Campbell, R. A.; Geysen, H. M. *Reactive Polymers* **1994**, *22*, 203–212.
13. Valerio, R. M.; Bray, A. M.; Campbell, R. A.; DiPasquale, A.; Margellis, C.; Rodda, S. J.; Geysen, H. M.; Maeji, N. M. *Int. J. Pept. Protein Res.* **1993**, *42*, 1–9.
14. Valerio, R. M.; Bray, A. M.; Maeji, N. J. *Int. J. Pept. Protein Res.* **1994**, *44*, 158–165, and references cited therein.
15. Moos, W. H.; Green, G. D.; Pavia, M. *Annu. Rep. Med. Chem.* **1993**, *28*, 315–324.
16. Bunin, B. A.; Plunkett, M. J.; Ellman, J. A. *Proc. Natl. Acad. Sci. U.S.A.* **1994**, *91*, 4708–4712.
17. Virgilio, A. A.; Ellman, J. A. *J. Am. Chem. Soc.* **1994**, *116*, 11580–11581.
18. Bray, A. M.; Chiefari, D. S.; Valerio, R. M.; Maeji, N. J. *Tetrahedron Lett.* **1995**, *36*, 5081–5084.
19. Houghten, R. A. *Proc. Natl. Acad. Sci. U.S.A.* **1985**, *82*, 5131–5135.
20. Houghten, R. A. U.S. Patent 4 631 211, 1986.
21. Jung, G.; Beck-Sickinger, A. *Angew. Chem. Int. Ed. Engl.* **1992**, *31*, 367–486, and references cited therein.
22. Beck-Sickinger, A. G.; Gaida, W.; Schnorrenberg, G.; Lang, R.; Jung, G. *Int. J. Pept. Protein Res.* **1990**, *36*, 522–530.
23. Dooley, C. T.; Chung, N. N.; Wilkes, B. C.; Schiller, P. W.; Bidlack, J. M.; Pasternak, G. W.; Houghten, R. A. *Science (Washington, D.C.)* **1994**, *266*, 2019–2021.
24. Meyers, H. V.; Dilley, G. J.; Durgin, T. L.; Powers, T. S.; Winssinger, N. A.; Zhu, H.; Pavia, M. R. *Mol. Diversity* **1995**, *1*, 13–20.
25. Pavia, M. R.; Cohen, M. P.; Dilley, G. J.; Dubuc, G. R.; Durgin, T. L.; Forman, F. W.; Hediger, M. E.; Milot, G.; Powers, T. S.; Sucholeiki, I.; Zhou, S.; Hangauer, D. G. *Bioorg. Med. Chem.* **1996**, *4*, 659–666.
26. DeWitt, S. H.; Kiely, J. S.; Stankovic, C. J.; Schroeder, M. C.; Cody, D. M. R.; Pavia, M. R. *Proc. Natl. Acad. Sci. U.S.A.* **1993**, *90*, 6909–6913.
27. DeWitt, S. H.; Kiely, J. S.; Pavia, M. R.; Schroeder, M. C.; Stankovic, C. J. U.S. Patent 5 324 483, 1994.

28. DeWitt, S. H.; Czarnik, A. W. *Acc. Chem. Res.* **1996**, *29*, 114–122.
29. DeWitt, S. H.; Schroeder, M. C.; Stankovic, C. J.; Czarnik, A. W. *Drug Dev. Res.* **1994**, *33*, 116–124, and references cited therein.
30. Fodor, S. P. A; Read, J. L.; Pirrung, M. C.; Stryer, L.; Lu, A. T.; Solas, D. *Science (Washington, D.C.)* **1991**, *251*, 767–773.
31. Pirrung, M. C.; Read, J. L.; Fodor, S. P. A.; Stryer, L. U.S. Patent 5 143 854, 1992.
32. Patchnornik, C. A.; Amit, B.; Woodward, R. B. *J. Am. Chem. Soc.* **1970**, *92*, 6333–6335.
33. Holmes, C. P.; Adams, C. L.; Fodor, S. P. A. In *Perspectives in Medicinal Chemistry*; Testa, B., Ed.; VCH: New York, NY, 1993; Chapter 31, pp 489–500.
34. Xiang, X. D.; Sun, X.; Briceno, G.; Lou, Y.; Wang, K. A.; Chang, H.; Wallace-Freedman, W. G.; Chen, S.; Schultz, P. G. *Science (Washington, D.C.)* **1995**, *268*, 1738–1740.
35. Cho, C. Y.; Moran, E. J.; Cherry, S. R.; Stephans, J. C.; Fodor, S. P. A.; Adams, C. L.; Sundaram, A.; Jacobs, J. W.; Schultz, P. G. *Science (Washington, D.C.)* **1993**, *261*, 1303–1305.
36. Zuckermann, R. N.; Kerr, J. M.; Siani, M. A.; Banville, S. C. *Int. J. Pept. Protein Res.* **1992**, *40*, 497–506.
37. Zuckermann, R. N.; Kerr, J. M.; Siani, M. A.; Banville, S. C.; Santi, D. V. *Proc. Natl. Acad. Sci. U.S.A.* **1992**, *89*, 4505–4509.
38. Zuckermann, R. N.; Martin, E. J.; Spellmeyer, D. C.; Stauber, G. B.; Shoemaker, K. R.; Kerr. J. M.; Figlozzi, G. M.; Goff, D. A.; Siani, M. A.; Simon, R. J.; Banville, S. C.; Brown, E. G.; Wang, L.; Richter, L. S.; Moos, W. H. *J. Med. Chem.* **1994**, *37*, 2678–2685.
39. Furka, A.; Sebastyen, F.; Asgedom, M.; Dibo, G. *Int. J. Peptide Res.* **1991**, *37*, 487–493.
40. Lebl, M.; Krchnak, V.; Sepetov, N. F.; Seligmann, B.; Strop, P.; Felder, S.; Lam, K. S. *Biopolymers* **1995**, *37*, 177–198.
41. Lashkari, D. A.; Huniche-Smith, S. P.; Norgren, R. M.; Davis, R. W.; Brennan, T. *Proc. Natl. Acad. Sci. U.S.A.* **1995**, *92*, 7912–7915.
42. O'Donnell, M. J.; Zhou, C.; Scott, W. L. *J. Am. Chem. Soc.* **1996**, *118*, 6070–6071.
43. Schwesinger, R.; Willardt, J.; Schlemper, H.; Keller, M.; Schmidt, D.; Fritz, H. *Chem. Ber.* **1994**, *127*, 2435–2454.
44. Terrett, N. K.; Bojanic, D.; Brown, D.; Bungay, P. J.; Gardner, M.; Gordon, D. W.; Mayers, C. J.; Steele, J. *Bioorg. Med. Chem. Lett.* **1995**, *5*, 917–922.
45. Peterson, M. L. Presented at CHI's Exploiting Molecular Diversity: Small Molecule Libraries for Drug Discovery, Coronado, CA, January 1996.
46. Peterson, M. L. Presented at Solid Phase Synthesis: Developing Small Molecule Libraries, Coronado, CA, February 1996.
47. Ohlmeyer, M. H. J.; Swanson, R. N.; Dillard, L. W.; Reader, J. C.; Asouline, G.; Kobayashi, R.; Wigler, M.; Still, W. C. *Proc. Natl. Acad. Sci. U.S.A.* **1993**, *90*, 10922–10926.
48. Borchardt, A.; Still, W. C. *J. Am. Chem. Soc.* **1994**, *116*, 373–374.
49. Chabala, J. C. *Curr. Opin. Biotechnol.* **1995**, *6*, 632–639.
50. Service, R. F. *Science (Washington, D.C.)* **1995**, *270*, 577.
51. Nikolaiev, V.; Stierandova, A.; Krchnak, V.; Seligmann, B.; Lam, K. S.; Salmon, S. E.; Lebl, M. *Peptide Res.* **1993**, *6*, 161–170.
52. Needles, M. C.; Jones, D. G.; Tate, E. H.; Heinkel, G. L.; Kochersperger, L. M.; Dower, W. J.; Barret, R. W.; Gallop, M. A. *Proc. Natl. Acad. Sci. U.S.A.* **1993**, *90*, 10700–10704.

11

Automated Approaches to Reaction Optimization

Jonathan S. Lindsey

Optimization of reaction conditions is essential to the practice of combinatorial chemistry, as high-quality chemical libraries are best prepared with clean, high-yielding reactions. Automation allows reaction optimization to be performed rapidly at the earliest stages of exploratory research. Here, I focus on the methods for automation of reaction optimization, rather than on specific chemistries.

The Need for Reaction Optimization

Reaction optimization is practiced in many areas of chemical research, including combinatorial chemistry. Reaction optimization, as I use the term here, is the search for appropriate conditions for carrying out a reaction (*see* the box). The search explores many parameters including solvent, catalyst or cocatalysts, reagents, concentrations of various species, addition sequence and timing, and reaction temperature. The set of parameters that describes how a reaction is carried out defines a "reaction space" (*1*), and reaction optimization explores as wide a variety of reaction spaces as possible to achieve the desired goal. Usually the objective is to obtain a higher yield of product, but in some cases one may want to find conditions that form less by-product, use less expensive materials, speed up the transformation, allow one to use milder conditions or less toxic reagents, avoid environmental disposal problems, or eliminate materials that present purification problems. Sometimes more than one of these objectives are important, but usually achieving high yields of product is paramount.

Despite its potential importance, reaction optimization is usually one of the last activities to take place when a new synthetic product is developed. New synthetic products are the result of a spectrum of research activities, generally starting

Reaction Optimization–A Definition

The definition of reaction optimization in the text is a narrow one, and is to be distinguished from two other uses of the term optimization in the chemical sciences. First, the term is often used to describe the development of fundamentally new methods and synthetic routes. For example, the development of solid-phase synthesis can be viewed as "optimizing" the synthesis of peptides, largely replacing previous solution techniques. This constitutes a correct but rather grand interpretation of the term optimization. We limit discussion here to optimization of reaction conditions for specific transformations, though the automation tools described here can certainly be used in support of the development of radically new synthetic routes. Second, optimization can refer to mathematical modeling of chemical processes. Chemical engineers often "optimize" reaction processes by computerized analyses of mathematical models, which can contain thousands of variables (2). Often these chemical processes are continuous flow processes in chemical plants involving numerous feedstocks, many different reactions, and a suite of products. Optimization issues in this domain involve, among other things, economics, feedstock availability, desired product mix, and environmental concerns. Here, we are concerned with specific chemical parameters for individual batch reactions where mathematical models are not available and identification of improved reaction conditions requires experimental investigation. The information gained about improved conditions for batch reactions certainly could be used in industrial-scale applications.

with broad-ranging exploratory research, moving on to more focused studies such as analog preparation, and ending with process optimization to refine and scale up a synthesis to large preparative levels (Figure 1). Reaction optimization is an integral part of process optimization, which is the optimization of the entire set of industrial operations that are involved in converting the starting materials into purified products. Reaction conditions that afford higher yields, cleaner reactions, or use less expensive materials will yield a more efficient process, and reaction optimization is therefore a central activity in preparing for industrial scale-up. But at earlier stages of the research process, reaction optimization has often been a much more casual endeavor. With the advent of combinatorial chemistry as part of exploratory research, however, it is not only desirable but necessary to optimize a reaction at the earliest possible point.

Traditionally, exploratory chemistry could tolerate low-yielding or messy reactions, because the goal was simply to obtain a small amount of a single target compound quickly. In combinatorial chemistry, however, one is concerned not only with a single compound but with the generation of a chemical library of many thousands of compounds. Libraries are typically generated by repetitive application of a few reaction types, with no purification between reactions (other than washing a solid-phase resin, which does not remove by-products attached to the resin). If the reactions employed are not clean and high-yielding, the quality of the library will

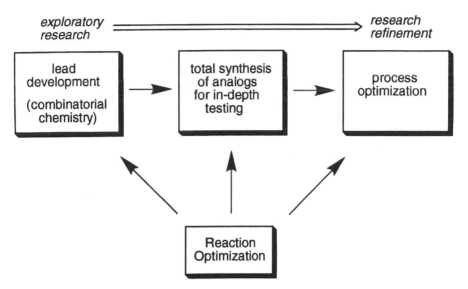

FIGURE 1. Optimization of reaction conditions is important at several stages of chemical research.

rapidly deteriorate as the number of synthetic steps increases. Reaction optimization is thus no longer a luxury at the exploratory research stage; it is essential for the generation of a high-quality combinatorial library.

Human Factors: Time and Tedium

The need to perform reaction optimization at the earliest stages of exploratory research presents a dilemma. On the one hand, thorough reaction optimization requires systematic investigation of the effects of a large number of variables on a reaction. On the other hand, repeatedly implementing one reaction with a slight change in one or more parameters to explore "reaction space" exhaustively can be extraordinarily time-consuming, and this is partly why reaction optimization is usually sidestepped in exploratory research. Furthermore, chemists often view reaction optimization as a tedious rather than creative activity in practicing their craft. When chemists do engage in reaction optimization, they typically examine a small subset of the possible reaction conditions rather than undertaking a more comprehensive optimization. Although such limited and unsystematic investigations can be quite productive, as they often identify some of the key parameters that deserve attention, for many purposes much more systematic optimization studies are needed.

Requirements for Automation Tools Used in Reaction Optimization

Reaction optimization requires the repetitive implementation of one reaction a large number of times with only slight changes in one or more parameters. Simple yet repetitive tasks are well-suited for automation, and those that additionally are dirty, dangerous, difficult, or considered demeaning are the most readily relinquished by humans (*3*). Automation at the bench scale has been applied to all stages of synthetic chemistry (*4–6*), systems for total synthesis of analogs have been constructed (*7*), and several automation systems have been built for the express purpose of reaction optimization (*8–20*). As our goal is to rapidly optimize chemical reaction conditions, the following points have become apparent:

1. Product isolation and detailed characterization are not necessary. This simplifies the design of the chemistry workstation, which need only carry out reaction initiation and monitoring (Figure 2). Analysis of reaction samples can be done on partially purified samples. This avoids the need to purify the product at preparative scale, which poses sample handling problems (particularly unexpected phase separations) that are among the most diffi-

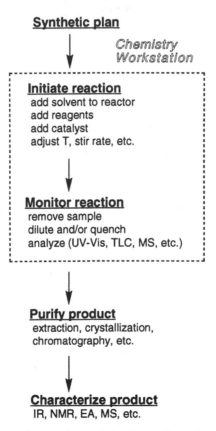

FIGURE 2. Automated systems for reaction optimization studies only need to perform the first two of four stages involved in chemical synthesis. (Reproduced with permission from reference 4. Copyright 1992 Elsevier.)

cult of all the techniques in synthetic chemistry to automate. The fact that these steps can be avoided makes automation of reaction optimization more feasible than the automation of total synthesis.

2. Only small samples are required for analytical detection and quantitation. Thus, microscale chemical reaction flasks (5–10 mL) can be used, minimizing the consumption of chemicals (particularly important given the large number of reactions that need to be run).

3. Speed is essential at the exploratory research stage. Thus, methods for performing large numbers of reactions in parallel are essential. Parallel experimentation in turn demands a host of other features, such as scheduling and resource management.

4. Many automation systems are "brittle", that is, they lack flexibility and are difficult to adapt to different applications. Most chemistry automation systems employ either flow or robotic designs (4–6). Though robotic systems have more moving parts than flow systems, the reprogrammability of robotic systems makes them more versatile in adapting to diverse chemistries.

Automated reaction optimization systems must be able to deliver both quality (precise accurate experimentation at the microscale level) and quantity (large numbers of experiments in a short period of time), and must be versatile to avoid the need for extensive re-tooling when different chemistries are investigated. Below, I discuss the basics of automated workstation design and show how these features are achieved.

Automated Chemistry Workstations

An automated chemistry workstation consists of both hardware and software. The hardware generally includes a reaction station equipped with a number of reaction flasks (fitted with stirring and temperature control), analytical instruments for monitoring reactions, and equipment for delivery of reagents and solvents (syringe pumps, and dispensers for solids delivery). In robotic systems, the robot generally serves only to move reagent samples and solvents from one site to another (for example, from reagent vials to reaction station or from reaction station to analytical instruments). The types of robots that have been used in laboratory chemistry include revolute robots, gantry robots, Cartesian robots, and cylindrical robots (21). The chief differences between these different types of robot are the physical layout of the work space and the motions the robotic arm is able to undergo. One popular robotic system allows different end effectors (a "hand" attached to the robotic arm) to be accessed and interchanged automatically. Automated workstations for synthetic chemistry, especially those dedicated to reaction optimization, have been extensively reviewed (4–6). In the remainder of this chapter, I focus on the automated chemistry workstations developed in my own research group.

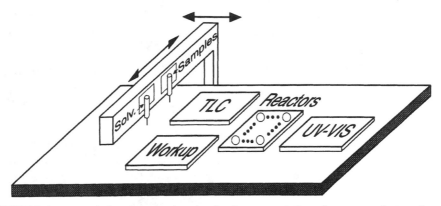

FIGURE 3. Automated chemistry workstation hardware consisting of a gantry robot equipped with syringes for solvent, reagent, and sample delivery, a multivessel microscale reaction station, a product workup station for preparing analytical samples, an automated TLC instrument, and a UV–visible absorption spectrometer. Each vessel in the reaction station is stirred magnetically and is temperature-controlled. (Reproduced with permission from reference 30. Copyright 1992 Elsevier.)

The workstation outlined in Figure 3 (*20*) uses a gantry robot for delivery of reagents, solvents, and samples, a 15-vessel reaction station for microscale chemistry, a workup station, a thin-layer chromatography (TLC) instrument (*22*), and a UV–visible absorption spectrometer. The reaction vessels are individually stirred and their temperatures individually controlled. The workup station does not perform preparative purifications but performs dilutions, neutralizations, or quenching procedures to prepare samples for analysis. This workstation handles liquids only (neat liquids or solutions). Equipment and techniques that address some solids-handling problems have been incorporated into other robotic systems (*10, 13, 23, 24*). General sample-handling techniques for optimizing solid-phase reactions, which are frequently used in combinatorial chemistry, have not yet been developed, however.

The overall architecture of the workstation, showing the interaction of the chemist with the hardware and the software, is diagrammed in Figure 4. Though hardware is the most visible part of an automated chemistry workstation, much of the success of a workstation hinges on the sophistication of the software. The software interface consists of six menu-driven experiment planning modules (*25*). The different modules are designed for different types of experimentation, including statistically designed experiments (*see* the section "Types of Experiments"). To plan an experiment, one enters a module in the experiment planner. A module provides a menu of all possible mechanical actions that can be performed; an experimental plan is composed by selecting operations from this menu to produce a list of mechanical operations and their start and finish times. The duration of each operation is determined automatically by referral to a set of timing formulae. Several other features, including data management structures, are available. A very significant feature of this system is the ability to perform experiments in parallel.

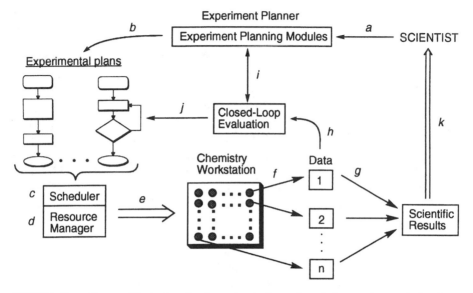

FIGURE 4. Flow diagram illustrating the throughput of experiments in an automated chemistry workstation. The hardware of the chemistry workstation is symbolized here as a multivessel reaction station, emphasizing its parallel experimentation capabilities. The steps shown are as follows: (a) The scientist works with the experiment planner to compose an experimental plan describing the operations to be performed. In some types of experiments (e.g., closed-loop), scientific objectives are stated (*see* (i) below for details). Information about available resources (chemicals and vessels) must also be provided. (b) From the experiment planning modules a plan or set of plans, consisting of a list of robotic operations, emerges. These may include conditionals whose output depends on experimental data. (c) The experimental plans are passed to a scheduler, where the plans are rendered in parallel to the extent possible. (d) A resource manager tabulates the total resource demands (chemicals and vessels) presented by the schedule, and determines which experiments are executable. (e) The executable experiments are passed to the automated chemistry workstation. (f) Data are generated from the analytical instruments and sensors. (g) The data from common experiments are combined in an output file. This file constitutes the results. When no decisions are made automatically about ongoing or planned experiments, this is referred to as open-loop experimentation. (h) Alternatively, the data are passed to the closed-loop evaluation unit of the experiment planning module. When decisions are made without user intervention about ongoing or planned experiments, this constitutes closed-loop experimentation. (i) The data are evaluated in the context of the scientific objective as stated in the experimental plan. For example, the simplex experiment requires an objective function to be stated. Examples include "maximize product yield", or "maximize the ratio of product to by-product for any product yield above 60%". (j) Depending on the results of the evaluation, ongoing experiments can be terminated or altered (e.g., the temperature can be increased or more reagent can be added), pending experiments can be expunged from the queue, or new experiments can be planned and implemented. (k) The results from open-loop and closed-loop experiments are available at all times for review by the scientist.

Scheduling and Resource Management in Parallel Experimentation

How can a workstation perform parallel experimentation? In the experimental plan shown in Figure 5, the experimental plan is displayed as a time line, with each robotic operation shown as a solid black mark. The robotic arm is idle at all other times while the chemical reaction takes place. The total duration of robotic activity in this plan is 36.4 min, only 7.8% of the duration of the experiment (464.2 min). This efficiency of robot utilization can be dramatically increased by interleaving several experiments (Figure 6). Within each plan, the individual operations are performed at fixed times relative to each other, but the start times of the different plans are staggered. This is a very simple but effective form of scheduling. For the example shown, the eight experiments are implemented in 844.0 min with a robot utilization of 34.5%. If done serially, the same eight experiments would take 3713.4 min. In this example, parallel scheduling has thus given a 4.4-fold enhancement of experimental throughput. Such scheduling generally gives efficiency enhancements of up to 10-fold (20), and when the workstation is run continuously rather than in batch mode, the efficiency is even further enhanced (by reducing the contribution of the relatively inefficient periods at the beginning and end of the schedule).

This scheduling approach is robust and has several noteworthy features (20). First, the scheduler used in this workstation can be used with sets of experimental plans, where each experiment in the set is different. When the set of experiments includes experiments of widely differing types (having different durations and densities of operations), the particular ordering of experiments can give considerably different schedules. For reaction optimization, the best schedule is the one that gives

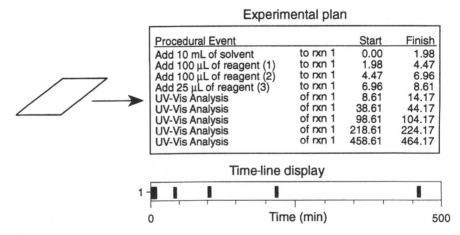

Experimental plan

Procedural Event		Start	Finish
Add 10 mL of solvent	to rxn 1	0.00	1.98
Add 100 µL of reagent (1)	to rxn 1	1.98	4.47
Add 100 µL of reagent (2)	to rxn 1	4.47	6.96
Add 25 µL of reagent (3)	to rxn 1	6.96	8.61
UV-Vis Analysis	of rxn 1	8.61	14.17
UV-Vis Analysis	of rxn 1	38.61	44.17
UV-Vis Analysis	of rxn 1	98.61	104.17
UV-Vis Analysis	of rxn 1	218.61	224.17
UV-Vis Analysis	of rxn 1	458.61	464.17

Time-line display

FIGURE 5. An experimental plan consists of a series of operations (procedural events) with associated start and finish times, and can be displayed as a list or as a time line. Activities requiring robot action are marked on the time line. Note that in this example, the first block in the time line encompasses the first five procedural events (additions of solvent and reagents, and one analysis procedure), which are performed in succession without pause.

the most rapid completion of all experiments. Heuristics can be used to order the experiments to achieve improved schedules. For example, a good heuristic is to order the experiments according to their lengths. This heuristic starts the longest experiment first and then schedules shorter and shorter experiments. In addition, we have recently developed a flexible scheduler that allows the start times of each operation to be varied to a specified degree (e.g., ±10%). The rationale for developing a flexible scheduler was the recognition that reaction optimization experiments rarely need to be performed with split-second precision. The experimental throughput of the flexible scheduler is up to twice that of rigid schedules, with robotic utilizations reaching 75% (*26*). Thus, the overall increase in experimental throughput offered by parallel scheduling is 10- to 20-fold. The discipline of scheduling is vast, and a variety of more elaborate schedulers can be used as workstations become more sophisticated (*27–29*). It should be possible to improve workstations of the design shown in Figure 3 still further, by 5–15-fold, using faster hardware components (*30*). This

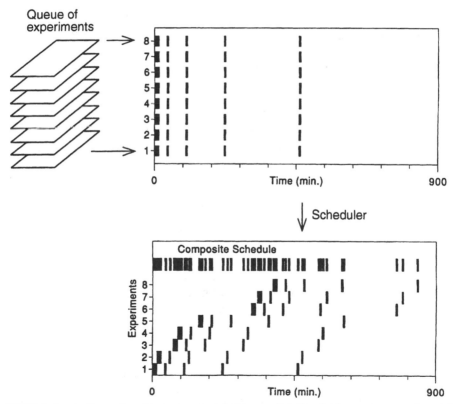

FIGURE 6. A set of experiments can be scheduled in parallel, thereby increasing experimental throughput. The scheduler offsets the start time of experiments, thereby interleaving the individual operations of the robot.

would enable over 100 reactions to be performed simultaneously. These next-generation workstations should be fast enough for use in exploratory research.

The scheduling that makes parallel experimentation possible generates other problems, however. The number of experiments running simultaneously is so large that keeping track of the resources needed becomes a significant problem. A resource management feature has been developed that projects the required resources (such as solvents, reagents, reaction flasks, and vials) for each planned schedule and compares them with a list of resources available. The schedule is then divided into executable experiments (for which resources are available) and experiments awaiting resources. The resource manager thus handles all of the resource bookkeeping, working in concert with the scheduler. As resources become available, waiting experiments are scheduled for completion, thereby enabling continuous operation of the workstation (*25*).

Types of Experiments

Open-Loop and Closed-Loop Experiments

Experiments can be categorized as "open-loop" or "closed-loop", depending on how the data they generate are used. In open-loop experiments, the chemist must examine the data generated by the workstation, decide what the next experiments should be, and compose a plan describing these experiments. No decision-making is performed by the workstation. In closed-loop experiments, experimentally generated data are analyzed automatically in the context of the experimental objective and new experiments are composed and performed automatically. This cycle is continued until the scientific objective is reached or resources are depleted. Closed-loop experimentation is quite attractive, as the workstation can modify its own actions on the basis of the data obtained, freeing the chemist for more creative activities (*17*). Such adaptive robotic systems lie at the heart of modern robotics research (*3*).

Statistically Designed Experiments

One attraction of automated chemistry is the opportunity to perform statistically designed experiments. Ideally, researchers should use statistical methods to design experiments that efficiently yield the desired quality of data, rather than use statistical methods retrospectively to determine the quality of data that have already been collected. For the purpose of reaction optimization, factorial design experiments and simplex experiments are most relevant, though other statistically designed methods can also be implemented.

Factorial Design Experiments. A factorial design experiment collects data at each point in a grid defined by one or more reaction parameters; the separation of the points and the extent of the overall grid is decided by the user. We shall refer to this user-defined grid as the search space, which is a subset of the reaction space. For example, consider an optimization study in which the reaction is performed at four

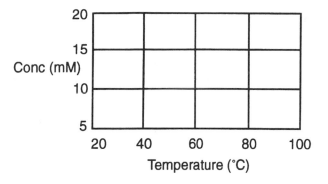

FIGURE 7. A 4 × 5 factorial design. Concentration and temperature form the two dimensions in this factorial design experiment, which comprises 20 experimental points.

concentrations ranging from 5 to 20 mM and five temperatures ranging from 20 to 100 °C. This is a 4 × 5 factorial design experiment, containing 20 experimental points (Figure 7). All combinations of the factors are examined, so 20 experiments must be done. In the jargon of this field, the example given constitutes a two-dimensional full factorial design; temperature and concentration are the two dimensions, and all combinations of factors are fully examined.

Factorial design experiments can provide a comprehensive characterization of a search space, but also generate a large number of experimental points to be investigated. Factorial design experiments can be expanded to cover any number of parameters, but the number of experiments that need to be performed expands dramatically with the increasing number of dimensions studied. Even with four dimensions and three data points per dimension, the number of experimental points would be $3^4 = 81$. In practice, various types of partial factorial design experiments are used to reduce the number of experiments required. These include star designs (in which only the experiments at the center of each face and the core of the search space are performed) and other patterns designed to capture a systematic but limited set of points. Several texts and introductory articles discuss factorial designs and their chemistry applications in more detail (*31–35*). For our purposes, one other point is noteworthy. In factorial design experiments, no data evaluation is done during the course of experimentation to guide the planning of future experiments. These are consequently always open-loop experiments and can be scheduled in parallel, accelerating comprehensive exploration of the search space.

Planning factorial design experiments is relatively simple. An experimental plan is composed using the experimental planner, and key operations are denoted as having variables that correspond to the dimensions of the search space. The specified dimensions and range of the search space are then melded automatically, yielding one experimental plan for each point to be examined in the search space (Figure 8). This set of experimental plans is then passed to the scheduler and resource manager for parallel implementation.

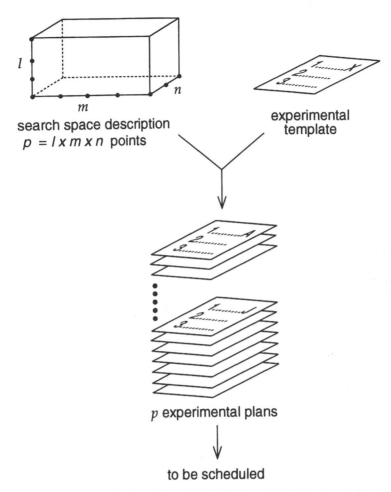

FIGURE 8. Implementation of a three-dimensional full factorial design, in combination with one experimental template, yields a set of p experimental plans. Factorial design experiments involve no decision-making (open-loop) and can be implemented in parallel.

Simplex Experiments. Simplex experiments are fundamentally different from factorial design experiments. The simplex algorithm is a type of evolutionary hill-climbing routine (*36*). It is evolutionary because the individual experiments are performed in succession, and it involves hill-climbing as the algorithm systematically projects the next experimental point in the direction of increasing response. Using a geographical analogy, the simplex algorithm causes movement from a low plain or valley (poor-response region) to the top of a peak (high-response region). An example of the simplex algorithm moving across a contour surface (as in a topographical map, or a reaction space showing product yield as a function of two variables) is

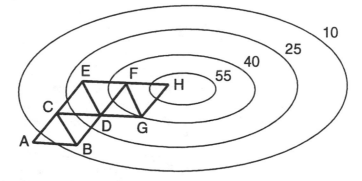

FIGURE 9. The original simplex algorithm illustrated on a two-dimensional surface, where the objective is to find the region of highest response. In a two-dimensional space, the results from three experimental points constitute a simplex. The initial simplex is defined by these points, A, B, and C. Experimental determination of the values (the responses) at these three points reveals that the worst point is A (i.e., among A, B, and C, A lies in the region of lowest response). The next simplex is formed by moving away from A and discarding this point. The second simplex is defined by points B, C, and D, and now the value of point D is determined by experimentation. Comparing the points in the current simplex (B, C, and D) reveals B as the worst point. The third simplex is formed by moving away from B, yielding a simplex composed of points C, D, and E. Point E is evaluated and another simplex move is performed. In this manner, the simplex evolves to the simplex formed by points F, G, and H. Subsequent simplex moves would revolve around the optimal region (not shown). The CMS algorithm would perform slightly differently than the original simplex algorithm, causing a slight change in the shape of the simplex as it moves. Upon reaching the region of optimal response, the simplex in the CMS algorithm would contract and give a finer-grained evaluation in this region. (Reproduced with permission from reference 40. Copyright 1992 Elsevier.)

shown in Figure 9. Although shown for two dimensions, the simplex algorithm can function in *n* dimensions. (A simplex is a geometrical object having *n* + 1 sides for an *n*-dimensional search space. Thus, in Figure 9, the simplex is the triangle that moves across the contour surface, and the simplex algorithm causes the movement of the simplex. In a three-dimensional space, the simplex would be a tetrahedron.)

The simplex algorithm has been known for several decades and has been applied to diverse optimization problems (*31–33, 37, 38*). The simplex algorithm has both strengths and weaknesses. Some of the weaknesses include oscillating on ridges, moving too slowly in regions of slowly changing response (as in a great plain that gently ascends to the base of a sharp peak, to continue the geographical analogy), moving in too-large steps about the region of optimal response, or losing its regular shape in certain types of search spaces. Many modifications to the simplex algorithm have been proposed to overcome these limitations. Betteridge et al. have evaluated 14 such modifications and have incorporated most of these into a "composite modified simplex" (CMS) algorithm (*39*). We have developed an experiment planning module that enables the CMS algorithm to be used in reaction optimizations (*40*). One nice feature of this experiment planner is that each of the modifications made by Betteridge et al. to the original simplex algorithm can be invoked individually in a software menu, allowing use of the complete CMS algorithm, the original simplex algorithm, or algorithms having intermediate features.

Experiments that use the CMS algorithm can be implemented in a straightforward fashion (Figure 10). Using the experiment planner, an experimental plan is composed, variables are identified, the range of the search space is defined, the objective function is specified (usually "maximize the yield of product", though more complex objectives can be stated), and start criteria (where the search should start) and stop criteria (maximum number of experiments, or minimum tolerable improvement in objective function) are specified. The experimental plan with designated variables constitutes a template that is used in each cycle. The simplex algorithm is then implemented, and one experiment is performed per cycle. With each cycle a new experimental point is projected, and this process is continued until the resources are expended or the stop criteria are satisfied. Versions of the simplex algorithm less elaborate than the CMS have been applied in other closed-loop automation systems (8, 14).

Comparison of Factorial Design and Simplex Experiments

How do simplex and factorial design experiments compare? In simplex experimentation, the experiments are performed sequentially (one at a time). This is unavoidable, as the parameter values of the next experimental point depend on the results of past and ongoing experiments. In factorial design experimentation, all experimental points are known in advance and there is no reason to perform experiments sequentially. Indeed, if the workstation has sufficient capacity, all experimental points in a factorial design experiment can be performed simultaneously. When a factorial design experiment is complete, the entire search space will be covered with a systematic set of points. If the space was finely gridded, the region of optimal response should be well defined. In contrast, a simplex experiment rarely covers the entire search space, instead focusing on a region of better response than the initial starting point. There is no guarantee that this is a global optimum; one can only begin to have confidence that a global optimum has been reached if several simplex searches started in different regions of the space give the same result. Upon reaching the region of optimal (or improved) response, the CMS algorithm will cause contraction and perform a fine-gridding of this region. In a sense, simplex and factorial design experiments are thus complementary (31–33, 40, 41). Simplex experiments can give an improved response without requiring a comprehensive exploration of the entire space, but the search must be performed one experiment at a time. Factorial design experiments are by nature unfocused, but yield at least some information about the whole of the space to be searched and can be performed in parallel as no decision-making is involved. On an automated chemistry workstation, simplex experiments are thus inefficient in time but efficient in chemical resources, while factorial design experiments have the opposite characteristics.

Factorial design and simplex experiments have been well known for several decades. These types of designed experiments have been applied sparingly in man-

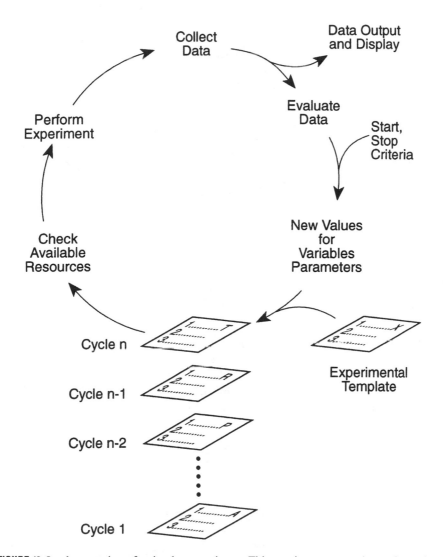

FIGURE 10. Implementation of a simplex experiment. This requires one experimental template, information about the search space, and start and stop criteria. Values describing specific points in the search space are inscribed in place of the variables in the experimental template. After required resources are checked against available resources, the experiment is performed, and the data are evaluated in the context of the objective function as stated in the experimental plan. The resulting evaluation is used to project new points for the next simplex move, which defines the input for another cycle of experimentation. (Reproduced with permission from reference 40. Copyright 1992 Elsevier.)

ual reaction optimization studies, however, as chemists usually prefer more intuitive and less extensive approaches. With the recent development of automated chemistry workstations, the important scientific objective of reaction optimization can be achieved without burdening the researcher with repetitive tasks. The focus on reaction optimization can be expected to intensify as a result of the requirement for clean reactions in combinatorial chemistry. A few reaction optimizations have already been performed with the workstations developed so far (*4–6, 8–14, 16, 19, 41*).

Major Challenges and New Directions in Automated Optimization

The demands of combinatorial chemistry can be expected to drive the development of more sophisticated workstations, statistical design tools, and integrated approaches for reaction optimization. The issues for future development are clear, and include the following major challenges:

1. Integrating diverse analytical instruments. Several different analytical instruments must be included in a single workstation to allow in-depth characterization of product distributions. New strategies will have to be developed to interpret the different modalities of data that will result, and to reconcile any apparent conflicts between different data streams.

2. Solids handling. The most pernicious aspect of automated chemistry involves solids-handling techniques. More reliable methods for handling solids will require categorization of different types of solids, methods for handling different types, and utilization of a variety of sensors to provide feedback on whether a given step has been successful. In addition, sample-handling techniques must be developed for optimizing reactions performed with reactants attached to a solid-phase support.

3. Enhancing flexibility without sacrificing efficiency. Efficiency and generality are often inversely related (*17*). Special-purpose workstations that are designed for one type of chemistry can be very efficient but tend to lack versatility, making it impossible to use them for other chemistries or other methodologies. The demands on the versatility of automation systems will expand as the range of chemistry optimization problems increases. Lack of versatility is a major obstacle to the widespread acceptance of laboratory automation systems for reaction optimization. It will be essential to design general-purpose automation systems that maintain high throughput without extensive manual reconfiguration. Modular hardware (and software) designs should help with this.

4. Improving safety. Most workstations at present lack even the most rudimentary safety features. Provisions to prevent programming dangerous procedures (e.g., heating a solvent far beyond its boiling point) must be developed. Sensors that detect dangerous situations that even a novice

would recognize if working manually (e.g., a runaway exotherm) must also be implemented. Databases of the properties of materials could be incorporated into the software, and might also be used to facilitate data interpretation as part of more sophisticated decision-making procedures.

5. Increasing the sophistication of searching procedures. New techniques that can be used to rapidly screen reaction methods would prove useful in reaction optimization. Factorial design and simplex experiments are quite useful but require extensive experimentation. The ideal methods would combine the focused evolutive nature of simplex experimentation with the systematic parallel nature of factorial design experiments.

Incorporation of some or all of these features into the next generation of automated chemistry workstations should provide rapid and efficient methods for reaction optimization. These workstations should be attractive to the wide range of chemists whose scientific objectives in one way or another lead them to need to develop improved conditions for a synthetic transformation.

References

1. Carlson, R. *Chemom. Intell. Lab. Syst.: Lab. Inf. Mgt.* **1992**, *14*, 41–56.
2. Grossmann, I. E.; Biegler, L. T. *Chemtech* **1995**, December, 27–35.
3. Critchlow, A. J. *Introduction to Robotics*; Macmillan: New York, 1985.
4. Lindsey, J. S. *Chemometrics and Intelligent Laboratory Systems: Laboratory Information Management* **1992**, *17*, 15–45.
5. Hardin, J. H.; Smietana, F. R. *Mol. Diversity* **1996**, *1*, 270–274.
6. DeWitt, S. H.; Czarnik, A. W. *Curr. Opin. Biotechnol.* **1995**, *6*, 640–645.
7. Sugawara, T.; Kato, S.; Okamoto, S. *J. Auto. Chem.* **1994**, *16*, 33–42.
8. Winicov, H.; Schainbaum, J.; Buckley, J.; Longino, G.; Hill, J.; Berkoff, C. E. *Anal. Chim. Acta* **1978**, *103*, 469–476.
9. Frisbee, A. R.; Nantz, M. H.; Kramer, G. W.; Fuchs, P. L. *J. Am. Chem. Soc.* **1984**, *106*, 7143–7145.
10. Legrand, M.; Bolla, P. *J. Auto. Chem.* **1985**, *7*, 31–37.
11. Porte, C.; Roussin, D.; Bondiou, J.-C.; Hodac, F.; Delacroix, A. *J. Auto. Chem.* **1987**, *9*, 166–173.
12. Lindsey, J. S.; Corkan, L. A.; Erb, D.; Powers, G. J. *Rev. Sci. Instr.* **1988**, *59*, 940–950.
13. Weglarz, T. E.; Morabito, P. L., Jr.; Garner, J. L. *Lab. Rob. Autom.* **1988**, *1*, 43–51.
14. Matsuda, R.; Ishibashi, M.; Takeda, Y. *Chem. Pharm. Bull.* **1988**, *36*, 3512–3518.
15. Kramer, G. W.; Fuchs, P. L. *Chemtech* **1989**, November, 682–688.
16. Catron, M. T.; Cunningham, L. J.; Huret, T. M.; Leach, J. T.; Wright, S. F. *Proc. Symp. Chemspec. USA'90, Spring Innovations, Manchester* **1990**, 5–10.
17. Corkan, A.; Lindsey, J. S. In *Advances in Laboratory Automation Robotics*; Strimaitis, J. R.; Helfrich, J. P., Eds.; Zymark Corp.: Hopkinton, MA, 1990; Vol. 6, pp 477–497.
18. Weglarz, T. E.; Atkin, S. C. In *Advances in Laboratory Automation Robotics*; Strimaitis, J. R.; Helfrich, J. P., Eds.; Zymark Corp.: Hopkinton, MA, 1990; Vol. 6, pp 435–461.
19. Josses, P.; Joux, B.; Barrier, R.; Desmurs, J. R.; Bulliot, H.; Ploquin, Y.; Metivier, P. In *Advances in Laboratory Automation Robotics*; Strimaitis, J. R.; Helfrich, J. P., Eds.; Zymark Corp.: Hopkinton, MA, 1990; Vol. 6, pp 463–475.

20. Corkan, L. A.; Lindsey, J. S. *Chemometrics and Intelligent Laboratory Systems: Laboratory Information Management* **1992**, *17*, 47–74.
21. Strimaitis, J. R. *J. Chem. Ed.* **1989**, *66*, A13–A17.
22. Corkan, L. A.; Haynes, E.; Kline, S.; Lindsey, J. S. In *New Trends in Radiopharmaceutical Synthesis, Quality Assurance and Regulatory Control*; Emran, A. M., Ed.; Plenum: New York, 1991; pp 355–370.
23. Kramer, G. W.; Fuchs, P. L. In *Advances in Laboratory Automation Robotics*; Strimaitis, J. R.; Hawk, G. L., Eds.; Zymark Corp.: Hopkinton, MA, 1988; Vol. 4, pp 339–359.
24. Caron, M.; Martin-Moreno, C.; Bondiou, J.-C.; Bourgogne, J.-P.; Porte, C.; Delacroix, A. *Bull. Soc. Chim. Fr.* **1991**, *128*, 684–696.
25. Kuo, P. Y.; Yang, K.; Corkan, L. A.; Lindsey, J. S. *Chemom. Intell. Lab. Syst.: Lab. Inf. Mgt.*, manuscript in preparation.
26. Aarts, R.; Lindsey, J. S.; Corkan, L. A.; Smith, S. *Clin. Chem.* **1995**, *41*, 1004–1010.
27. French, S. *Sequencing and Scheduling*; John Wiley and Sons Inc.: New York, NY, 1982.
28. Morton, T. E.; Pentico, D. W. *Heuristic Scheduling Systems*; John Wiley and Sons Inc.: New York, NY, 1993.
29. *Intelligent Scheduling*; Zweben, M.; Fox, M., Eds.; Morgan Kaufmann: Palo Alto, CA, 1993.
30. Lindsey, J. S.; Corkan, L. A. *Chemometrics and Intelligent Laboratory Systems: Laboratory Information Management* **1993**, *21*, 139–150.
31. Bayne, C. K.; Rubin, I. B. *Practical Experimental Designs and Optimization Methods for Chemists*; VCH: Deerfield Beach, FL, 1986.
32. Brereton, R. G. *Chemometrics: Applications of Mathematics and Statistics to Laboratory Systems*; Ellis Horwood: New York, 1990.
33. Massart, D. L.; Vandeginste, B. G. M.; Deming, S. N.; Michotte, Y.; Kaufman, L. *Chemometrics: A Textbook*; Elsevier: Amsterdam, 1988.
34. Carlson, R.; Nordahl, A. *Top. Curr. Chem.* **1993**, *166*, 1–64.
35. Deming, S. N. *Chemtech* **1992**, October, 604–607.
36. Spendley, W.; Hext, G. R.; Himsworth, F. R. *Technometrics* **1962**, *4*, 441–461.
37. Deming, S. N.; Morgan, S. L. *Anal. Chim. Acta* **1983**, *150*, 183–198.
38. Betteridge, D.; Wade, A. P.; Howard, A. G. *Talanta* **1985**, *32*, 709–722.
39. Betteridge, D.; Wade, A. P.; Howard, A. G. *Talanta* **1985**, *32*, 723–734.
40. Plouvier, J.-C.; Corkan, L. A.; Lindsey, J. S. *Chemometrics and Intelligent Laboratory Systems: Laboratory Information Management* **1992**, *17*, 75–94.
41. Corkan, L. A.; Plouvier, J.-C.; Lindsey, J. S. *Chemometrics and Intelligent Laboratory Systems: Laboratory Information Management* **1992**, *17*, 95–105.

Application of Automated Parallel Synthesis

Adnan M. M. Mjalli

With the significant advances in the automation of compound screening, individual scientists are able to screen thousands of compounds per day. This high-throughput screening capacity far exceeds the synthetic output of traditional medicinal chemists. To bridge the gap, several hardware and software tools have been developed. Relatively few applications of these tools to the automated parallel synthesis of non-peptide-based molecules have been published. In this chapter, I therefore describe a number of combinatorial strategies involving automated high-speed parallel synthesis that have been established at Ontogen, before describing the few syntheses that have been reported by other workers.

Requirements for High-Speed Parallel Synthesis

High-speed parallel synthesis of organic compounds can be used throughout the drug discovery process, from the discovery of active lead compounds to the clinical evaluation stage, as shown in Figure 1 (*1*).

For the successful implementation of high-speed parallel synthesis in drug discovery, the following issues need to be considered:

1. Compound purity: Pure compounds are needed to allow rapid structure–activity relationship (SAR) evaluation directly from biological assay results.

2. Compound quantity: Milligram quantities are required for numerous chemical and biological analyses and possibly for further purification.

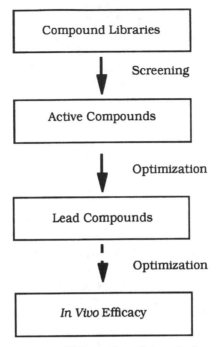

FIGURE 1. The drug discovery process. High-speed parallel synthesis can be used at each of these three steps. (Reproduced with permission from reference 1. Copyright 1995 Network Science.)

3. Compound identification: Exact compound characterization, including structural identity, is needed in order to optimize drug candidates.

4. Structural diversity: Diversity within a chemical library is important to ensure a successful hit rate across a wide range of biological targets.

5. Method of synthesis: A highly efficient method for the synthesis of all required compounds is needed. This method should result in the synthesis of compounds that can be optimized by SAR and are diverse, pharmaceutically attractive (with low molecular weights and high metabolic stability) and novel (to ensure substantial patent protection).

6. Automation technology: The technology should be applicable to a wide range of chemical reactions (conducted under extremes of temperature and an inert atmosphere, for example) in order to synthesize most, if not all, proposed chemical structures.

High-speed parallel synthesis can be used to generate libraries of compounds with previously identified pharmacophores such as the 1,4-benzodiazepine pharmacophore. A highly efficient methodology has allowed for the manual synthesis and evaluation of tens of thousands of 1,4-benzodiazepine-based compounds (2).

Alternatively, an investigator may wish to modulate new biological targets such as protein tyrosine phosphatases (PTPases) (*3, 4*), in which case a targeted library with novel pharmacophores is needed. The following information should be considered in the design of such libraries:

1. The biology of the target;

2. The nature of known substrates or ligands;

3. The mechanism of the target–substrate interaction; and

4. Related literature information, including three-dimensional structural information.

Stages of Library Optimization and Preparation

There are three stages in the preparation of any given library.

1. Feasibility and Basic Research: Chemistry generated in-house and any available literature information are evaluated with the presumption that a wide range of starting material (inputs) is readily available and can be used in the synthesis of compounds.

2. Optimization and Process Development: The method of synthesis undergoes full optimization of reaction condition parameters, such as the solvent, reaction time, temperature, and number of steps to be used. The method should allow for a wide range of building block inputs, so that significant structural diversity is generated. In many cases, the optimized protocol allows for the synthesis of large amounts of compounds (up to several kilograms), which may be required for further in vivo testing.

3. Automated Synthesis and Production: The fully optimized protocol can be executed to produce chemical libraries in a high-speed manner using automated technology.

The OntoBLOCK System

The "OntoBLOCK System" (*5*) consists of a synthesis robot for resin and reagent delivery into arrays of reaction vessels held in the reaction block, a shaker station to shake the reaction block after addition of reagents, a rinse and wash station to remove excess reagents, and a rinse and cleavage station to cleave the compounds from the polymer into 96-well microtiter plates (Figure 2).

The following sections describe libraries that have been synthesized in a high speed fashion using the OntoBLOCK System, Diversomer technology (see Chapter 10), and other custom robotic instruments.

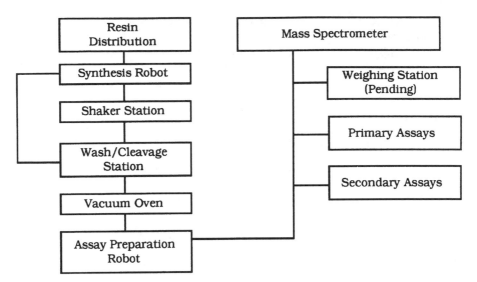

FIGURE 2. The OntoBLOCK System. The synthetic robot controls three stations (for resin distribution, shaking, and washing, rinsing, and cleavage). (Reproduced with permission from reference 5. Copyright 1996 Zymark Corporation.)

Libraries Produced Using the OntoBLOCK System

Tetrasubstituted Imidazoles (Method 1)

Compounds with a five-membered ring system represent a large class of biologically active molecules that are used in various therapeutic areas (*6*). In particular, compounds containing imidazole groups have attracted a great deal of attention because of their ability to act as a hydrogen bond donor and acceptor as well as their high affinity for metals such as zinc, iron, and magnesium (*7–10*). In peptides, the replacement of an amide bond with an imidazole ring has been shown to enhance the pharmacokinetics and bioavailability of some protease inhibitors (*11*). Therefore, a high-speed protocol for the synthesis of highly functionalized imidazoles (for example, the tetrasubstituted imidazole shown below) is of considerable value.

There are four different substituents (R_1, R_2, R_3, and R_4) around the imidazole pharmacophore. Literature searches indicate that trisubstituted imidazoles can be synthesized in solution from the condensation of diaryl dione ($R_3(CO)_2R_4$) with an

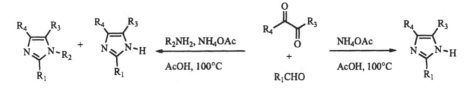

SCHEME I. Synthesis of tri- and tetrasubstituted imidazoles in solution. (Reproduced with permission from reference 16. Copyright 1996 Elsevier Science Ltd.)

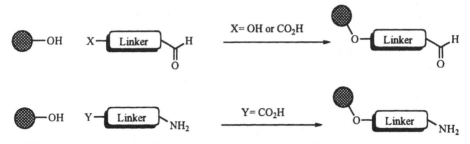

SCHEME II. Attachment of aldehyde and amine linkers onto solid support for the solid-phase synthesis of substituted imidazoles. The aldehyde can be attached by an ester or ether linkage, whereas the amine can only be attached via an ester.

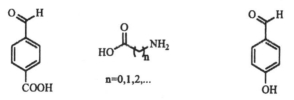

SCHEME III. Examples of aldehyde and amine linkers used for the solid-phase synthesis of substituted imidazoles.

aldehyde (R_1CHO) and ammonium acetate in acetic acid (*12*). If an additional component such as R_2NH_2 is used in the reaction, mixtures of tri- and tetrasubstituted imidazoles are obtained (Scheme I).

The optimization of this method on solid support relies on a number of critical factors, including chemoselectivity, temperature, resin stability, linker strategy, and reagent concentration. It has been observed that the reaction in AcOH does not proceed at temperatures below 60 °C. For certain diones, the reaction proceeds well in organic solvents such as dimethylsulfoxide (DMSO) and dimethylformamide (DMF) using *para*-toluenesulfonic acid as a catalyst. We have found that the reaction proceeds smoothly on Wang resin (*13*) when the reacting aldehydes or amines are attached to the polymer via ester (*see* reference 14 for a review) or ether (*15*) linkages (Scheme II). Examples of these linkers are shown in Scheme III.

The concentration of each reactant in the reaction mixture was found to be crucial for complete conversion. The optimization results are summarized in Table I.

TABLE I. Optimization of Reactant Concentrations for Imidazole Substitution Reactions

[Dione] (M)	[R₄NH₂] (M)	% Conversion
0.289	0.289	25
0.434	0.434	50
0.578	0.578	77
0.723	0.723	99
0.867	0.867	99

For this optimization, the reaction time was fixed at 30 h, the temperature was held at 100 °C, and the concentration of ammonium acetate was 50 mM. It was also observed that increasing the amount of NH_4OAc resulted in the formation of appreciable amounts of the corresponding trisubstituted NH-imidazoles.

In the next study, the reaction time was optimized and the optimal concentration of each component fine-tuned. These results are summarized in Table II. The final optimized synthesis of highly functionalized imidazoles is extremely efficient and suitable for use with a wide range of substrates (*16*). For example, aldehyde resin **3** was prepared and reacted as outlined in Scheme IV. The reaction of Wang resin **1** with carbon tetrabromide in DMF provided polymer **2**, which was reacted with 4-hydroxybenzaldehyde in the presence of triethylamine in DMF to yield resin-bound aldehyde **3**. Compound **3** was reacted with a series of diones and amines (Table III) to produce the corresponding polymer-bound imidazoles **4**. Cleavage with 20% trifluoroacetic acid (TFA) in dichloromethane (CH_2Cl_2) afforded the desired tetrasubstituted imidazoles **5** (Scheme IV).

In a typical synthesis, resin **3** was distributed among the reaction vessels as a suspension in 75:25 CH_2Cl_2:CH_3CN. The reaction block was placed on the synthesis station, and a single dione, single amine, and solution of ammonium acetate in AcOH were delivered to each vessel by the synthesis robot. The reaction block was transferred from the synthesis station to a preheated oven at 100 °C and heated for 15 h. The blocks were then allowed to cool slowly and were transferred to a wash station in which the solvent was drained. All 80 wells were washed five times (each wash involved the use of DMF, then CH_2Cl_2, and then methanol), with a final wash with CH_2Cl_2 before addition of 20% TFA in CH_2Cl_2 (0.6 mL). The reaction block was then transferred to the shaking station, where it was agitated for 20 min. The wells

TABLE II. Reaction Time Optimization for Imidazole Substitution Reactions

[R₄NH₂] (M)	[Dione] (M)	Time (h)	% Conversion
0.723	0.723	30	99
0.723	0.723	15	99
0.723	0.723	<15	<99
0.867	0.867	10	99
1.167	1.167	4	99

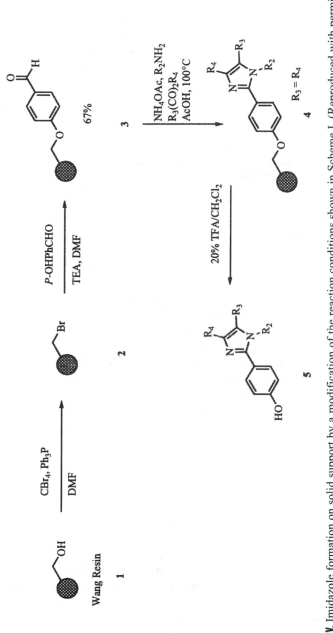

SCHEME IV. Imidazole formation on solid support by a modification of the reaction conditions shown in Scheme I. (Reproduced with permission from reference 16. Copyright 1996 Elsevier Science Ltd.)

TABLE III. Amines and Diones Used in the Imidazole Synthesis

Amines	Diones
n-Butylamine	4,4'-Dimethoxybenzil
n-Hexylamine	3,3'-Dimethoxybenzil
n-Octylamine	4,4'-Dimethylbenzil
2-Fluorophenethylamine	2,2'-Dichlorobenzil
3-Fluorophenethylamine	4,4'-Difluorobenzil
Phenethylamine	Benzil
p-Methylphenylamine	5,5'-Dibromosalicil
Benzylamine	Furil
Ammonia	
Aniline	

NOTE: Each amine was reacted with the diones individually to produce unique single compounds in each well.

were drained into 96-well microtiter plates and rinsed with CH_2Cl_2 (0.6 mL) to ensure complete recovery of the cleaved compounds. The microtiter plate was then placed in a vacuum oven and the solvent was removed at 23 °C and 1 mm Hg. Several daughter plates were prepared (using DMSO as the solvent) for use in mass spectral analysis and biological assays. Eighty distinct compounds were synthesized, and the success of the synthesis was immediately confirmed by high-speed mass spectral analysis (100% success rate) and proton NMR analysis (>90% purity).

All of the aldehydes and amines (including aliphatic and aromatic compounds) were successfully incorporated using this method. The method is limited to diaryl diones, however. Substituted (electron-donating or electron-withdrawing) diones react very smoothly, but aliphatic diones do not. When ammonium acetate was used, the trisubstituted imidazoles were obtained in >95% yield. Tetrasubstituted imidazoles were obtained in >95% yield when a primary amine was used in addition to ammonium acetate. The potential library size using this method is the product of the number of aldehydes, primary amines, and diones available. Ontogen has produced 5000 compounds using this method.

Pentasubstituted Pyrroles

Synthetic Rationale

The synthesis of a pyrrole library (17) represents an extension of our interest in establishing highly efficient methods for the preparation of five-membered ring heterocycles. The 1,3-dipolar cycloaddition of münchnones 7 to various alkynes, 8, has been reported (Scheme V) (18–21). Münchnone 7 is usually generated from acid 9, which is prepared in several synthetic steps from natural amino acids.

We envisioned that the selective hydrolysis of the RNHCO bond in compound 10 would provide acid 9. It has been shown that the condensation between an acid

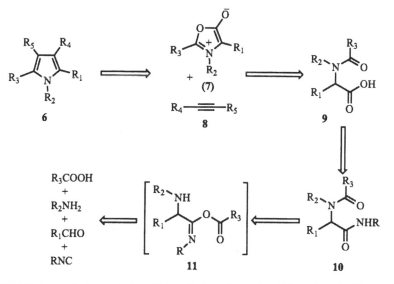

SCHEME V. Retrosynthetic analysis of tetrasubstituted pyrroles. (Reproduced with permission from reference 17. Copyright 1996 Elsevier Science Ltd.)

(R_3COOH), an amine (R_2NH_2), an aldehyde (R_1CHO), and an isocyanide (RNC) provides the N-acyl-N-alkyl-α-aminoamide derivative **10** (*22*). This reaction proceeds via formation of the imine $R_1C=NR_2$, followed by protonation and finally addition of the isocyanide RNC and the acid R_3COOH to give intermediate **11**. This intermediate undergoes acyl transfer to provide **10**. It is well documented that a wide range of sterically and electronically diverse acids, amines, aldehydes, and isocyanides can be used successfully in this four-component condensation reaction (*22*). Our strategy is aimed at taking advantage of this reaction to introduce four substituents in one chemical step. As there are hundreds of commercially available aldehydes, carboxylic acids, and amines, the synthesis of millions of diverse compounds can be achieved. Furthermore, peptide **10** can be readily converted to compounds that are pharmaceutically more useful, such as imidazoles, pyrroles, and oxazoles. The synthesis of a huge number of diverse pyrroles and other small heterocycles is realized in a manner that would be difficult and perhaps impossible to achieve with conventional approaches.

Thus the pyrrole library synthesis involves:

1. The synthesis of pure four-component condensation products on solid support.

2. The use of a convertible isocyanide in the synthesis of N-acyl-N-alkyl-α-aminoamides **10**. This isocyanide will provide an amide that can then be converted to the corresponding acid under mild conditions.

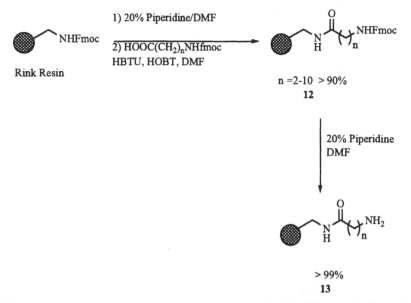

SCHEME VI. Synthesis of the starting amine for pyrrole synthesis on Wang resin. (Reproduced with permission from reference 17. Copyright 1996 Elsevier Science Ltd.)

3. The conversion of N-acyl-N-alkyl-α-amino acids **9** to highly substituted pyrroles on solid support.

Linker Synthesis

The preparation of this library on solid support requires the attachment of one of the inputs (aldehyde, amine, or acid) to the polymer. For example, the Fmoc-protected amine linker HOOC(CH$_2$)$_n$NHFmoc was attached to Rink resin (*23*) using a standard procedure (2-(1H-benzotriazol-1-yl)-1,1,3,3-tetramethyluronium hexafluorophosphate (HBTU) and 1-hydroxybenzotriazole hydrate (HOBT)) (*24, 25*) in DMF to give amide **12**, which was deprotected with 20% piperidine in DMF to afford the desired amine-bound polymer **13** in 99% yield (Scheme VI).

Isocyanide Synthesis

Another necessary component in this pyrrole synthesis is a convertible isocyanide. An ideal isocyanide would be readily available, highly reactive in the four-component condensation reaction, and easily convertible to other derivatives such as acids, esters, amides, and ureas. Initial experiments indicated that BnNC (*26*) reacts with a variety of aldehydes, acids, and amines to provide the corresponding N-acyl-N-alkyl-α-aminobenzylamides. These amide derivatives can be treated with t-Boc$_2$O-DMAP (dimethylaminopyridine) in tetrahydrofuran (THF) followed by LiOH hydrolysis in a 1:1 THF:water mixture to give the corresponding N-acyl-N-alkyl-α-

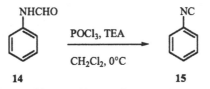

SCHEME VII. Synthesis of phenyl isocyanide, another reagent necessary for the synthesis of highly substituted pyrroles.

amino acids. Unfortunately, both reactions were very slow ($t_{1/2} > 7$ days), with low overall yield (10%). However, the rate and the yield of these reactions improved drastically when BnNC was replaced with PhNC. Phenyl isocyanide, **15**, was prepared in >80% yield by the reaction of formanilide, **14**, with POCl$_3$ and triethylamine (TEA) in CH$_2$Cl$_2$ (Scheme VII) (*27*). Both isocyanides are prepared in one step from commercially available formamides.

Pyrrole Synthesis

Reaction of amine-bound polymer **13** with a series of aldehydes (R$_1$CHO), acids (R$_2$COOH), and phenyl isocyanide in a 1:1:1 mixture of MeOH:CHCl$_3$:pyridine at 65 °C provided the corresponding Ugi product, **16**, attached to the polymer in good overall yield (50–70%) (Scheme VIII). No reaction was observed at room temperature. Pyridine was essential to stabilize the acid-sensitive isocyanide during the course of the reaction. Phenylamides **16** were treated with *t*-Boc$_2$O and DMAP to give *N*-acyl-carbamates **17** with >95% yield. Compounds **17** were hydrolyzed to the corresponding carboxylic acids **18** using aqueous LiOH in THF. Compounds **19** were cleaved from the polymer in quantitative yield using 10% TFA in CH$_2$Cl$_2$ (Scheme VIII).

For the final steps of pyrrole synthesis, polymer-bound acids **18** were subjected to neat acetic anhydride or isobutyl chloroformate and triethylamine in toluene followed by the addition of a series of acetylenic esters to provide pentasubstituted pyrroles **21** (Scheme IX). The reaction proceeds via the formation of münchnone **20**. Münchnone **20** undergoes 1,3-dipolar cycloaddition with a variety of alkynes to yield pyrroles **21**. Subsequent cleavage of the pyrroles from solid support was accomplished using 20% TFA in CH$_2$Cl$_2$ to provide pyrroles **21** and **22** as a mixture of isomers in a ratio of ~4:1. The desired pentasubstituted pyrroles were obtained in good overall yield (35–72% over eight steps). This reflects a process in which each synthetic transformation occurred on average with >85% yield. Some of the compounds prepared by this method are listed in Table IV.

The regiochemistry of these isomers was assigned using nuclear Overhauser enhancement spectroscopy experiments. The ratio of cycloaddition is consistent with results obtained with solution chemistry (*28*). In conclusion, this method of synthesis represents an efficient approach to the preparation of highly functionalized pyrroles. Large numbers of diverse compounds can be produced. Ontogen has produced 2000 compounds using this method.

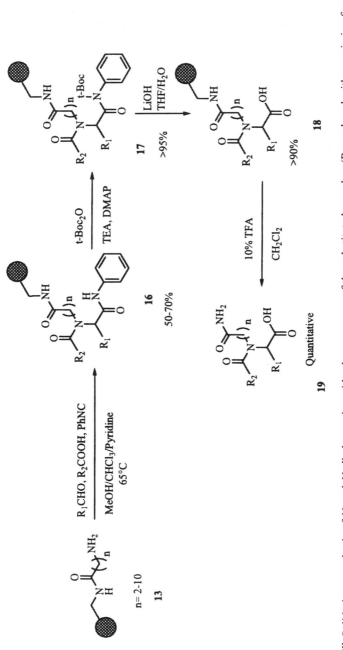

SCHEME VIII. Solid-phase synthesis of *N*-acyl-*N*-alkyl-α-aminoacids, the precursors of the substituted pyrroles. (Reproduced with permission from reference 17. Copyright 1996 Elsevier Science Ltd.)

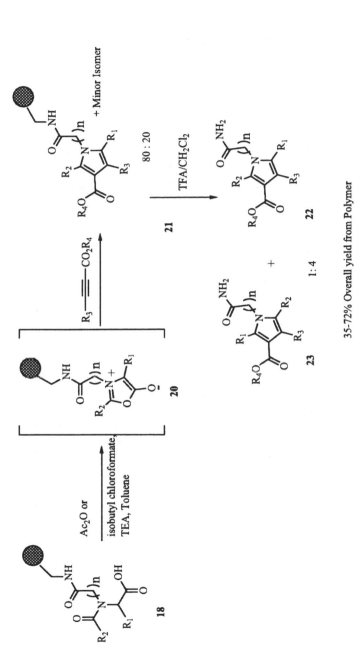

SCHEME IX. The synthesis of pentasubstituted pyrroles on a solid support. (Reproduced with permission from reference 17. Copyright 1996 Elsevier Science Ltd.)

TABLE IV. Pentasubstituted Pyrrole Derivatives Synthesized on Polymer

n	R_1	R_2	R_3	R_4	% Yield
1	Et	4-Br-C_6H_4	CO_2Me	Me	46
2	Et	4-Br-C_6H_4	CO_2H	H	49
2	Et	4-Br-C_6H_4	Et	Et	35
2	i-Pr	$PhCH_2$	CO_2Me	Me	72
2	n-Pr	Ph	CO_2Me	Me	46
2	i-Pr	4-MeO-C_6H_4	CO_2Me	Me	45
2	n-Bu	4-CF_3-C_6H_4	CO_2Me	Me	40

NOTE: The table shows the four variable groups on the final products of the pentasubstituted pyrrole synthesis, and *n* gives the length of the linker (*see* structures in Schemes VIII and IX).

SOURCE: Reprinted with permission from reference 17. Copyright 1996 Elsevier Science Ltd.

Tetrasubstituted Imidazoles (Method 2)

It has been reported that *N*-alkyl-*N*-acyl-α-aminoketones react with ammonium acetate in acetic acid at 100 °C to provide the corresponding imidazoles (*29, 30*). The use of α-ketoaldehydes in a four-component condensation reaction would result in the synthesis of *N*-alkyl-*N*-acyl-α-aminoketone libraries (*31, 32*). These ketone derivatives could then be converted into highly substituted imidazoles with diversity similar to that discussed in the previous section.

Phenylglyoxal is the only commercially available aromatic α-ketoaldehyde of this nature. A convenient method to generate a wide range of substituted aryl-ketoaldehydes was required. It has been reported that aryl bromomethyl ketones can be converted to their corresponding aldehydes by treatment with diethylhydroxy-amine in methanol in quantitative yield (*33–35*). The compounds are used directly after the reaction without any further purification. More than 120 ketoaldehydes can be easily synthesized using this procedure (Scheme X).

Resin for the synthesis was supplied by reacting Wang resin with HOOC-$(CH_2)_n$NHCHO in the presence of diisopropylcarbodiimide (DIC), DMAP, and TEA (*36*) to afford formamide **24** in excellent yield (Scheme XI). The amide was

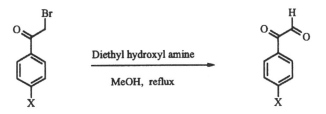

SCHEME X. Synthesis of aryl ketoaldehydes, to be used in a four-component reaction for the synthesis of tetrasubstituted imidazoles.

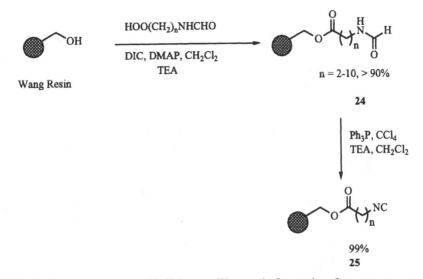

SCHEME XI. Synthesis of isocyanide linkers on Wang resin for use in a four-component reaction for the synthesis of tetrasubstituted imidazoles. (Reproduced with permission from reference 37. Copyright 1996 Elsevier Science Ltd.)

converted to the corresponding isocyanide, **25**, using triphenylphosphine and carbon tetrachloride in the presence of TEA (99%).

Polymer-bound isocyanide **25** was reacted with a series of aromatic ketoaldehydes, amines, and carboxylic acids to provide the corresponding polymer-bound keto diamides, **26**. Ammonium acetate was added and the reactants were heated in acetic acid at 100 °C to provide tetrasubstituted imidazoles **27** attached to the polymer. The desired imidazoles, **28**, were cleaved from the polymer in 44–56% overall yield by treatment with 10% TFA in CH_2Cl_2 (*37*) (Scheme XII).

Examples of imidazoles synthesized by this method are shown in Table V. This method of synthesis yields a large number of novel compounds that cannot be accessed by Method 1. Ontogen has produced 5000 compounds using this method.

NH-Acyl-α-Aminoamides

In the course of our search for novel, small-molecule inhibitors of protein tyrosine phosphatases, a highly efficient method for the synthesis of *N*-acyl-α-aminoamide derivatives, **30**, was needed.

Protein tyrosine phosphatases (PTPases) are a family of enzymes important in signal transduction pathways, including those that control cell growth and differentiation (*3, 4*). These enzymes catalyze the hydrolysis of a phosphorylated tyrosine via the formation of a transient covalent thiophosphate linkage between the cysteine moiety in the enzyme active site and the phosphate group of tyrosine (Scheme XIII) (*38, 39*).

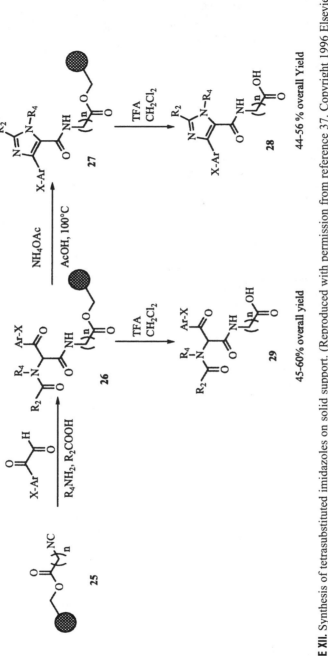

Reproduced with permission from reference 37. Copyright 1996 Elsevier Science Ltd.

SCHEME XII. Synthesis of tetrasubstituted imidazoles on solid support. (Reproduced with permission from reference 37. Copyright 1996 Elsevier Science Ltd.)

TABLE V. Tetrasubstituted Imidazoles Synthesized on Polymer

n	X	R_2	R_4	% Yield
10	H	Ph	i-C_4H_9	45
10	H	4-F-Ph	i-C_4H_9	35
10	H	$PhCH_2$	i-C_4H_9	47
10	H	Ph	4-MeO-Ph	43
10	MeO	Ph	i-C_4H_9	51
2	H	Ph	i-C_4H_9	44
10	F	Ph	i-C_4H_9	49

NOTE: The table shows the variable groups on **28**, the final products of the tetrasubstituted imidazole synthesis, and *n* gives the length of the linker (*see* structures in Schemes XI and XII). X is the identity of the variable position of the arylketoaldehyde (*see* Scheme X).

SOURCE: Reprinted with permission from reference 37. Copyright 1996 Elsevier Science Ltd.

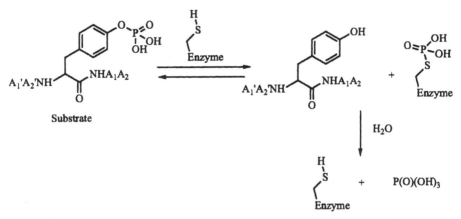

SCHEME XIII. The mechanism of hydrolysis of phosphorylated tyrosine substrates by PTPases.

Our strategy was to design a focused library of small molecules from which substrate-based inhibitors could be identified. The library design was based upon the interaction between PTPases and their substrates, related literature information, and the X-ray structures of some PTPases. The three variables in the library (Scheme XIV) were the tyrosine phosphate group in box 1, the A_1A_2 in box 2, and the $A_1'A_2'$ in box 3. The R_1, R_2, and R_3 groups can be thought of as representing the tyrosine phosphate group or the peptide backbone (A_1A_2, $A_1'A_2'$).

Having defined this library pharmacophore, we turned our attention to the establishment of a highly efficient method for the high-speed synthesis of compounds of formula **30**. This method of synthesis must allow for independent variations of each of the substituents R_1, R_2, and R_3. From the four-component condensation reaction discussed earlier, recall that the reaction between an acid R_1COOH, an aldehyde R_2CHO, an isocyanide R_3NC, and an amine R_4NH_2 generates an *N*-alkyl-*N*-acyl-α-aminoamide of formula **31**. It was anticipated that if the amine com-

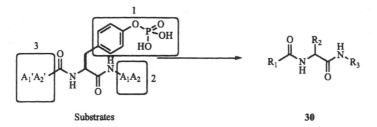

Substrates   **30**

SCHEME XIV. The pharmacophore for the PTPase inhibitors.

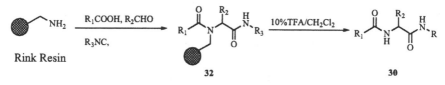

Rink Resin
                                                    32   **30**

SCHEME XV. The synthesis of NH-*N*-acyl-α-aminoamides on Rink resin.

ponent R$_4$NH$_2$ was replaced by ammonia, the corresponding NH-*N*-acyl-α-aminoamides **30** could be synthesized.

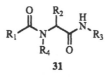

31

When the reaction was carried out with ammonia, several unidentified products were formed in addition to the desired compound **30**. This problem was circumvented by substituting Rink amine resin for ammonia. This provided a solid support, and yielded pure, polymer-bound, four-component condensation product **32** (Scheme XV) (*40*). The desired NH-*N*-acyl-α-aminoamides, **30**, were cleaved from the polymer upon treatment with 10% TFA in CH$_2$Cl$_2$. These compounds were synthesized in good yield with high purity.

Using the OntoBLOCK system described in Method 1, 800 members of this library were synthesized and screened against several protein tyrosine phosphatases (PTPases), and several classes of potent and selective inhibitors of various PTPases were identified. For example, compounds of general formula **33** and bearing a cinnamate group were shown to exhibit low micromolar inhibitory activity against HePTP (*40*), a phosphatase specific to hematopoeitic cells and implicated in acute leukemia (*41, 42*). Some of these inhibitors are shown in Table VI.

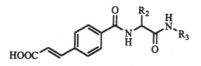

33

TABLE VI. Inhibition of HePTP by Cinnamic Acid Derivatives

R_2	R_3	IC_{50} (μM)
n-Hexyl	n-Butyl	9
n-Hexyl	Tetrabutyl	7.5
n-Hexyl	Cyclohexyl	9
n-Hexyl	Benzyl	6.0
n-Hexyl	CH$_2$COOH	7.5
n-Hexyl	CH$_2$CO$_2$Me	7.2
n-Hexyl	CH$_2$CO$_2$Et	10
Phenyl	n-Butyl	6.7
Phenyl	tert-butyl	4
Phenyl	Cyclohexyl	6.2
Phenyl	Benzyl	3.9
Phenyl	CH$_2$COOH	10.4
Phenyl	CH$_2$CO$_2$Me	20.2
Phenyl	CH$_2$COOEt	9.6
Methyl	Benzyl	7.2
Ethyl	Benzyl	15
n-Propyl	Benzyl	6.1
n-Butyl	Benzyl	6.3

SOURCE: Reprinted from reference 40. Copyright 1995 American Chemical Society.

α-Hydroxyphosphonates

Phosphonate-containing compounds have been shown to have a wide range of biological activities and have been used as enzyme inhibitors for HIV protease (*43*) and renin (*44, 45*). We therefore initiated the development of a highly efficient method for the synthesis of compounds of general formula 38 (Scheme XVI). This combinatorial method was to have two different inputs, R_1 and R_2. Atom X could be an oxygen, sulfur, or nitrogen. We envisioned that attachment of the H-phosphonate group to the polymer, as in 35, would provide a facile entry to compounds of general formula 38 with X = O. Reaction of Wang resin with 2-chloro-4*H*-1,3,2-benzo-dioxaphosphorin-4-one (*46*) in CH$_2$Cl$_2$ and pyridine provided 34, which was hydrolyzed with aqueous NaHCO$_3$ and triethylamine to provide the H-phosphonate 35. (The stock solution of Et$_3$N and NaHCO$_3$ was prepared by adding 1.0 mol of Et$_3$N and 1.0 mol of NaHCO$_3$ to 1 L of H$_2$O. During addition to the resin, the pH of the mixture was kept below 8.5.) Polymer 35 was synthesized in multigram quantities and was used as a building block. Polymer 35 was esterified to 36 through the reaction with pivaloyl chloride followed by displacement with a series of alcohols. Other nucleophiles such as amines and thiols can also be used in this reaction, in which case the X in the final product 38 would be N or S, respectively. Resin 36 was deprotonated with 1,8-diazabicyclo[5.4.0]undec-7-ene (DBU) and reacted with a series of aldehydes to provide the polymer-bound α-hydroxyphosphonate 37. Treatment with 10% TFA in CH$_2$Cl$_2$ produced the desired compounds 38 in good overall yield with > 85% purity.

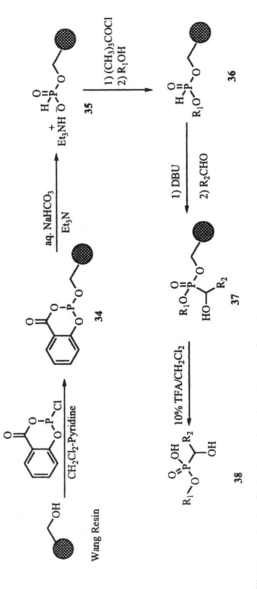

SCHEME XVI. The synthesis of α-hydroxyphosphonates on polymer.

TABLE VII. α-Hydroxyphosphonate Derivatives Synthesized on Polymer

R_1	R_2	% Yield
H	p-F-Ph	86
Et	p-F-Ph	90
PhCH$_2$	n-Pr	87
PhCH$_2$	Ph	95
PhCH$_2$	p-MeO-Ph	77
PhCH$_2$	p-F-Ph	92
PhCH$_2$	1-Naphthyl	81
PhCH$_2$	3-Thiophene	82
2-CF$_3$-PhCH(CH$_3$)	Ph	79

This strategy represents a highly efficient method for the synthesis of phosphonates, thiophosphonates, aminophosphonates, and other related structures. More than 1000 compounds of general formula **38** were synthesized using the Onto-BLOCK System. Some of these compounds are listed in Table VII.

Libraries Produced Using the Diversomer Technology

The following libraries have been synthesized using the Diversomer technology developed at Parke-Davis.

1,4-Benzodiazepine-2,5-dione Library

The 1,4-benzodiazepines have a seven-membered ring, and several of them exhibit a wide range of biological activities (*2*). Milligram quantities of 40 distinct benzodiazepines have been synthesized using the Diversomer pin approach (*47*) (for details see Chapter 10). Five Boc-protected amino acids were attached to Merrifield resin via an ester bond to provide polymer **39** (Scheme XVII). Deprotection of the Boc group using 50% TFA in CH$_2$Cl$_2$ gave amines **40**, which were reacted with eight 2-aminobenzophenone imine derivatives at 60 °C for 24 h to provide imines **41**. Compounds **41** cyclized upon cleavage to provide the desired diazepines **42**, which were isolated in 9–63% yield. Some of these compounds are listed in Table VIII.

This process is an efficient combinatorial method with only two chemical inputs: an amino acid and a benzophenine. Even though the synthesis of only 40 compounds has been reported at this time, one can expect that many more compounds will be synthesized using this methodology.

Hydantoin Library

Eight polymer-bound amino acids of formula **43** were each reacted with five different isocyanates in a two-dimensional array to produce ureas **44**. These were cyclized to provide hydantoins **45** (*47*) upon treatment with 6 M HCl at 85–100 °C

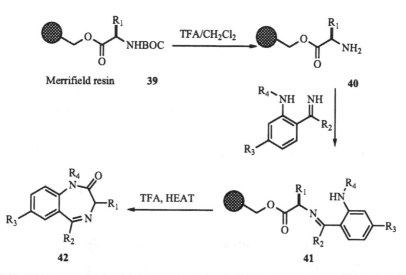

SCHEME XVII. The synthesis of 1,4-benzodiazepine-2,5-diones on polymer. (Reproduced with permission from reference 47. Copyright 1993 National Academy of Sciences.)

TABLE VIII. 1,4-Benzodiazepine-2,5-dione Derivatives Synthesized on Polymer

R_1	R_2	R_3	R_4	% Yield
Me	Ph	H	H	40
Me	Ph	Cl	H	56
Me	4-MeOPh	H	H	34
H	Ph	H	H	44
H	Ph	Cl	Me	20
Benzyl	Ph	H	H	52
3-CH$_2$indolyl	Ph	NO$_2$	H	23
n-Propyl	Ph	H	H	31
n-Propyl	2-Thienyl	H	H	37
n-Propyl	Ph	NO$_2$	H	9

SOURCE: Reprinted with permission from reference 47. Copyright 1993 National Academy of Sciences.

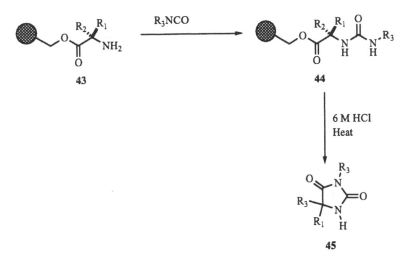

SCHEME XVIII. The synthesis of hydantoins on polymer. (Reproduced with permission from reference 47. Copyright 1993 National Academy of Sciences.)

(Scheme XVIII). These compounds were synthesized and fully characterized in 4–81% yield.

Antibacterial quinolone-based libraries were also synthesized using the Diversomer technology. A total of eight different quinolones were prepared on Wang resin in 4–24% yield with high purity (*48–50*).

Libraries Produced using Custom Robotic Instruments

Researchers at Chiron have reported the synthesis and screening of 5000 oligo-*N*-(substituted) glycine derivatives or "peptoids" (Chart I) using a custom robotic instrument (*51*). Peptoids have been shown to be more resistant to hydrolysis by proteases than natural peptides, and so are considered to be more suitable drug candidates (*52*).

The derivative of Wang resin **46** was reacted with bromoacetic acid in DMF and DIC to provide polymer **47**, which was treated with a series of primary amines (R$_2$NH$_2$) to give **48**. The acylation process with BrCH$_2$COOH was repeated twice. Each acylation was followed with the nucleophilic addition of amines R$_3$NH$_2$ and R$_4$NH$_2$ on the first and second times, respectively. This process yielded trimer peptoid **49**. TFA cleavage afforded the desired peptoids **50** in excellent overall yield (Scheme XIX).

The library, which consisted of 18 pools with 204 peptoid trimers in each mixture, was screened against α1-adrenergic and μ-opiate receptors (*53*). Iterative deconvolution led to the discovery of a potent, selective antagonist of the α1-adren-

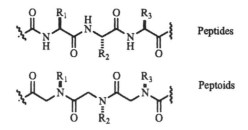

CHART I. The structures of peptides and peptoids.

ergic receptor (CHIR 2279, with a K_i of 5 nM) and the μ-opiate receptor (CHIR 4531, with a K_i of 56 nM) (Chart II).

A combinatorial library consisting of 30,752 compounds has been synthesized on solid support using the Advanced Chemtech 396 multiple peptide synthesizer (54). The chemistry involves amide bond formation reactions to create trimeric compounds using amino acid building blocks. Thirty-two monomeric natural and unnatural amino acids (L and D forms) were used to produce 31 mixtures of 992 compounds each. These mixtures were screened against dog spleen endothelin receptors type A (ETA) (55). Iterative deconvolution led to the discovery of several low-nanomolar endothelin antagonists.

The Advanced Chemtech 357 multiple peptide synthesizer was used to produce unnatural di- and tripeptide-based libraries on Wang resin. This synthesis strategy was based on the formation of imines, which are produced from the reaction of the appropriate polymer-bound amines with benzophenones. This step was followed by alkylation and imine hydrolysis using aqueous HCl in tetrahydrofuran. The terminal polymer-bound amine was N-acylated with Fmoc-Cl. All peptides were transesterified and cleaved from the polymer as the corresponding O-allyl esters (56).

A liquid-dispensing robot was used for the solution-phase preparation of a combinatorial library consisting of 2500 2-aminothiazole-based compounds (57). This chemistry involves a practical and effective condensation reaction between N-substituted thiourea and α-bromoketones to provide a wide range of 2-aminothiazoles in good yields with high purity.

A Zymark R robotics system was used to produce 50 compounds of quinuclidines as squalene synthetase inhibitors (58). Solution-phase automated equipment was also used to create 8000 aminimides, 5000 diaminides, 9000 aminophenoles, and 4400 benzimidazoles (59).

In addition, the Bohdan automated system was used to create 50,000-compound solution-phase libraries of aldols, alcohols, amides, thiocarbonates, thiocarbamates, esters, ureas, ketones, sulfonamides, sulfones, urethanes, and hydrazones (59).

It is expected that custom robotic instruments, with their ability to handle more samples and more diverse reaction conditions, will be useful in the generation of more complex combinatorial libraries.

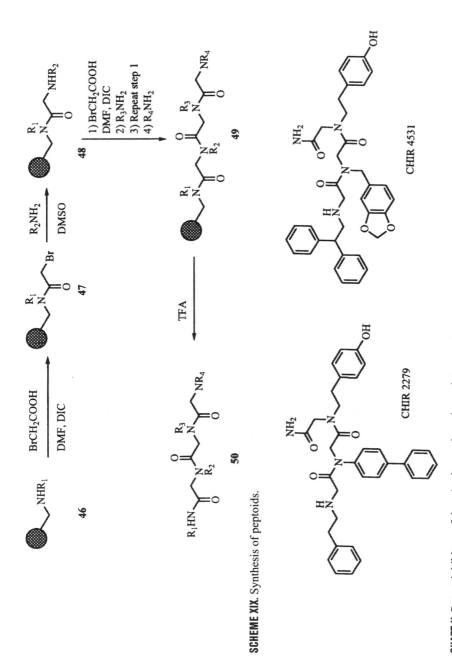

SCHEME XIX. Synthesis of peptoids.

CHART II. Potent inhibitors of the α1-adrenergic and μ-opiate receptors.

Summary

Automated high-speed parallel synthesis can result in the synthesis of thousands of compounds per experiment, and thus satisfies the demands of high-throughput screening currently unmet by the output of traditional medicinal chemistry. These combinatorial syntheses can yield highly functionalized, small-molecule compounds such as tetrasubstituted imidazoles, trisubstituted oxazoles, pentasubstituted pyrroles, α-hydroxyphosphonates, NH–acyl-α-aminoamides, benzodiazepines, hydantoins, and peptoids and, when combined with high-throughput screening, result in the discovery of pharmaceutically attractive and biologically active small molecules. These techniques are likely to shorten drug development time lines and reduce the discovery costs for new therapeutics.

References

1. Mjalli, A. M. M.; Toyonaga, B. E. *Net. Sci.* **1995**, *1*, accessible via WWW at URL *http://www.awod.com/netsci*.
2. Thompson, L. A.; Ellman, J. A. *Chem. Rev.* **1996**, *96*, 555–600, and references therein.
3. Hunter, T. *Biochem. Soc. Trans.* **1996**, *42*, 307–327.
4. Chan, A. C.; Desai, D. M.; Weiss, A. *Annu. Rev. Immunol.* **1994**, *12*, 555–593.
5. Cargill, J. F.; Maiefski, R. R.; Toyonaga, B. E. *International Symposium on Laboratory Automation & Robotics Proceedings (ISLAR '95)*; Zymark Corporation: Boston, MA, 1996; pp 221–234.
6. Lednicer, D.; Mitscher, L. A. *Inorganic Chemistry of Drug Synthesis*; Wiley-Interscience: New York, 1977; pp 226–253.
7. Hill, D. T.; Girard, G. R.; Weinstock, J.; Edwards, R. M.; Weidley, E. F.; Ohlstein, E.; Peishoff, C. E.; Baker, E.; Aiyar, N. *Biorg. Med. Chem. Lett.* **1995**, *5*, 19–24.
8. Judd, D. B.; Dowle, M. D.; Middlemiss, D.; Scopes, D. E. C.; Ross, B. C.; Jack, T. I.; Pass, M.; Tranquillini, E.; Hobson, J. E.; Panchanl, R. A.; Stuart, P. G.; Paton, J. M. S.; Hubbard, T.; Hilditch, A.; Drew, G. M.; Robertson, M. J.; Clark, K. L.; Travers, A.; Hunt, A. A. E.; Polley, J.; Eddershaw, P. J.; Bayliss, M. K.; Manchee, G. R.; Donnelly, M. D.; Walker, D. G.; Richards, S. A. *J. Med. Chem.* **1994**, *37*, 3108–3120.
9. Thompson, S. K.; Murthy, K. H. M.; Zhao, B.; Winborne, E.; Green, D. W.; Fisher, S. M.; DesJarlais, R. L.; Tomaszek, T. A., Jr.; Meek, T. D.; Gleason, J. G.; Abdel-Meguid, S. S. *J. Med. Chem.* **1994**, *37*, 3100–3107.
10. Rotstein, D. M.; Kertesz, D. J.; Walker, K. A. M.; Swinney, D. C. *J. Med. Chem.* **1992**, *35*, 2818–2825.
11. Abdel-Meguid, S. S.; Metcalf, B.W; Carr, T. J.; Demarsh, P.; DesJarlais, R. L.; Fisher, S.; Green, D. W.; Ivanoff, L.; Lambert, D. M.; Murthy, K. H. M.; Petteway, S. R., Jr.; Pitts, W. J.; Tomaszek, T. A., Jr.; Winborne, E.; Zhao, B.; Dreyer, G. B.; Meek. T. D. *Biochemistry* **1994**, *33*, 11671–11677.
12. Krieg, B.; Manecke, G. Z. *Naturforschg* **1967**, *22b*, 132–141.
13. Wang, S. S. *J. Am. Chem. Soc.* **1973**, *95*, 1328–1333.
14. Fields, G. B.; Noble, R. L. *Int. J. Peptide Res.* **1990**, *35*, 161–214.
15. Richter, L. S.; Gadek, T. R. *Tetrahedron Lett.* **1994**, *35*, 4705–4756.
16. Sarshar, S.; Siev, D.; Mjalli, A. M. M. *Tetrahedron Lett.* **1996**, *6*, 853–838.
17. Mjalli, A. M. M.; Sarshar, S.; Baiga, T. J. *Tetrahedron Lett.* **1996**, *17*, 2943–2946.

18. Potts, K. T. In *1,3-Dipolar Cycloaddition Chemistry*; Padwa, A., Ed.; Wiley: New York, 1984; pp 1–82.
19. Padwa, A.; Burgess, E. M.; Gingrich, H. L.; Rouch, D. M. *J. Org. Chem.* **1982**, *47*, 786–791.
20. Brunn, E.; Funke, E.; Gotthardt, H.; Huigsgen, R. *Chem. Ber.* **1971**, *104*, 1562–1572.
21. Huigsgen, R.; Gotthardt, H.; Bayer, H. O. *Chem. Ber.* **1970**, *103*, 2368–2387.
22. Ugi, I.; Dömling, A.; Hörl, W. *Endeavour* **1994**, *18*, 115–122.
23. Rink, H. *Tetrahedron Lett.* **1987**, *28*, 3787–3790.
24. Knorr, R.; Trzeciak, A.; Bannwarth, W.; Gillessen, D. *Tetrahedron Lett.* **1987**, *30*, 1927–1938.
25. Dourtoglou, V.; Gross, B. *Synthesis* **1984**, 572–574.
26. Flynn, D. L.; Zelle, R. E.; Grieco, P. A. *J. Org. Chem.* **1983**, *48*, 2424–2426, and references therein.
27. Walborsky, H. M.; Ronman, P. *J. Org. Chem.* **1978**, *43*, 731–734.
28. Coppola, B. P.; Noe, M. C.; Schwartz, D. J.; Abdon, R. L. II; Trost, B. M. *Tetrahedron Lett.* **1994**, *50*, 93–116.
29. Evans, D. A.; Lundy, K. M. *J. Am. Chem. Soc.* **1992**, *114*, 1495–1496.
30. Schneiders, P.; Heinze, J.; Baumgartel, H. *Chem. Ber.* **1973**, *106*, 2415–2417.
31. Bossio, R.; Marcaccini, S.; Pepino, R. *Synthesis* **1994**, 765–766.
32. Bossio, R.; Marcaccini, S.; Pepino, R. *Liebigs Ann. Chem.* **1991**, 1107–1108.
33. Kornblum, N.; Powers, J.; Anderson, G. J.; Jones, W. J.; Larson, H. O.; Levand, O.; Weaver, W. *J. Am. Chem. Soc.* **1957**, *79*, 6562–6563.
34. Gunn, V. E.; Anselme, J. P. *J. Org. Chem.* **1977**, *42*, 754–755.
35. Mikol, G. J.; Russell, G. A. *Org. Syntheses, Coll.* **1973**, *5*, 937–940.
36. Mathias, L. J. *Synthesis* **1979**, 561–571.
37. Zhang, C.; Moran, E. J.; Woiwode, T. F.; Short, K. M.; Mjalli, A. M. M. *Tetrahedron Lett.* **1996**, *6*, 751–755.
38. Guan, K.; Dixon, E. J. *J. Biol. Chem.* **1991**, *266*, 17026–17030.
39. Cho, H.; Krishnaraj, R.; Kitas, E.; Bannwarth, W.; Walsh, C. T.; Anderson, K. S. *J. Am. Chem. Soc.* **1992**, *114*, 7296–7298.
40. Cao, X.; Moran, E. J.; Siev, D.; Lio, A.; Ohashi, C.; Mjalli, A. M. M. *Bioorg. Med. Chem. Lett.* **1995**, *24*, 2953–2958.
41. Zanke, B.; Suzuki, H.; Kishihara, K.; Mizzen L.; Minden, M.; Pawson, A.; Mak, T. W. *Eur. J. Immunol.* **1992**, *22*, 235–239.
42. Zanke, B.; Squire, J.; Griesser, H.; Henry, M.; Suzuki, H.; Patterson, B.; Minden, M.; Mak, T. W. *Leukemia* **1994**, *8*, 236–244.
43. Stowasser, B.; Budt, K-H.; Jian-Qi, L.; Peyman, A.; Ruppert, D. *Tetrahedron Lett.* **1989**, *33*, 6625–6628.
44. Patel, D. V.; Rielly-Gauvin, K.; Ryono, D. E. *Tetrahedron Lett.* **1990**, *31*, 5587–5590.
45. Patel, D. V.; Rielly-Gauvin, K.; Ryono, D. E. *Tetrahedron Lett.* **1990**, *31*, 5591–5594.
46. Marugg, J. E.; Tromp, M.; Kuyl-Yeheskiely, E.; van der Marel, G. A.; van Boom, J. H. *Tetrahedron Lett.* **1986**, *27*, 2661–2664.
47. DeWitt, S. H.; Kiely, J. S.; Stankovic, C. J.; Schrornder, M. C.; Cody, D. M. R.; Pavia, M. R. *Proc. Natl. Acad. Sci. U.S.A.* **1993**, *90*, 6909–6913.
48. MacDonald, A. A.; DeWitt, S. H.; Hogan, E. M.; Ramage, R. *Tetrahedron Lett.* **1996**, *37*, 4815–4818.
49. DeWitt, S. H.; Czarnik, A. W. *Acc. Chem. Res.* **1996**, *29*, 114–122.
50. MacDonald, A. A.; DeWitt, S. H.; Ramage, R. *Chimia.* **1996**, *50*, 266–271.
51. Zuckermann, R. N.; Martin, E. J.; Spellmeyer, D. C.; Stauber, G. B.; Shoemaker, J. M. K.; Figliozzi, J. M.; Goff, D. A.; Siani, M. A.; Simon, R. J.; Banville, E. G. B.; Wang, L.; Richter, L. S.; Moos, W. H. *J. Med. Chem.* **1994**, *37*, 2678–2685.

52. Miller, S. M.; Simon, R. J.; Zuckermann, R. N.; Kerr, J. M.; Moos, W. H. *Bioorg. Med. Chem. Lett.* **1994**, *4*, 2657–2662.
53. Potenza, M. N.; Graminski, G. F; Lerner, M. R. *Anal. Biochem.* **1992**, *206*, 315–322.
54. Terrett, N. K.; Bojanic, D.; Brown, D.; Bungay, P. J.; Gardner, M.; Gordon, D. W.; Mayers, C. J.; Steele, J. *Tetrahedron Lett.* **1996**, *5*, 917–922.
55. Doherty, A. M. *J. Med. Chem.* **1992**, *35*, 1493–1498.
56. O'Donnell, M. J.; Zhou, C.; Scott, W. L. *J. Am. Chem. Soc.* **1996**, *118*, 6070–6071.
57. Bailey, N.; Dean, A. W.; Judd, D. B.; Middlemiss, D.; Storer, R.; Watson, S. P. *Bioorg. Med. Chem. Lett.* **1996**, *6*, 1409–1414.
58. Main, B. G.; Rudge, D. A.; Strimaitis, J. R.; Hawk, G. L. (Eds.) International Symposium on Laboratory Automation and Robotics (ISLAR), Zymark Corp., Hopkinton, MA.
59. Balkenehohl, F.; Bussche-Hunnefeld, C. von dem; Lansky, A.; Zechel, C. *Angew. Chem. Int. Ed. Engl.* **1996**, *35*, 2288–2337, and references therein.

Information Management and Biological Applications

13

Information Management

Steven M. Muskal

Combinatorial chemistry methodology has significant acceleration and cost-containment potential throughout the compound development process. But before this methodology can realize its full potential, the information generated needs to be adequately captured, organized, distributed, and interpreted. Through adequate information management, combinatorial work flow can be optimized. Once optimized, combinatorial methodology will have a greater impact on reducing the time and costs associated with bringing new chemical substances to market.

Introduction: Business Issues Behind Information Management

As with any business, chemical, agrochemical, and pharmaceutical companies constantly look to maximize their return on investment. Whether their products are reagents, insecticides, or drugs, these companies invest huge sums of money to discover, produce, and market chemical substances. Unfortunately, return on investment significantly decreases as costs, regulations and competitive pressures increase. As return decreases, so does business viability.

In the pharmaceutical industry, for example, it can take more than 10 years to "hand-craft" and evaluate thousands of compounds (*1, 2*), at an average cost of $7500 per compound (*3*), to arrive at a single marketable drug. And despite noticeable advances in technology in the last 20 years, between 1970 and 1993 the average time required to develop a new chemical entity (NCE) from first synthesis to first market actually rose from 7.7 to 11.8 years (*4*). Another staggering study estimated that only 4% of the 512 NCEs launched during 1979–88 achieved sufficient sales to break even (*5*). In today's business climate, the pressure to bring better, more effective chemical substances to market for less money and in less time is so intense that virtually every pharmaceutical company is scrutinizing virtually every aspect of every process throughout compound discovery and development.

357

Even after a pharmaceutical-related compound reaches market, its remaining years of patent protection are frequently challenged with more effective and more marketable drugs. Fully aware of this ever-shrinking window of return, many have been looking to new technologies to decrease the time and cost required to bring new chemical substances to market. To many, combinatorial chemistry is one such technology with great acceleration and cost-containment potential (6–16).

For any set of new technologies to reach its full potential, it is essential that the technologies be adopted and utilized. In many respects, the philosophy of this book has been to provide the fundamentals of combinatorial chemistry so as to further propagate some of the most exciting technological advances of our time. The purpose of this chapter is to highlight the issues associated with combinatorial information management.

The very premise behind combinatorial chemistry is the cost-effective and rapid exploration of compound space during lead discovery and optimization. Without the capture, organization, distribution, and interpretation of combinatorial information, the dream of revolutionizing the chemical, agrochemical, and pharmaceutical industries with combinatorial technologies might not be realized.

Few dispute the importance of maintaining some record of singly produced molecules along with associated physical and biological data. These molecular memoirs not only minimize duplication of effort, but also insure and protect intellectual property. Furthermore, with careful study, this history can suggest future directions for research. Establishing that the maintenance of such records in electronic form is essential to cost-effective research and development might, however, require a more detailed discussion beyond the scope of this text. For now, let us accept the premise that electronic storage and searching of published and proprietary information throughout the research and development process can greatly facilitate company, project, and individual work flow. Efficient work flow translates directly into reduced costs, enhanced productivity, and hence a greater return on investment.

Similarly, one can argue that electronic storage of the data associated with combinatorial libraries will be essential to cost-effective research and development. While the per molecule costs associated with producing a combinatorial library (time and reagents, and assay and hardware costs) can be significantly less than those associated with "one-at-a-time" methods (Table I), if not contained and exploited, these costs might inhibit organizational adoption of new combinatorial programs. Though combinatorial programs will be given certain leeway initially, unless they can produce more successful compounds per dollar and per day, these programs will lose funding to other promising or traditional methods.

Organizations measure the performance of technology against their strategic objectives. Management is constantly faced with justifying the cost of new and ongoing methodologies like combinatorial synthesis. Hard evidence such as reagent dollars, planning duration, building duration, and screening effectiveness are some of the factors that can be used in these calculations. Electronic records of a developing or ongoing combinatorial program have become increasingly important in the evaluation of and the budgeting for a combinatorial effort.

TABLE I. Cost of Producing 11,200 Discrete 1,4-Benzodiazepines

Item	Cost (in dollars)
Methodology development	$200,000
Reagents	$100,000
Disposal	$5,000–10,000
Total cost	>$300,000
Cost per compound	~$30

NOTE: Methodology development cost is based on three "academic"-person years. Industrial costs are expected to be significantly greater.

SOURCE: Data are taken from reference 38.

Just as management requires information for intelligent decision-making, so do individual researchers. If a synthesis is to be successful, the researcher must first ask what has been done before, which building blocks are compatible with a given methodology, which building blocks are commercially available, and what are the expected products of a given sequence of synthesis steps.

The next sections describe the fundamentals of information management. After defining information and some of its sources throughout the compound development process, chemical information management is discussed in more detail. Then, after highlighting relational database technology, current developments in the Internet and Intranet are discussed. Finally, putting everything in context, combinatorial work flow is discussed with emphasis on the areas that require sound information management.

The Nature of Information

What Is Information?

Information is organized data in use. Like technology, information only has value when applied to a particular problem (*17*). Furthermore, information has limited value if it is inaccessible or difficult to use. Proper information management will transform data into information and leverage information into knowledge. The quest for efficient and effective information management has evolved into a whole philosophy and discipline. For our discussion, let us simply state the six key elements to information management:

- Creation
- Capture
- Organization
- Distribution
- Interpretation
- Commercialization

Information is created in every step of every process. But because information is data in use, the first step to full utility begins with information capture. Information is best captured at its source. Who or what created the data, how the data was created, when the data was created, where the data was created, and why the data was created are all fundamental elements in initial data capture. It is important to recognize, however, that if information capture is highly burdensome and involves painful and laborious data entry, it is often avoided. It is for this reason that mechanisms which facilitate rapid and efficient data entry should be implemented from the start.

One common trap that data-capture systems fall into is that their user interfaces are often designed to accommodate an internal data structure or model, with little regard for human interaction. For data that is not automatically collected or machine-read, graphical user interfaces (GUIs) that enable intuitive data entry are of utmost importance. It is easier to design GUIs that fit natural work flow than it is to change work flow to fit a GUI. This is especially true as the number of people using a data-capture system increases.

User interface design is as much an art as a science. The art of user interface design involves the marriage between the requirements of human task-flow and internal program structure. These interfaces not only serve as the primary mechanism to populate internal data structures, but also facilitate distribution and interpretation of information. GUI developers need to understand and work within the confines of human nature as well as software program control. Good GUI developers are tough to come by and typically command a premium.

Organizing data is fundamental to information management. The nature of the process from which information is captured, as well as the current and future use of the captured information, should be considered prior to its organization. Equally important is the technology that supports the efficient, robust, and controlled organization of information. The section on database technology will highlight many of the requirements for database management systems.

The distribution of information within a group or an entire organization is one of the most important dividends paid once information is captured and organized. Critical to the success of any information management strategy are clear, well defined channels of information delivery. These channels require an infrastructure that supports authorization and flexible reporting. Efficient, high-bandwidth distribution of information depends on supporting GUIs that are even more forgiving and easy to use than those used in data capture. Recent developments in Internet and Intranet technology have revolutionized the way in which information is delivered and distributed, opening new doors to nonspecialists typically unfamiliar with the details or history behind the information presented. The sections on the Internet and Intranet detail these developments further.

Perhaps the greatest dividend from effective information management results when knowledge can be derived from the managed information. Since problem-solving efficiency increases with increasing knowledge, our effectiveness as problem-solvers tends to increase with enhanced information access, visualization, and analysis. Problems find solutions when trends are spotted and redundancy is realized. Interpretation is as likely to lead individuals and entire organizations toward success as toward failure, so its importance should never be understated.

Information management can be self-funding, provided that the managed information is commercialized. Commercialization is about buying and selling. People typically think of commercialization of goods and trade, but often overlook the commercialization of information. Information is a form of wealth and power. It is a resource equal in status to other resources like capital, labor, and technology. And like other resources, there is always a cost to gain access. It is for this reason that numerous companies have designed their business models around the commercialization of information.

Without serious incentives, members within any organization are typically unwilling to share information. Fundamental to effective information management, however, is the infrastructure, technology, and desire to work through these "politics" of information sharing. Davenport et al. (*18*) nicely describe the politics of information in the context of five governmental models.

- Technocratic utopianism—Information assets are categorized and modeled. Great reliance is placed on emerging technologies.

- Anarchy—Information is managed by individuals. No information management policy is in place.

- Feudalism—Information is managed by business units or functions. These groups define their own information needs and report only limited information to the overall corporation.

- Monarchy—Information is categorized and organized by the leaders of a firm, who may or may not be willing to share the information.

- Federalism—Information management is based on consensus and negotiation of the key information elements. Reporting structures are agreed upon and shared throughout the organization.

After studying 25 companies, Davenport found that feudalism and utopianism were the most common, yet the least effective. In general, utopianism, anarchy, and feudalism are less effective, whereas monarchy and federalism are the most effective. Making these models explicit, choosing one and sticking to it, will pay off considerably in the end. Maintaining multiple models is confusing and consumes enormous resources.

Information is not free. There are real costs associated with its creation, capture, organization, distribution, and commercialization. Human nature suggests that information will not be shared freely unless individuals, groups, and organizations have incentives to do so.

Published and Proprietary Information

Every phase of drug development, including compound discovery, development, clinical evaluation, regulatory filing, and postmarketing surveillance, depends on some kind of information retrieval. Initially, such retrieval begins with the appraisal of scientific literature and a comprehensive patent survey. Intellectual protection is fundamental to the chemical substance business, and so understanding existing public information is of critical importance. For most corporations, this information is used to determine whether there is a niche to fill in the marketplace and to gauge existing and potential competition. When this published information is coupled with proprietary information, the combination can stimulate new ideas and create new business opportunities.

Numerous on-line searching services provide published information relevant to chemical substance creation and utility. Maizell (*19*) details numerous sources of published chemical information, with an emphasis on on-line services. Popular services are offered from vendors such as Chemical Abstracts Service (CAS), Knight-Ridder, Beilstein, and Derwent (see the appendix for more information).

Because the chemical substance business depends on novelty, the most valuable information is proprietary to an organization. From white-boards to reports to laboratory notebooks, this unpublished information has incredible value. The more information that is captured, the more it is put to use, and hence the greater its value. It is for this reason that most organizations have established or are evolving systems to capture information resulting from day-to-day operations (see the section "Operational Databases and Data Warehousing"). With the changing patent laws and the trend toward electronic investigational new drug (IND) and new drug application (NDA) filings, electronic tracking of compounds through their developmental lifetimes can become a distinct and strategic advantage.

By archiving and providing access to every element of chemical synthesis, biological evaluation (early discovery), process refinement, and scale-up (postdiscovery), immeasurable savings can be accrued by simply avoiding "reinvention of the wheel", or duplication of effort. Imagine, for example, a molecular biologist spending a few months to characterize, clone, and purify a biological target. If another scientist in the same company located in a different laboratory (say across the ocean) just completed the exercise, weeks to months of duplicated effort could be avoided were the information communicated between the respective groups. This potential for duplication of effort can grow exponentially as companies grow to employ many hundreds to thousands of scientists. If the costs associated with these inefficiencies are not contained, they can directly affect the overall viability of the business.

Minimizing duplication of effort begins with effective project management. Even with clear and well-defined objectives, however, the information associated with a particular project becomes localized unless reports are disseminated between projects and teams. Localized information often loses strategic value. Corporate repositories are one means of facilitating and controlling the dissemination of information between disparate projects.

Chemical Information

The electronic management of chemical information typically parallels the processes used in its creation. Historically, molecules have been delicately handcrafted and characterized "one-at-a-time". As a result, most chemical-information management systems were designed to store and retrieve one molecule and its associated data "one-molecule-at-a-time". As organizations grew, and more compounds were synthesized and characterized, these chemical information systems had to support the searching and display (either on computer screens or paper) of larger compound collections. Relational database systems soon became the obvious choice for storage of numerical and textual data collected and derived from chemical substances as they progressed through the developmental cycle.

The management of chemical structural information, in particular, has added another level of complexity. Chemical substances, formulations, and the reactions used to produce each are pictures that chemists have been using for communication for hundreds of years. These language-independent pictures describe not only the identity of the compound but how it was produced. Long before computers were even developed, notecards were used to archive this valuable data, with typically one card per compound.

As the number of compounds grew, these notecards became particularly cumbersome, even with sophisticated, now archaic, index systems. When computer systems became affordable, chemical information was among the first data to be stored in electronic form. Despite International Union of Pure and Applied Chemistry (IUPAC) standards, the naming of chemical substances and the textual descriptions of their transformations can vary quite wildly. Remarkably, this single fact increased (and still increases) the complexity of retrieving stored compounds and associated information.

A picture of a chemical structure (or structures) can resolve naming ambiguities, provided a database system's access control language can support a means of search and retrieval by structure, substructure, or similar structure. Early in the evolution of the computer industry, time-shared systems enabled on-line vendors like CAS, Beilstein, and Derwent to grow and profit immensely by selling access to published chemical information. Other software vendors such as MDL, Tripos, Oxford Molecular, Daylight, and Chemical Design (see the appendix) arose to provide software and databases to support proprietary chemical information management internal to an organization. Today, numerous systems are available to store and search chemical structures, each using its own proprietary method to search by structure.

Relational database systems like Oracle, Sybase, and Informix were initially developed to support business processes. Because the development of chemical substances is a business process, it should come as no surprise that relational systems have found great application in chemical information management at the corporate, project, and even personal level.

With combinatorial methodology, populations or collections of molecules are produced together. Devising methods to electronically store these collections, whether they are produced as mixtures or as arrays of discretes, has become an important and largely unsolved problem. Locating libraries and groups of molecules within libraries (subcollections) is becoming just as important as the earlier task of locating individual molecules within the archives of molecules produced individually. Associating physical and biological performance with the respective samples has become crippling to most present-day systems. And for most, sifting through the enormous volume of information generated by high-throughput approaches has compromised any ability to thoroughly interpret the data.

Library Representation

To adequately manage library information, we must first establish a formal working definition of a library. Consider the following definition:

> A library is a physical, trackable collection of samples resulting from a definable set of processes or reaction steps.

This definition encompasses libraries produced as mixtures or as arrays of discretes. It also includes libraries produced biologically, by fermentation, and even by enzymatic degradation. Another, more mathematical way of stating the same thing is as follows:

> Library = [sample 1, sample 2, sample 3, ...]—a collection of sample(s)
>
> and
>
> {procedure 1, procedure 2, procedure 3, ...}—a set of procedures(s).

In this way, a library can be described as a collection of samples (products), where each sample may be a discrete entity, a mixture, or a pool. The procedure(s) used to create or manufacture the samples within the library have an implied order and contain information that relates to specific samples (e.g., building block A1 incubated with building block B2 in pin (x, y) at temperature T for H hours to produce sample$_i$).

When samples contain one or more chemical substances, consider the following, where "generic" is a picture describing one or more chemical structures within a sample (see Figure 1):

> Sample$_i$ = (generic$_a$, generic$_b$, generic$_c$, ...)—a pool of generic(s).

Questions about samples within a library are highly analogous to questions about substructures within individual molecules. This is why it is often very convenient to represent the expected products within a library with one or more pictures. The generic structure in Figure 1 is one way to describe the structures in a sample with a single picture. The generic structure in Figure 1 can be used to represent a

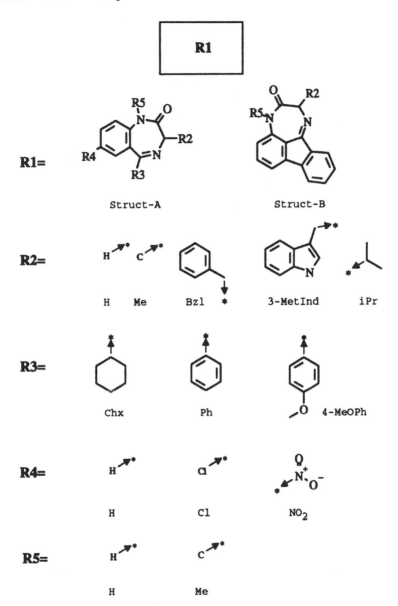

FIGURE 1. A generic structure. Unlike a Markush structure, this structure is unambiguous (i.e., it does not contain functional groups like "alkyl" or "aryl"); in this case it describes exactly 100 specific structures. A generic structure consists of a root and explicitly defined R-group components with their specified connection points (denoted with asterisks). In this example, the root is defined as "R1", where R1 is explicitly defined as one of two possible benzodiazepine classes, Struct-A and Struct-B. In Struct-A, R-groups R2, R3, R4, and R5 are used. In Struct-B, only R-groups R2 and R5 are used. If one were to enumerate each specific structure from this generic structure, there would be 90 Struct-A examples and 10 Struct-B examples, totaling 100 structures.

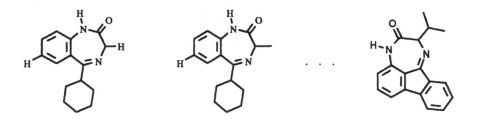

Struct-A.H.Chx.H.H   Struct-A.Me.Chx.H.H   Struct-B.iPr.H

FIGURE 2. Examples of specific structures enumerated from the generic structure in Figure 1. Associated with each specific structure is a name generated from the names beneath each R-component in the generic structure seen in Figure 1.

physical sample of a mixture of 100 compounds (90 of structural class Struct-A and 10 of structural class Struct-B), or as a tool to enumerate (Figure 2) or spell out the 100 specific structures that might be samples themselves.

Finally, using these formalisms, a sublibrary might be viewed as a collection within a library:

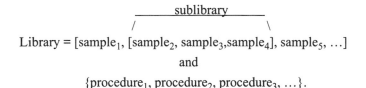

Representing a combinatorial library with the process used to produce it (e.g., in reaction format) has a clear advantage of more closely describing what the chemist actually did. By describing the "recipe", and detailing all the "ingredients", the picture of a library might be supplemented with a description such as that in Figure 3. This type of library representation can serve as a step-by-step procedure that other chemists can follow to reconstruct either the specific library or, if the synthesis procedure can tolerate a different set of building blocks, another library. Furthermore, by supplementing this representation with other more detailed information like temperature gradients, starting material tolerances, and washing cycles, this type of picture can be used to feed robotic systems capable of manufacturing molecules far faster than human chemists.

The process representation of a library seen in Figures 3 and 4 might appear to be more useful than the product-based representation seen in Figure 1. Keep in mind, however, that applying combinatorial methodology toward lead discovery and optimization always requires some form of assessment of properties. In the case of pharmaceutical research, this assessment is typically of biological consequence. The association of chemical samples with their respective physical and biological

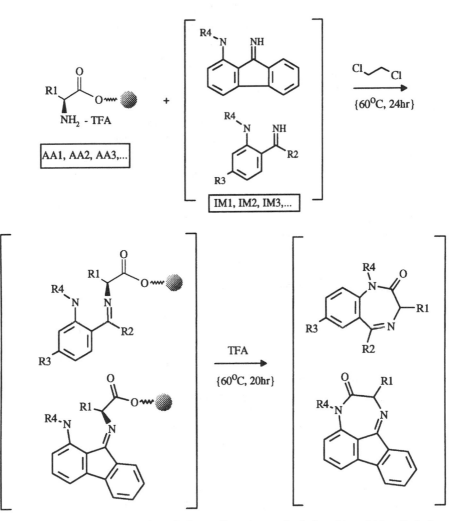

FIGURE 3. A process representation of a benzodiazepine synthesis from bi- and tricyclic imines (*34, 35*). Note that the R-group labels differ slightly from those used in Figure 1, and the explicit R-group members have not been detailed as in Figure 1. Lists of building block amino acids [AA1, AA2, AA3, ...] and 2-aminobenzophenonimines [IM1, IM2, IM3, ...] have been included; in a real process representation, these would be actual names of individual building blocks. Specific procedural details have been left out, but are significant when others wish to reproduce or automate the production of the library. This reaction scheme is one convenient means of representing a library.

properties is therefore paramount. This is why a product-based representation is so important.

Can the product-based representation be generated from the process representation? In many cases, the answer is yes. In some cases, especially as the complexity of the chemistry grows, the answer is not so clear. Consider the reaction methodology developed by Murphy et al. (20) (Figure 4). Upon first inspection, it might appear straightforward to generate all products from this scheme given the type of building blocks used. It should become clear upon closer inspection, however, that a direct association cannot be made between every reactant atom and the corresponding atom in the final product, so the transformation (which most synthetic chemists can follow) will cause some difficulty to a computer. Furthermore, not all chemists draw their reaction transformations in a standard way, often skipping trivial yet fully understood steps. For a computer program to properly generate the expected final products from a series of transformations and lists of building blocks, an exhaustive "entry mechanism" is necessary. But requiring chemists to specify seemingly obvious information might be too burdensome and thus threaten the adoption of the procedure. It is for this reason that explicit generic structure representation for library products (as seen in Figure 1) might be necessary.

Searching Library Information for Overlap

Just as locating all or part of a molecule within a set of individually produced molecules became important as compound inventories grew, so will the location of all or part of a library within the electronic archives. As we have seen in the search and retrieval of individual molecules, ambiguous naming strategies (in this case of libraries) has challenged information retrieval. Perhaps for this reason alone, locating a library (or sublibrary) by locating one or more samples within it might become of critical importance. Because the samples of a library are typically produced (or acquired) together as a collection, they share certain characteristics. For example, some of the samples might contain a certain feature as a result of an impurity, a new polymer support, or a new linker strategy. It is for this reason that new search operations need to be derived to facilitate collection-based querying.

The ability to search in this manner is very important to resolve questions of "library overlap", which have real cost implications. Should, for example, a researcher produce a particular library when there is a chance that another researcher has already produced a very similar library? And even more quantifiable, should a collection of compounds be acquired or shared between collaborators when perhaps the majority have already been produced and currently exist in inventory?

When discussing overlap, it is important to distinguish between sample overlap and structural overlap. The distinction between sample and structure is subtle, but important. Here, a sample as described above can be a single compound or a mixture of compounds. A sample is tangible, testable, and trackable. A structure, on the other hand, is within a sample. In the case of mixtures, a structure is but one entity within a pool of entities. A structure within a mixture is not individually

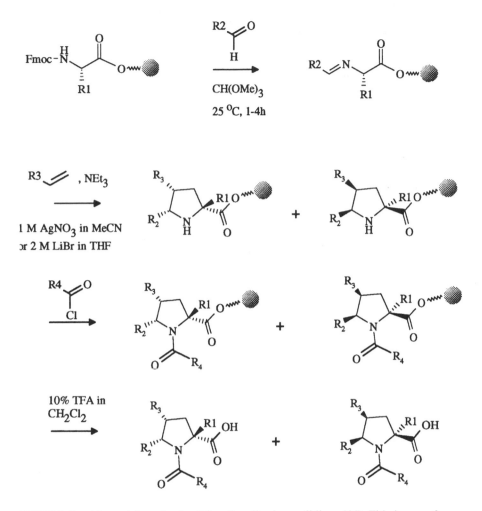

FIGURE 4. Combinatorial synthesis of functionalized pyrrolidines (*20*). This is one of many examples that illustrate the difficulty of enumerating expected products when given a reaction transformation and starting materials. As an exercise, try locating the contributions from the amino acids, aldehydes, olefins, and acid chlorides in the pyrrolidine products. The imine formation, olefin cycloaddition, and stereochemistry preference throughout the synthesis each add elements of complexity that challenge computer programs that generate (enumerate) products from starting materials.

testable (i.e., the mixture is tested as a whole). The consequences of this difference can be seen in the different operations that are possible when comparing samples or structures, as shown in Tables II and III.

Representing a library with one or more generic structures has a distinct advantage in that it facilitates obtaining answers to some of the questions that might be asked. We have seen how generic structures can be used to represent mixtures

TABLE II. Useful Operations When Comparing Samples

		Hit	
		Specific$_h$	*Generic$_h$*
Query	*Specific$_q$*	Exact: [Specific$_q$ = Specific$_h$] (i.e., discretes are the same)	N.A. (discrete ≠ mixture)
		Similar [Sim(Specific$_q$, Specific$_h$) ≥ Sim Val] (i.e., discretes are similar)	
	Generic$_q$	N.A. (mixture ≠ discrete)	Exact: [Generic$_q$ = Generic$_h$] (i.e., discretes are the same)

NOTE: A sample is a physical, trackable entity that might contain a pure compound, a mixture of related compounds, or a pool of disparate compounds. "Query" (left) is what a user would pose to a computer system, while "Hit" (top) is the expected return value(s). "Generic" is a generic structure representing a mixture of chemical substances. "Specific" is a single structure representing one discrete chemical substance. In sample comparisons, one is really asking which of the samples is the same. Therefore, one should neither see discrete molecule hits with mixture queries (lower left quadrant), nor mixture hits with discrete molecule queries (upper right quadrant).

Abbreviation: N.A., not applicable.

and how they can be used as tools to generate submixtures as well as discrete molecules. However, because combinatorial methodologies often produce mixtures of compounds not easily described with generic structures, or can produce unknown or side-products, other library representations will be necessary.

Information capture is one of the most critical aspects of information management. Registering library information needs to be fluid and intuitive, yet must also pay dividends on the time spent archiving the information. Pictures that are not easily searchable, or not easily resolved into products that can be associated with a property or activity, can be useful to those doing the same type of work but can be restrictive downstream. A picture is worth a thousand words, but, unless the words are in a language that is understood or translatable, the words lose utility and simply become noise.

Database Technology

Database Management Systems

All database management systems (DBMSs) store and manipulate information. A good DBMS must reliably manage large quantities of data in a multiuser environment and support concurrent data access. All of this should be managed while delivering high performance to the people and applications using the database. A good

TABLE III. Useful Operations When Comparing Structures Within Samples

		Hit	
		$Specific_h$	$Generic_h$
Query	$Specific_q$	Exact: [$Specific_q = Specific_h$] SubStructure: [$Specific_q$ within $Specific_h$] Similar: [$Sim(Specific_q, Specific_h) \geq$ Sim Val]	Contained: [$Specific_q = (Specific_h$ in $Generic_h)$] SubStructure: [$Specific_q$ within $(Specific_h$ in $Generic_h)$] Similar: [$Sim(Specific_q, (Specific_h$ in $Generic_h)) \geq$ Sim Val]
	$Generic_q$	Exact: [$(Specific_q$ in $Generic_q) = Specific_h$] Contained: [$(Specific_q$ in $Generic_q)$ within $Specific_h$] Similar: [$Sim((Specific_q$ in $Generic_q), Specific_h) \geq$ SimVal]	Exact: [$Generic_q = Generic_h$] Specifics in common: [$(Specific_q$ in $Generic_q) = (Specific_h$ in $Generic_h)$] Similar: [$Sim((Specific_q$ in $Generic_q), (Specific_h$ in $Generic_h)) \geq$ SimVal]

NOTE: As in Table II, "Query" (left) is what a user would pose to a computer system, while "Hit" (top) is the expected return value(s). "Generic" is a generic structure representing a mixture of chemical substances. "Specific" is a single structure representing one discrete chemical substance. In structure comparisons, operations operate on structures within samples. Here, one is really asking questions such as which of the structures within a given sample are found in another sample, which of the structures within a given sample are found within structures (i.e., as substructures) in another sample, and which structures in a given sample are similar to those in another sample. Most commercially available chemical information systems adequately address the operations in the upper left quadrant (see MDL, Daylight, Tripos, Chemical Design, and Oxford Molecular), but the operations in the three other quadrants have yet to be adequately implemented.

DBMS must also be secure to unauthorized access and provide efficient solutions for failure recovery. More specifically, a DBMS:

- provides a repository for the storage of data;

- provides concurrent user access for information creation, reporting, and update;

- provides efficient security mechanisms to restrict activities on sensitive data;

- provides mechanisms to ensure data integrity;

- provides a data access language that conforms to industry standards; and

- enables the segmentation of operations between one or more servers and many clients that have no shared storage capability.

Databases are the foundations for information storage and retrieval. As databases may contain hundreds of gigabytes to terabytes of data, a DBMS needs to fully control space usage within one or more hardware devices. A multiuser database needs to provide support for large numbers of concurrent users, each potentially executing different database applications on the same set of data. The DBMS must minimize data contention, and guarantee data concurrency.

Most users become frustrated very quickly with applications that perform poorly. A DBMS must therefore maintain the functions just described without slowing down the speed of processing. In some companies, database applications work continuously, 24 hours a day, with no down time. This means that normal systems operations such as database backup and partial computer system failures cannot interrupt database usage. A good DBMS must be capable of controlling the selective availability of data, either at the database level, subdatabase level, or both.

A database system should adhere to the industry-accepted standards set forth for the data access language, operating systems, user interfaces, and network communication protocols. By complying with these standards, the database system is truly an "open" system that protects customers' investments.

To protect against unauthorized database access and use, a DBMS must provide fail-safe security features that allow the limitation and monitoring of data access. To support a large number of users, the DBMS should provide security features that make it easy to manage even the most complex design for data access.

A DBMS needs to enforce data integrity or "business rules" that dictate the standards for acceptable data. By allowing the DBMS to define and enforce such rules at the database level, the drawbacks of having to code and manage such checks within database applications are eliminated.

To take full advantage of a given computer system or network, the DBMS must allow for the separation of processing between the database server and the client application programs. All responsibilities of shared data management can be processed by the computer running the DBMS, while the workstations running the database applications can concentrate on the interpretation and display of data.

As computer environments are connected via networks, a DBMS must be able to combine the data physically located on different computers into one logical database that can be accessed by all network users. A distributed database must supply the same degree of user transparency and data consistency as nondistributed systems, yet still support the advantages of local database management.

Relational Databases

Database management systems have evolved from hierarchical to network to relational models. Today, the most widely accepted database model is the relational model. Therefore, we will focus on relational database management as its application is universal throughout compound discovery and development processes. The relational approach to database management is based on a mathematical model with foundations in relational algebra and calculus. An informal, working definition of a relational database management system (RDBMS) might include the following:

- All information is represented in table form; and

- Three major operations including *selection, projection* and *join* are used to specify different views of data.

The father of the relational model, E. F. Codd, developed a detailed list of criteria that a relational model must meet. A comprehensive explanation of "Codd's rules" is beyond the scope of this text, but, as a brief summary of Codd's test for relational systems, the RDBMS should:

1. represent all information in the database as tables;

2. keep the logical representation of data independent from its physical storage characteristics;

3. use one high-level language for structuring, querying, and changing information in the database (really, any number of database languages fit this bill, though in practice SQL (structured query language) is the language of RDBMS);

4. support the main relational operations (selection, projection, and join) as well as set operations such as union, intersection, difference ,and division;

5. support alternative ways for users to look at data in tables;

6. provide a method for differentiating between unknown values (nulls) and zero or blank; and

7. support mechanisms for integrity, authorization, transactions, and recovery.

Numerous vendors supply relational database technology meeting Codd's rules. Some of the more popular vendors include Oracle, Sybase, and Informix (see the appendix).

Operational Databases and Data Warehousing

Most information systems developed over the past decade store data in a form that supports the operations of a group, department, or organization. The data are usually organized or databased in a manner that is suitable for access and modification by the respective operational system. This database is referred to as an operational database. Examples include data collection systems, inventory systems, and document (report) tracking systems.

By nature, operational databases are unsuitable for decision support because the data is usually normalized (i.e., stored without any redundancies, often using codes rather than descriptions). To most end-users, searching or querying normalized data is counterintuitive and cumbersome. And because operational databases typically reside on the computer systems that are dedicated to the capture of the operational data, decision support often suffers from the diversion of computing power to these other tasks.

A "data warehouse" is a higher level database that is dedicated to providing data in a form suitable for access by end-users. Data is structured in a form that is optimized toward the types of questions people typically ask. The data is often denormalized, introducing redundancies to avoid having to refer to several disparate tables of files to satisfy certain types of queries. Codes used in operational databases for efficiency are pretranslated so that they make more sense to end-users.

In a data warehouse, data is cleaned or scrubbed to remove flawed data that could result in an erroneous analysis by the end-user. A data warehouse typically contains precalculated summaries of information and might contain historical data (legacy data) not collected by the current operational system. In general, a warehouse database is more accurate, refined, and larger than an operational database. In short, operational databases are designed to efficiently capture information, and warehouses are designed to enable efficient exploitation of the information.

The Internet and World Wide Web

The Internet consists of a group of computers connected together using standard protocols to exchange information. Today's Internet is made up of a loose collection of interconnecting commercial and noncommercial networks, including on-line information services. Servers are scattered around the world, linked to the Internet on a variety of high, and low, capacity paths. In addition to electronic mail and file exchange, the Internet supports "Web browsing". The World Wide Web (WWW) refers to those servers connected to the Internet that offer graphical pages of information. When connected to one of those servers, a screen of information (graphics and text) can appear with hyperlinks, which facilitate navigation from one page to the next.

The WWW has been around since 1992. The first prototype was developed in 1990 at CERN, the European laboratory for particle physics. In 1993, the alpha version of MOSAIC (the first graphical web browser) was developed, and roughly 130 Web servers were in use. In 1995, it was estimated that over 13,000 Web servers were displaying over 10 million Web pages.

In just a few short years, the World Wide Web has become a part of daily life. Every day, millions of people all over the world browse the Web for news, product information, entertainment, and even hard data. Perhaps the greatest criticism of the WWW, besides frequent "traffic jams", is that there is an incredible amount of unedited information that anyone with a personal computer and modem can submit.

Indeed, the ease with which companies and individuals can "publish" information on the Internet is changing the whole idea of what it means to publish.

The Internet has become a reasonably reliable communications channel, though there have been occasional breakdowns and security problems. On November 2, 1988, for example, thousands of computers connected to the network began to slow down. Robert Morris, Jr., a 23-year-old graduate student at Cornell University, was later convicted of violating the 1986 Computer Fraud and Abuse Act, a federal offense, for designing and unleashing a "worm", a mischievous computer program that spreads from one computer to another, "replicating" as it goes. Morris was sentenced to three years of probation, fined $10,000, and required to work over 400 hours of community service.

Despite these occasional security problems, the Internet provides millions of people with a means of exchanging electronic mail, documents, programs, and other data. In principle, the same cost is involved in requesting data from a server a mile away as in requesting data from the other side of the world. As the foundation of the Internet consists of a collection of leased lines, or special-purpose telephone lines dedicated to calls between two or more sites, the Internet can support an enormous volume of traffic quite inexpensively.

Most businesses use a T-1 line to connect to the Internet. Subscribers pay a local phone company a monthly charge for the T-1 line, which moves their data to the nearest Internet access point, and a flat rate (~$20,000 a year) to connect the company to the Internet. In the United States, there are currently a handful of companies that provide these Internet connections. The yearly charge is based on the capacity of the connection (i.e., the size of the "on ramp") and covers all Internet usage whether constant or infrequent, and whether the traffic travels a few miles or a few thousand. Recently, these Internet providers have been offering "home access" subscriptions to personal computer users at a low monthly rate. On-line services like America On-line (AOL) and CompuServe have also been providing their subscribers with Internet access. The sum of these Internet payments funds the entire Internet network.

Numerous scientific resources can be located on the Web. In addition to the information provided in the appendix, the following URLs might be of interest.

- **http://www.rpi.edu:80/dept/chem/cheminfo/chemres.html**
  The Rensselaer "Internet Chemistry Resources list" is a selective list of resources related to chemistry and associated fields. Resources are organized according to either Internet service or subject.

- **http://galaxy.einet.net:8000/galaxy/Science/Chemistry.html**
  This is the chemistry site within Galaxy's guide of worldwide information and services (provided by TradeWave Corporation). Topics include analytical chemistry, electrochemistry/spectroscopy, biochemistry, inorganic chemistry, organic chemistry, and physical chemistry.

- **http://www.ch.ic.ac.uk/infobahn/Paper23/**
  Chemical information resources on the World Wide Web.

- **http://info.cas.org/ON-LINE/CATALOG/descript.html**
  The STN Database Catalog provides a concise, yet comprehensive background of each database on STN International.

- **http://www.ch.ic.ac.uk/chemime/**
  The Chemical MIME Home Page. This page provides a concise entry point for various information sources relating to chemical MIME.

- **http://chem.leeds.ac.uk/papers/html/chem-com/chem-com.html**
  "Chemical applications of the World-Wide-Web system."

Intranets

Businesses of every kind have found that the Internet's World Wide Web is a great way to get information to customers, partners, or investors. Now, they are finding that using the same technology internally (on Intranets) can improve information capture and exchange dramatically within an organization. Intranets exploit the same inexpensive approach as the Web for capture and delivery of electronic information, using the same Web browsers developed for the Internet. Numerous Web browsers (e.g., Netscape Navigator, Lynx, NCSA Mosaic and Spyglass Enhanced Mosaics) are available. All sorts of documents including internal phone books, procedure manuals, training materials and requisition forms can be converted to HTML format (Hypertext Markup Language), centralized and delivered cross-platform (e.g., to both PCs and Macintoshes) through Web browsers. By centralizing this information, changes and addenda can be made centrally and delivered upon request. In addition, using "fire walls", people can still venture out onto the Internet and utilize the information services on the WWW without allowing unauthorized users to enter the company's private Intranet.

Just as the Internet connected millions of computers around the world together, Intranets are connecting islands of information within corporations and sparking unprecedented collaborations. For many organizations, Intranets have broken down the walls within corporations. By presenting information in the same way to every computer, Intranets can bring together computers, software, and databases that dot the corporate landscape into a single system that enables employees the opportunity to locate information wherever it resides. This universal reach has great potential to dramatically enhance company, project, and individual work flows.

Sales of software to run Intranet servers in 1995 were $476 million and are expected to exceed $4 billion in 1997 (*21*). Estimates for 1998 exceed $8 billion, four times the size of the Internet server business. Indeed, as Sun CEO Scott G. McNealy said, "Intranets are huge."

Intranets are not the magic bullet for every software ill, however. Today, nothing on the Web can replace complex business programs that have been refined over many years. Some companies may still opt for the unimpeachable security of conventional programs. But for most organizations, Intranets represent a sound mechanism for company-wide capture and distribution of information.

Combinatorial Work Flow

Combinatorial research is a collaborative effort, requiring the input of several disciplines within the organization. Now that we have discussed the business and technologies behind information management, let us turn to the activities or work flow typical of most combinatorial programs. The four major stages in combinatorial work flow are:

- planning—methodology development, library design, and reaction rehearsal;
- library synthesis—organic synthesis, complete with quality control (QC);
- screening—evaluation of properties (such as biological or physical); and
- interpretation—information extraction and assimilation.

These four stages in combinatorial work flow are depicted in Figure 5. With efficient, unobtrusive mechanisms of archiving, information can be siphoned from the process and utilized in subsequent research efforts.

Planning

The planning of a combinatorial library synthesis is similar to the planning of a single compound synthesis, but with an added dimension of complexity. As the paradigm shifts from single compound synthesis to the synthesis of a collection or population of molecules, each issue underlying conceptual design and synthesis strategy takes on this new dimension.

For example, during retrosynthetic planning of single molecules, chemists begin with a target and work backwards to readily available starting materials. When planning a combinatorial synthesis, the analogous "retrocombinatorial analysis" (*14*) requires not only the traditional retrosynthetic dissections (Figure 6), but must also answers questions about building block availability for different modes of synthesis, and the scope and reliability of the necessary reactions. Figure 7 shows information flow throughout the planning process.

Databases of published and proprietary reactions are extremely useful in exploring the different possible transformations (i.e., those that are sufficiently robust, have reasonable yields, and can accommodate multiple components), as are databases of available (commercial or inventory) building blocks. The more closely integrated these databases of transformations and available reagents are, the more efficient the retrocombinatorial analysis.

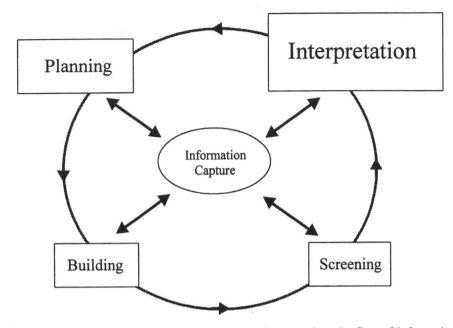

FIGURE 5. Sample work flow. Combinatorial chemistry requires the flow of information through each of the stages (planning, building, screening, and interpretation). Central to this paradigm is information capture. The time spent on each task varies from months to years for interpretation, to months for planning, weeks for building and screening, and days for information capture.

Throughout the planning process, trial synthetic runs are often necessary to work out or "rehearse" a combinatorial methodology. These efforts produce enormous quantities of synthetic methodology information that, when verified, can be shared with other project team members and eventually the entire organization. Efficient capture of this information in electronic form can represent one big step toward electronic laboratory notebooks.

Another very important element in the planning process is building block cost and availability. Indeed, one of the largest bottlenecks in the construction of combinatorial libraries is the selection, location, and purchase or synthesis of basic building blocks. Chemical information systems that provide access to up-to-date inventory and commercially available building blocks can be very useful tools for round-up and ordering.

Once the availability and cost of building blocks have been assessed, a critical activity in the design of combinatorial libraries is the rational selection of building blocks from the available choices of compounds of a particular class. Consider the synthesis of the 1,4-benzodiazepine library discussed in Figure 8. This three-component synthesis, involving two 2-aminobenzophenones, 12 amino acids, and 8 alkylating agents, produced 192 structurally diverse 1,4-benzodiazepines. While the number of commercially available 2-aminobenzophenones is relatively small, the

"Traditional" Retro-Synthetic Analysis

$$R_{eagent_M} \cdots R_{eagent_1} \longleftarrow P_{recursor_N} \leftrightarrow P_{recursor_1} \longleftarrow T_{arget}$$

1) What is the most "efficient" route?
2) How can I maximize purity/yield?
3) Where can I obtain reagents and precursors?

Vs.

Retro-Combinatorial Analysis

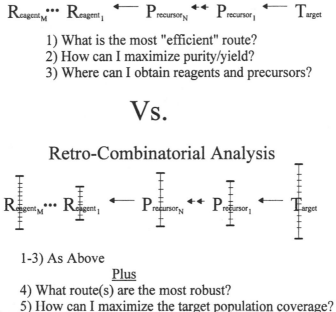

1-3) As Above

Plus

4) What route(s) are the most robust?
5) How can I maximize the target population coverage?
6) Where can I obtain the cheapest/most diverse reagents?

FIGURE 6. A comparison of "traditional" retrosynthetic analysis and retrocombinatorial analysis. Another dimension is added in retrocombinatorial analysis as the paradigm shifts from synthesizing single targets to synthesizing a population of targets. Populations of targets (and hence populations of precursors and reagents) are denoted by hash marks.

number of commercially available Fmoc-protected amino acids is quite large. And, relatively speaking, the number of commercially available alkylating agents is enormous. The question then remains, which reagents should a chemist choose when confronted with lists of hundreds or thousands?

Rational building block choices are both necessary and crucial. Typically, when selecting building blocks the researcher is attempting to:

- squeeze or expand the library into a manageable format given the hardware, manpower, and reagent-cost constraints;

- remove overt structural redundancy, i.e., similar or identical members that are either in the same or different libraries; and

- improve the survey of "compound space", so that it is broader for lead finding or more focused for optimization.

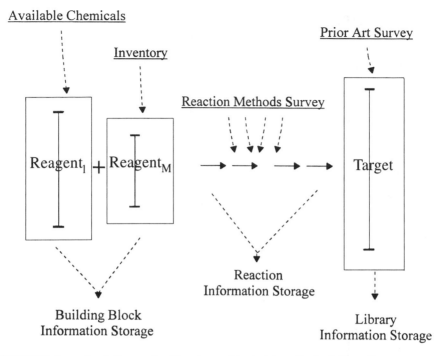

FIGURE 7. Information inflow and outflow in combinatorial synthesis. Information sources useful throughout the planning of a combinatorial library are shown above the reaction scheme. Databases of reaction methodology, reagent availability, and relevant prior art will prove extremely useful in planning combinatorial synthesis. Data capture is shown below the reaction scheme.

The compelling nature of these problems has driven many computational chemists into the "diversity assessment" playing field. But even before sophisticated methods of diversity assessment are considered, chemists can do a substantial amount of building block selection simply based on inspection. Reducing large lists of possible starting materials with molecular weight constraints and with simple structure-based queries are just some of the simple "search and reduce" techniques that can be applied to building block selection.

Building block selections (or additions) based on instinct, intuition, and experience are not, however, always easy to formalize in terms of structure- or database-based queries. In these cases, brute-force inspection may be required. Graphical interfaces that display large numbers of structures, perhaps in a grid format and grouped (e.g., based on structural similarity), can become useful devices for rapid survey and exploration.

In some cases, even after building block lists are pruned on the basis of synthetic compatibility, a researcher may still face the daunting task of having to select from large lists of possibilities. In principle, one could computationally generate

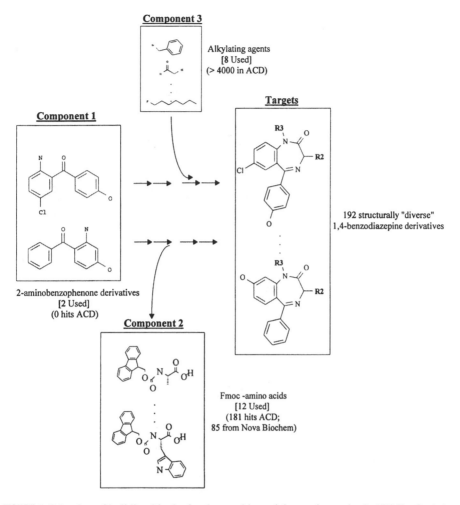

FIGURE 8. Selection of building blocks for the combinatorial organic synthesis (COS) of a 1,4-benzodiazepine library (*36, 37*). This three-component synthesis, using two 2-aminobenzophenones, 12 amino acids, and 8 alkylating agents, produced a library of 192 structurally diverse 1,4-benzodiazepine derivatives containing a variety of chemical functionalities including amide, carboxylic acid, amine, phenol, and indole groups.

Abbreviation: ACD, Available Chemicals Database

every possible combination of building blocks (i.e., generate a "virtual library"), score each molecule by its likelihood of being active, and select those building blocks used to build the highest scoring molecules (Figure 9). In some cases, however, the combinations of building blocks might result in enormous libraries even after feasibility pruning. To exhaustively build and computationally evaluate these enormous virtual libraries will be prohibitively time consuming even for today's

Feasible Starting Materials

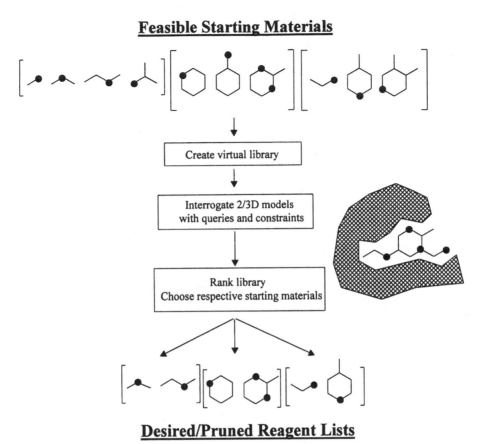

FIGURE 9. Constrained reagent selection. Given a synthetic methodology and an opportunity to select from feasible reagent lists, a process that builds a virtual library of 2D or 3D models and interrogates each structure with queries (e.g., calculated physicochemical properties and conformational flexibility) can assist the reagent selection process provided the members in the virtual library retain information about their respective building blocks.

most powerful computers. Furthermore, there is no guarantee that such computational work is even justified, especially if the computational assessment of each molecule is average at best. To address some of these issues, two general approaches have been reported in the literature (see Chapter 2 for more details). One technique generates and evaluates "on the fly" (*22*), and the other "optimally" selects building blocks on the basis of their maximal dissimilarity (*23*).

The Sheridan and Kearsley approach (*22*) uses a genetic algorithm to explore and evolve a subset of computer-generated molecules from an initial set of building blocks (derived from a feasible set of starting materials). By using a mathematical model to assess each molecule's likelihood of possessing a particular biological activity, this approach "evolves" a virtual library by building and computationally assessing molecules as it runs. The process still requires a great deal of computing power, and depends heavily on the "selection function" used to evaluate each molecule.

The other general approach, developed by Martin et al. (*23*), concentrates on selecting maximally dissimilar building blocks from the feasible set of building blocks prior to the generation of any complete molecules. The primary goal of this technique is to design a library of maximal diversity, and Martin et al. have worked with the premise that, if the building blocks are diverse, so too will be the library. While numerous, well-established, chemically and biologically significant descriptors are used in this elegant D-optimal design strategy, some argue that full molecular diversity is still not achieved. Others argue that the designed libraries, while diverse, are not necessarily biologically focused. Perhaps a third general approach, combining the two strategies, will evolve to better address the issues associated with computer-assisted library design.

In addition to selecting from existing or commercially available building blocks, much effort is dedicated to the design and synthesis of novel building blocks. Nature spent millions of years evolving a set of 20 common amino acids that (together with a few prosthetic groups) can be used to construct a large variety of functionally diverse proteins (Figure 10). Analogously, chemists have been creating building blocks (many of which are now commercially available) that can be used to construct a large variety of functionally diverse, therapeutically viable, and hence marketable, compounds. The process of designing or obtaining these building blocks can be of critical importance to the success of a planned library.

Library Synthesis

Once a synthetic plan or strategy is developed, the actual synthesis of a combinatorial library proceeds relatively quickly. Indeed, while it can take from months to years to work out the chemistry for a combinatorial synthesis, it often only takes from days to weeks to build the library. If the process is well defined, and perhaps even programmed into an automated system, this information can be used to electronically generate expected products, and thus trackable information, without requiring burdensome user entry.

Depending on the modes of extracting information (*6–9, 24–30*), however, resynthesis or decoding may be necessary. The storage and retrieval of reaction histories are essential for expedient resynthesis, even more so because emphasis is placed on synthesizing populations of molecules rather than a single molecule. In mixture synthesis, varying reaction yields will often result in the production of an excess of one library member, and little or none of another. This uncertainty in product representation is tolerated only if the synthetic process is easily reproduced.

Whenever a process is automated, controls must be added to ensure that the process is running properly. Few things are more frustrating to research organizations than the extra work caused by false hits in primary screens. Although it is impossible to completely eliminate false hits, appropriate controls and data management systems can go a long way toward minimizing their impact. One strategy for

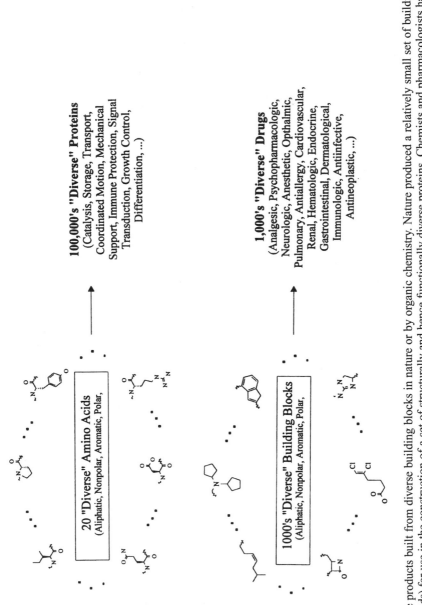

FIGURE 10. Diverse products built from diverse building blocks in nature or by organic chemistry. Nature produced a relatively small set of building blocks (amino acids) for use in the construction of a set of structurally and hence functionally diverse proteins. Chemists and pharmacologists have also "evolved" a set of building blocks to construct structurally and functionally diverse compounds. Combinatorial programs have put much effort into the design and development of building blocks that can be used to construct functionally diverse libraries.

information verification and quality control is to couple systems that generate information about the expected products (such as their molecular weights) with systems that analyze or evaluate tracked samples to see whether the products have the expected properties.

The possibility of a fully automated laboratory with robots capable of preparing, assaying, and analyzing compound libraries lies in the not-too-distant future. The intense desire to minimize human involvement in highly repetitive tasks has already led to the increased use of robotics. Many of the software systems controlling these robots, the synthetic procedure, and sample locations contain what is analogous to project-level (i.e., localized) data. To maximize the fruits of laboratory automation, it will be necessary to integrate these data (both controlling and acquired) for easy access at the project and corporate levels (*see* the section "Information Capture").

Screening

For the past 50 years, high-, medium-, and low-throughput screening efforts have been used to expedite compound discovery. Advances in molecular biology, cloning, and the use of receptors (*31*) have intensified the thirst for new compound supplies, a thirst that combinatorial programs have helped quench. Together with natural products, acquired compounds, and those available as a result of traditional organic synthesis, the compounds generated by combinatorial organic synthesis (COS) will feed this range of screening technologies in the pursuit of profitable compounds.

Not every company is equipped with the hardware to screen their entire sample collection every time a new target is identified, and not every assay is conducive to high-throughput screening (HTS). A successful combinatorial program must, however, have at least the capability for low- to medium-throughput screening.

Many larger companies with well-established high-throughput screening efforts already submit combinatorial libraries to these centralized units, which are capable of assaying hundreds of thousands of compounds in a month. However, almost all companies pursuing COS utilize low- to medium-throughput screening. The variation in the nature of these screening technologies means that the association of combinatorial library data with bioassay data needs to be accommodated at both the project and corporate level of archiving.

The results of screening are generally used to reveal and tune compound effectiveness. High-throughput screens can reveal unexpected leads (lead finding), and low- to medium-throughput screens can be used to tune, or optimize leads (lead optimization). The results of these in vitro, cellular, and tissue-based assays can be used as a cost-effective means of gauging potential utility in animals and eventually humans.

On numerous occasions, companies have "rediscovered" compounds in the compound stores that have been dismissed in one project, but demonstrate great promise in other projects. Just as many companies routinely screen their entire sample collections to ferret out new uses for compounds that have already been synthesized and characterized. In the process, many capture and retain even the most "negative" data in the hope that someone, someday will benefit from it. Combinatorial libraries elevate this concept to another level, however. Some feel that "negative" data derived from combinatorial library evaluation is of little value because of the uncertainty associated with the production of the library. The level of uncertainty is often greater for libraries of mixtures than for libraries of individual compounds produced in a spatially addressable format.

The ultimate goal of all discovery research is, of course, to find new therapeutic agents that are more effective and have fewer undesirable effects than those currently in use, or to find therapeutic agents for conditions for which no cure currently exists. Although it is unlikely that any new drug will be found directly by high-throughput screening, the information gathered during the HTS process can be crucial in optimizing a promising compound. Thus there are three major goals for any HTS project: first, to locate the small percentage of compounds that are active against a biological target; second, to accumulate information that will help researchers better understand the structural basis for activity; and third, to help prioritize which hits warrant downstream optimization based on target selectivity.

Interpretation

Perhaps the most time-consuming aspect of COS lies in interpretation. In a sense, interpretation is necessary during all phases of combinatorial work. From deconvolution and decoding (i.e., information extraction) to library profiling (screening) and future library design, powerful interpretation tools will be the hallmark of a successful combinatorial program.

Clearly, sophisticated methods of statistical analysis and data visualization will become even more critical to research and development as more compounds are produced and assayed. Without such tools, the value of the data that is collected will not be fully exploited. Traditional tools used for the analysis of structure–activity relationships (SARs) can accommodate certain library work when the number of compounds is manageable (i.e., when using focused libraries of analogs). With larger libraries, however, more sophisticated applications that can automatically learn relationships between structures and activity data will become increasingly important.

The tools used for data analysis will need to become more accessible to the combinatorial chemist. These tools, currently used by specialists (i.e., computational chemists), will be essential to allow each researcher and research manager to navigate through, learn from, and exploit the mountain of data produced by COS and screening methodologies.

Information Capture

Information can be captured for later use at each stage during the construction and testing of a combinatorial library. The development of methodologies and selection of building blocks in the early planning stages, procedural and quality controls in the building stages, physical and biological property evaluation and association in the screening stages, and performance reports and quantitative SARs developed in the interpretation stages each contribute an enormous amount of information.

Unfortunately, some of the bottlenecks associated with the process inhibit information capture. As was stated earlier, information capture must pay interest and dividends on registry efforts. One strategy when using this "capture then exploit" philosophy begins with a layered approach to information capture, in which information is extracted at many different steps (using automatic capture and generation) throughout the planning and development of a library. The layered approach begins with data collection by the individual. The information then flows to the group and then to the organization. Frequent steps of quality control and peer review are necessary for this model to work. The remaining sections detail the three-layered approach to information capture, in which personal capture is followed by project and then corporate capture.

Personal Information Capture

Personal information capture begins with individual contributions. As one example, consider the enormous effort spent developing a new, more robust synthetic methodology (*11, 32, 33*). Before others will be willing to use a new synthetic strategy, they must have confidence that it is reproducible and reliable. By nature and by training, scientists do not publish their results until they have them verified. If the scientist treats the electronic storage of a synthesis as a sort of publication, the results can be used to facilitate peer review, and this data can be transferred into project-level repositories. The information pool will gain strength from the numbers of individuals who are creating and capturing information. Provided consistent and efficient data entry mechanisms are in place for personal information capture, and that everyone contributes, the repositories of information will grow in size and value.

But as described earlier in this chapter, information is a form of wealth that is not readily shared. Therefore incentives need to be put in place that not only encourage personal capture, but also reward it. One example is to provide a report of the number of novel methodologies, compounds, libraries, building blocks, and receptors that an individual has electronically stored. These reports can detail other qualifying information such as activity of the compounds, and similarity or dissimilarity of a method to other project and corporate entries. These "bank statements" can be used for personal accounting as well as for performance review. Their main purpose, however, is to motivate and provide an incentive for individuals to contribute to what will eventually become corporate-wide information.

Project Information Capture

Once individual contributions are published within a group or project team, they can be scrutinized and refined by the peer review of team members. This process can remove inaccuracies and suggest new directions for research. Once team members become confident of the accuracy of the information, it can be published in a manner similar to that used for individual publication. As before, incentives can be put in place to motivate project-level publication.

Corporate Information Capture

Just as the project benefits from the contribution of and exchange between individual researchers, the company benefits from exchange between individual projects. The most dramatic example of this is when one project's analogue becomes another project's lead. With centralized information capture, information exchange between projects is greatly facilitated.

Every major chemical, agrochemical, and pharmaceutical company now has some kind of central registry system in place. Small- to medium-sized companies either have, or have recognized the importance of a central archival system. Most high-throughput screening efforts are centralized, depositing information into relational systems. Low- and medium-throughput screening data, while first archived at the project level, soon find their way into these corporate relational archives. Associating chemical and biological data collected throughout an organization can have unexpected and often highly desirable results.

The number of compounds produced in combinatorial programs adds another dimension to corporate information capture. In fact, using combinatorial techniques, some companies have produced more chemical and biological data in a year than they have in their entire history! Although some argue that combinatorial data is of a different quality to that generated using more traditional methods, hidden gems may await serendipitous rediscovery if the information from these programs is captured, organized, and made accessible.

Conclusions

For every compound that reaches market, a company may synthesize 5,000–10,000 compounds. Using combinatorial techniques, the same company may synthesize 500,000–1,000,000 compounds for every one marketed. To achieve increased productivity, the company must produce more novel and effective compounds per dollar and per day. If combinatorial chemists can quickly plan their libraries and successfully select from huge arrays of potential building blocks, they will be capable of building more successful libraries in a more timely fashion. If project teams can learn from other project teams, an organization can collectively design new, more successful libraries. And if libraries can better interrogate assays of greater biological significance, organizations might increase the number of candidates to be moved forward, thus increasing the probability of developing a more successful and more profitable compound. If the planning, building, screening, and interpretation of a library and the capture, organization, and distribution of the resulting data can be coordinated, combinatorial chemistry will live up to its promise to help minimize the time and cost associated with bringing new chemical substances to market.

Appendix: Summary of Tools

Numerous scientific software applications are described (alphabetically) in the following Web site: **http://www.awod.com/netsci/academicsw.html**. Another useful Web site is: **http://www.und.ac.za/und/nprg/ppl/chem.html**.

CAS

Home page: **http://www.cas.org/**
Chemical Abstracts Service (CAS), based in Columbus, Ohio, is a division of the American Chemical Society. CAS is a not-for-profit organization founded in 1907.

Chemical Abstracts Service
2540 Olentangy River Road
Columbus, Ohio 43210, USA

Phone: 1-614-447-3600 (main switchboard)
Fax: 1-614-447-3713
E-mail: help@cas.org

Products

- STN International
- SciFinder
- CD-ROM products
- Printed products
- CAS client services
- CAS/STN customer support and training
- CAS document detective service
- Chemical Patents Plus: WWW access to U.S. Patents 1971–

Chemical Design

Chemical Design was founded in 1983 by two Oxford University chemists—Keith Davies and his sister Mary Bogdiukiewicz.

Chemical Design Ltd.	(USA)
Roundway House, Cromwell Park	Suite 120
Chipping Norton, Oxfordshire	200 Route 17 South
OX7 5SR, UK	Mahwah, NJ 07430, USA

Phone: +44 (0)1608 644000   Phone: (201) 529-3323
Fax: +44 (0)1608 642244   Fax: (201) 529-2443
E-Mail: CHEMDESIGN@applelink.apple.com (in Europe)
            CHEMX@applelink.apple.com (in North America)

Products

Chemical Design's product information is displayed on NETSCI's site (**http:// www.awod.com/netsci**).

Software

Chem-X/DIVERSE is an entry-level software system that includes ten standard Chem-X modules supporting:

- Library registration and design
- R-group selection
- Pharmacophore diversity for lead generation
- Pharmacophore searching for lead explosion
- Property-based lead optimization

Optional Chem-X modules can be added to Chem-X/DIVERSE:

- Client-Server
- Oracle interface
- ChemRXS for library registration using reactions
- MDL SDFile reading/writing
- Inventory upgrade

Databases

- Chapman & Hall's Dictionary of Drugs
- Chapman & Hall's Dictionary of Natural Products
- Chapman & Hall's Dictionary of Fine Chemicals
- Derwent's World Drug Index
- BioByte's Medchem Masterfile
- SPECS & BioSPECS
- Maybridge
- National Cancer Institute's NCI
- InfoChem's ChemSelect
- InfoChem's ChemSynth
- InfoChem's ChemReact

ComGenex

ComGenex is a Hungarian-based company with laboratory services and software products.

(Europe)	(USA)
ComGenex Ltd.	ComGenex USA Inc.
Hollan Erno u. 5	Princeton Corporate Center
H-1136	5 Independence Way
Budapest, Hungary	Princeton, NJ 08540, USA
Phone: +36-1-1124-874	Phone: (609) 520-0599
Fax: +36-1-1322-574	Fax: (609) 520-0897
E-mail: df@cdk-cgx.hu	E-mail: 102254.2601@compuserve.com

Daylight

Home page: **http://www.daylight.com**
Daylight, founded in 1986, is a privately held company that provides toolkits, application software, and databases. Daylight is currently based in Santa Fe, New Mexico, USA.

Products

- Toolkits (object-oriented programming libraries)
- SMILES—Chemical objects and algorithms
- Depict—2D and 3D objects and display
- SMARTS—substructure description and recognition
- Fingerprint—chemical structure characterization
- Monomer—object support for combinatorial mixtures
- Thor—object-oriented interface for all database operations
- Merlin—object-oriented interface for high-speed searching
- X-Widgetsx—high-level X-Windows user interfaces
- Program Object Toolkit

Applications (end-user programs; x=X-Windows)

- Thorx—user interface for Thor databases over any TCP/IP network
- Merlinx—a spreadsheet-like interface to Thor databases for high-performance searching and display of data and structures
- Thor Manager—Thor & Merlin database management
- Printing Package—printing support for PostScript devices
- Clustering Package—efficient, nonparametric methods for clustering large chemical databases by molecular structure
- Rubicon—A rule-based distance geometry program providing very flexible generation of sensible 3D conformations
- PCmodelsx—hydrophobicity and polarizability prediction

Databases available in DAYLIGHT's Thor/Merlin format

- SPRESI—3.4 million substances, and all their data
- MedChem—Over 28,000 structures with measured LogP values
- WDI—World Drug Index (by Derwent publications)
- TSCA—Environmental Protection Agency's Toxic Substances Control Act
- Maybridge—Organic Chemical Catalog
- ACD—Available Chemicals Directory (by MDL)

ISI

Home page: **http://www.isinet.com**
The Institute for Scientific Information (ISI) is a database publishing company. ISI was founded in 1958 by Eugene Garfield. The ISI database covers over 16,000 international journals, books, and proceedings in the sciences, social sciences, and arts and humanities, indexing complete bibliographic data, cited references, and author abstracts. These products and services include current awareness data, retrospective citation indexes, customized alerting services, chemical information products, even specially structured sets of bibliometric data. They are available in a variety of media: print, diskette, CD-ROM, magnetic tape, on-line, and via the Internet.

Products

- Index Chemicus
- Current Chemical Reactions
- Reaction Citation Index
- Chemistry Citation Index
- Biochemistry & Biophysics Citation Index
- Science Citation Index
- Current Contents/Physical, Chemical & Earth Sciences

Knight-Ridder

Home Page: **http://www.dialog.com**
Knight-Ridder Information is based in Mountain View, California, USA, with 51 regional offices and affiliate companies in 37 countries to serve the diversified needs of a global customer base. Knight-Ridder Information is a subsidiary of Knight-Ridder, Inc., of Miami, Florida, one of the world's largest newspaper, electronic publishing, and business information companies.

Knight-Ridder Information, Inc.
2440 El Camino Real
Mountain View, CA 94040 USA

Phone: (415) 254-7000
Fax: (415) 254-7070

Products
See **http://www.corp.dialog.com/dialog/whatsnew/whats new.html**

- DataStar
- DIALOG
- DIALOG Alert for Current Awareness
- ERA Service
- KR BusinessBase

- KR GlobalReach
- KR OnDisc
- KR ProBase
- KR ScienceBase
- KR SourceOne

MDL

Home page: **http://www.mdli.com**
Founded in 1978, Molecular Design Limited, now MDL Information Systems Inc., has become one of the world's leading suppliers of chemical information management software, chemical databases, and related services. MDL is a publicly traded company with 1996 revenues exceeding $60 million. MDL is based in San Leandro, California, with sales offices throughout the world:

MDL Information Systems, Inc.
14600 Catalina Street
San Leandro, CA 94577, USA

(510) 895-1313
(800) 635-0064
Fax: (510) 483-4738

Combinatorial Chemistry and High Throughput Screening (HTS) Products

- Project Library
- Central Library
- MDL SCREEN

Client/Server Products

- ISIS/Draw, a client-based chemical drawing package.
- ISIS/Base, a client-based database management application.
- ISIS/Host, a server-based application that gives ISIS/Base access to data stored in flat, binary, relational, text, chemical, reaction, and 2D and 3D chemical structure databases.

Mainframe Products

- MACCS-II (Molecular ACCess System)
- REACCS (Reaction ACCess System)

Database Products

- Available Chemicals Directory (ACD)

- Comprehensive Medicinal Chemistry-3D (CMC-3D)
- Comprehensive Heterocyclic Chemistry (CHC)
- Chemical Products Information (CPI) File
- The ChemInform Reaction Library (ChemInform RXL)
- Current Synthetic Methodology (CSM)
- MDL Drug Data Report (MDDR)
- Metabolite
- National Cancer Institute Database (NCI-3D)
- ORGSYN
- The OHS Safety Series:
  - OHS MSDS ON DISC
  - OHS Cornerstone MSDS Database
  - OHS MSDS Reference Database
  - OHS MSDS Inventory Match Databases
  - OHS On-Line Services
  - OHS Fast FAX
- The Reference Library of Synthetic Methodology
- REACCS-JSM
- SPORE

Oracle

Home page: **http://www.oracle.com**
Oracle Corporation, established 18 years ago, is the world's largest vendor of database software and information management services, with 1995 revenues of almost $3 billion. Oracle has offices throughout the world, but is based in California:

Oracle Corporation
500 Oracle Parkway
Redwood Shores, CA 94065, USA

For a detailed listing of Oracle's products see: **http://www.oracle.com/products**

Oxford Molecular

Home page: **http://www.oxmol.co.uk**
Founded in 1989, Oxford Molecular Group PLC is a leading developer and marketer of computer-aided chemistry and bioinformatics software.

Oxford Molecular Ltd.
The Medawar Centre
Oxford Science Park
Oxford OX4 4GA UK

Phone: +44 1865 784600
Fax: +44 1865 784601

Products

See **http://www.oxmol.co.uk/PRODUCTS/prod_top.html**

- AbM (HP, SGI): Modeling of antibody structures from sequence data.
- AMBER (HP, SGI, IBM): Molecular dynamics simulations for biomolecules
- Anaconda (SGI, IBM): Interactive molecular similarity system for visual comparison of surface properties and pharmacophore discovery.
- Asp (HP, SGI, IBM): Calculation of quantitative molecular similarity based on the shape, and electrostatic and lipophilic properties of molecules.
- Cameleon (HP, SGI, IBM): An integrated protein sequence alignment and analysis tool that enables protein sequence and structure information to be correlated visually.
- Cobra (HP, SGI, IBM): Comprehensive system for automated conformational analysis and 3D structure generation using artificial intelligence techniques.
- Iditis (HP, SGI): Instant access to protein motifs, substructures, and interactions in the Brookhaven data bank.
- Iditis Architect (HP, SGI): The ability to add proprietary Protein Data Bank (PDB) files to Iditis Data.
- Iditis Data (HP, SGI): The most recent PDB release in relational database format.
- Nemesis (PC, Macintosh): Interactive molecular modeling system.
- ProSeries (SGI, IBM): Suite of software tools.
- RS3: Relational database management of chemical structures.
- Tsar (HP, SGI, IBM): Integrated package for interactive investigation of QSARs.
- Vamp (HP, SGI, IBM): Robust semiempirical molecular orbital package for rapid calculation of optimized geometries, molecular orbitals, and molecular properties.

Sybase

Home page: **http://www.sybase.com**
Sybase is the sixth largest software company in the world, with headquarters in Emeryville, California. Sybase focuses on four major market segments: on-line transaction processing, data warehousing, mass deployment, and on-line electronic commerce. Sybase markets products and services under two brands, Sybase and Powersoft.

For a detailed listing of Sybase's products see **http://www.sybase.com**

Tripos

Home page: **http://www.webcom.com/~tripos2**

Tripos, Inc., was founded in 1979 to support the use of computers in molecular design, visualization, and analysis. Its patented CoMFA technology is used for QSAR applications. Tripos has branched into fields of analytical information processing (NMR) and chemical information management (databases).

For a detailed listing of Tripos' products, see **http://www.webcom.com/ ~tripos2**

References

1. Dimasi, J. A. *J. Health Econ.* **1991**, *10(2)*, 107–142.
2. Halliday R. G.; Walker, S. R.; Lurnley, C. E. *J. Pharm. Med.* **1992**, *2*, 139–154.
3. Chabala, J. C. Presented at the Pharmaceutical Manufactures Association, Drug Discovery Management Subsection, Philadelphia, PA, September 1993.
4. MacFarlane, F. G.; Kunze, Z. M.; Drasdo, A. L.; Lumley, C. E.; Walker, S. R. *J. Pharm. Med.* **1995**, *5*, 17–31.
5. Lehman Brothers. *PharmaPipelines: Implications of Structural Change for Returns on Pharmaceutical R&D.* Lehman Brothers: London, 1994.
6. Geysen, H. M.; Meloen, R. H.; Barteling, S. J. *Proc. Natl. Acad. Sci. U.S.A.* **1984**, *81*, 3998–4002.
7. Geysen, H. M.; Rodda, S. J.; Mason, T. J.; Tribbick, G.; Schoofs, P. G. *J. Immunol. Methods* **1987**, *102*, 259–274
8. Geysen, H. M.; Mason, T. J. *Bioorg. Med. Chem. Lett.* **1993**, *3*, 397–404.
9. Fodor, S. P. A; Read, J. L; Pirrung, M. C.; Stryer, L.; Lu, A. T.; Solas, D. *Science (Washington D.C.)* **1991**, *251*, 767–773.
10. Lam, K. S.; Salmon, S. E.; Hersh, E. M.; Hruby, V. J.; Kazmierski, W. M.; Knapp, R. J. *Nature (London)* **1991**, *354*, 84–86.
11. Pavia, M. R.; Sawyer, T. K.; Moos, W. H. *Bioorg. Med. Chem. Lett.* **1993**, *3*, 387–396.
12. Zuckermann, R. N. *Curr. Opin. Struct. Biol.* **1993**, *3*, 580–584.
13. Gallop, M. A.; Barrett, R. W.; Dower, W. J.; Fodor, S. P. A.; Gordon, E. M. *J. Med. Chem.* **1994**, *37*, 1233–1251.
14. Gordon, E. M.; Barrett, R. W.; Dower, W. J.; Fodor, S. P. A.; Gallop, M. A. *J. Med. Chem.* **1994**, *37*, 1385–1401.
15. Gordon, E. M.; Gallop, M. A.; Patel, D. V. *Acc. Chem. Res.* **1996**, *29*, 144–154.
16. Terrett, N. K; Gardner, M.; Gordon, D. W.; Kobylecki, R. J.; Steele, J. *Tetrahedron* **1995**, *51*, 8135–8173.
17. McGee, J. V.; Prusak, L. *Managing Information Strategically*; The Ernst & Young Information Management Series; John Wiley & Sons: New York, 1993.
18. Davenport, T.; Eccles, R.; Prusak, L. "Information Politics" *Sloan Management Review* **1992**, *34(1)*, 53–65.
19. Maizell, R. E. *How to Find Chemical Information—A Guide for Practicing Chemists, Educators, and Students*; John Wiley & Sons: New York, 1987.
20. Murphy, M. M.; Schullek, J. R.; Gordon, E. M.; Gallop, M. A. *J. Am. Chem. Soc.* **1995**, *117*, 7029–7030
21. Cortese, A. "Here Comes the Intranet" *Business Week* **1996**, *February 26*, 77.
22. Sheridan, R. P.; Kearsley, S. K. *J. Chem. Inf. Comput. Sci.* **1995**, *35*, 310–320.
23. Martin, E. J.; Blaney, J. M.; Siani, M. A.; Spellmeyer, D. C.; Wong, A. K.; Moos, W. H. *J. Med. Chem.* **1995**, *38*, 1431–1436.
24. Houghten, R. A.; Pinilla, C.; Blondelle, S. E.; Appel, J. R.; Dooley, C. T.; Cuervo, J. *Nature (London)* **1991**, *354*, 84–86.
25. Houghten, R. A. *Gene* **1993**, *137*, 7–11.

26. Blake, J.; Litzi-Davis, L. *Bioconj. Chem.* **1992**, *3*, 510–513.
27. Needels, M. C.; Jones, D. G.; Tate, E. H.; Heinkel, G. L.; Kochersperger, L. M.; Dower, W. J.; Barrett, R. W.; Gallop, M. A. *Proc. Natl. Acad. Sci. U.S.A.* **1993**, *90*, 10700–10704.
28. Nikolaiev, V.; Stierandova, A.; Krchnak, V.; Seligmann, B.; Lam, K. S.; Salmon, S. E.; Lebl, M. *Peptide Res.* **1993**, *6*, 161–170.
29. Ohlmeyer, M. H. J.; Swanson, R. N.; Dillard, L. W.; Reader, J. C.; Asouline, G.; Kobayashi, R.; Wigler, M.; Still, W. C. *Proc. Natl. Acad. Sci. U.S.A.* **1993**, *90*, 10922–10926.
30. Simon, R. J.; Martin, E. J.; Miller, S. M.; Zuckermann, R. N.; Blaney, J. M.; Moos, W. H. In *Techniques in Protein Chemistry V*; Crabb, J. W.; Jones, W. A., Eds.; Academic: San Diego, 1994; pp 533–539.
31. Williams, M. W. *Med. Res. Rev.* **1991**, *11*, 147–184.
32. Leznoff, C. C. *Acc. Chem. Res.* **1978**, *11*, 327–333.
33. Thompson, L. A.; Ellman, J. A. *Chem. Rev.* **1996**, *96*, 555–600.
34. Cody, D. R.; DeWitt, S. H.; Hodges, J. C.; Kiely, J. S; Moos, W. H.; Pavia, M. R.; Roth, B. D.; Schroeder, M. C.; Stankovic, C. J. U.S. Patent 5 324 483, 1994.
35. DeWitt, S. H.; Kiely, J. S.; Stankovic, C. J.; Schroeder, M. C.; Cody, D. M. R.; Pavia, M. R. *Proc. Natl. Acad. Sci. U.S.A.* **1993**, *90*, 6909–6913.
36. Bunin, B. A.; Ellman, J. A. *J. Am. Chem. Soc.* **1992**, *114*, 10997–10998.
37. Bunin, B. A.; Plunkett, M. J.; Ellman, J. A. *Proc. Natl. Acad. Sci. U.S.A.* **1994**, *91*, 4708–4712.
38. Bunin, B. A.; Plunkett, M. J.; Ellman, J. A. *Methods Enzymol.* **1996**, *267*, 448–465.

14

Screening of Combinatorial Libraries

F. F. Craig

Drug screening today involves the testing of thousands to millions of samples in a biological assay to identify compounds that affect the target molecule or cellular process of interest. These compounds may serve either as tools to elucidate the role of the target in cellular processes or as a lead for further optimization, perhaps eventually becoming a pharmaceutical product. In recent years, an explosion in the numbers of compounds, generated mainly by combinatorial chemistry approaches, has influenced screening in many areas, including bioassay formats, screen design, screen automation, and lead optimization. This chapter reviews the effects of combinatorial chemistry on the screening process and discusses future trends in this area.

The Screening Process

Definition and Types of High-Throughput Screening

High-throughput screening operates chiefly as an empirical approach that involves the testing of many samples (tens of thousands to millions of samples per year) against a number of specific biological targets of interest (*1*). The samples can include natural products (broths from bacterial or fungal cultures, or extracts from plants or marine animals) or solutions of single or pooled compounds derived from manual or automated methods of synthesis including combinatorial chemistry. High-throughput screening can be:

1. random, involving the testing of large libraries of samples in no given order against a specific target; or

2. directed, comprising the testing of small libraries of compounds preselected for their potential to modulate the activities of a specific target.

Lead Generation and Optimization

A screening critical path defines the sequence of operations as active hits progress from the high-throughput screen to subsequent identification of lead compounds (Figure 1). If primary screening is performed in replicates rather than single samples, a retest stage may not be required, and confirmed hit compounds can progress directly into the secondary screen stage. The screening function has now been centralized in most companies, although in a few companies screening is still performed within therapeutic research groups. Screening groups must interface effectively with the research project groups to design the screens, agree on the critical path for screening, and transfer hit compound structures. The secondary screens may be performed in either a screening or research group, depending on the management structure of the program.

Each target usually requires a different range of secondary screens. Secondary screens are designed to produce selectivity, toxicity, or potency information on each

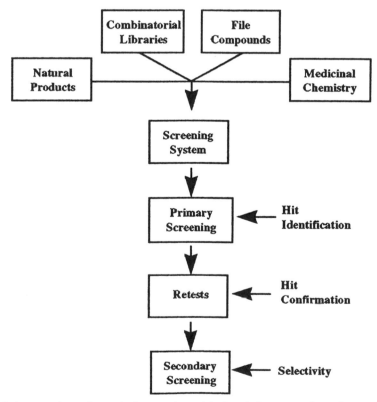

FIGURE 1. An overview of a typical screening critical path for processing of potential lead compounds. Chemicals from one of four sources are entered into a manual or automated screening system, and hits are identified in a primary screen, confirmed by retesting, and refined in secondary screens.

molecular hit, and to allow further progression toward a promising lead compound possessing the correct and desired properties. A recent trend is systems-based research, in which a large number of related targets (e.g., receptors of the same molecular class) are screened simultaneously. This approach can thus give selectivity and potency information at the primary screening stage and may thus accelerate the screening process.

Analogs of a lead compound may be synthesized using manual or automated chemistry or collected from a company's pre-existing chemical collection. The analogs are tested to determine the impact of various structural modifications of the lead compound and to obtain a structure–activity relationship (SAR). Combinatorial approaches have made a significant impact on this process. Directed libraries are often made on the basis of key structures of a lead compound or the naturally occurring ligand of a related receptor. An optimized lead may then proceed to further pharmacological study and, perhaps, even become an early development candidate (*2*).

Biological Targets

Major pharmaceutical companies currently screen against therapeutic targets for one of two reasons. First, they attempt to identify a drug lead against a known target that is strongly implicated in the pathogenesis of a disease. Second, they employ a screen when molecular tools are needed to further dissect the therapeutic target to ascertain if it is genuinely involved in the disease process. Pharmaceutical firms perform research on therapeutic targets in the areas of both infectious and non-infectious disease.

Anti-infectious-disease studies involve identifying inhibitors of the following specific pathogens (the disease caused is shown in parentheses):

- Pathogenic bacteria such as *Mycobacterium tuberculosis* (tuberculosis).

- Viruses such as human acquired immunodeficiency virus (AIDS), herpes simplex (herpes or cold sores), and influenza ("flu").

- Fungi such as *Candida albicans* (thrush).

- Other disease-causing parasites such as trypanosomes (African sleeping sickness), Plasmodium (malaria), and prions (spongiform encephalopathies).

For human noninfectious disease, there are many classes of targets involved. These are summarized in Table I.

Biological Screen Design

There are currently a variety of biological screens, including molecular, subcellular, and cellular screens. Molecular assays can involve measuring binding (e.g., of receptor and ligand, antibody and antigen, or DNA-binding protein and DNA) or

TABLE I. A Summary of Classes of Targets and Their Respective Involvement in (Noninfectious) Therapeutic Areas

Class of Target	Therapeutic Area
G-protein-coupled receptors	Central nervous system, inflammation, metabolic disease
Intracellular receptors	Cancer, metabolic disease
Cytokine receptors	Cancer, cardiovascular, inflammation
Transcription factors	Various areas including osteoporosis and inflammation
Tyrosine kinase receptors	Cancer, cardiovascular, inflammation
Kinases	Various areas including cancer and metabolic disease
SH2 and SH3 domains	Various areas including diabetes and cancer
Ion channels	Cardiovascular, central nervous system
Proteases	Cardiovascular, arthritis and osteoporosis

enzymatic function (e.g., substrate turnover or modification of a target molecule). Very often, binding or enzyme assays can be performed using subcellular components such as membrane preparations or cellular fractions. The more complex cellular assays may involve measurement of cellular events such as secretion, triggering of intracellular ion fluxes, generation of signaling molecules, or stimulation of gene expression.

Each different class of biological screen has advantages and disadvantages. Molecular and subcellular screens are often easier to design and maintain, and the compound hits identified usually act on the target molecule of interest. However, it is not uncommon to find that hits identified from molecular screens do not affect the natural process within cells, where the target molecule interacts with other proteins in a dynamic manner and must act within a specific cellular compartment. The hit compound may also be nonselective, toxic, or unable to cross membranes—properties that cannot be uncovered in molecular assays. Cellular screens offer the advantages of a more natural physiological context and an ability to identify hits that affect multiple targets controlling the cellular process of interest. Cellular screens can, however, be more difficult to construct, and identified hits may need to be retested in several other screens to ascertain which exact target is affected.

Genomics initiatives have also affected screening approaches. When a specific DNA sequence is found to be implicated in a disease, there is very often little or no information on the function of the encoded protein, and the ligand is not known. In this instance, biochemical screens are of limited use, but cell-based screens may provide a solution, as they can detect agonists that trigger signals from the target molecule. A prime example would be to screen a cell line expressing an orphan G-protein-coupled receptor for agonists, looking for the generation of second messengers known to be produced by G-protein-coupled receptors such as Ca^{2+} or cAMP (3). Also, novel tools such as "promiscuous" G-proteins appear to direct many G-protein-coupled receptor signals down the phospholipase C/Ca^{2+} pathway and may offer an improved way to build orphan G-protein-coupled receptor screens by ensuring a uniform readout (4). Once agonists are identified, antagonists can then be derived, and functional analysis of the target molecule in its natural context can be accelerated using the hits or molecular tools identified from the screening campaign. In general, most major pharmaceutical companies will use a combination of molecular and cellular screens based on scientific experience and the type of target.

Labels used for screening are either direct (e.g., radioisotopic, fluorescent) or indirect (e.g., enzymes that catalyze the production of colored, fluorescent, or luminescent signals). The trend in the industry is to move away from radiolabels wherever possible to nonradioactive labeling systems (5). Radioactivity has problems of storage, safety, and disposal, all of which have been exacerbated in recent years because of the greater throughputs allowed by screen automation.

Common features required by any type of screen include robustness (i.e., tolerance to solvents such as dimethyl sulfoxide (DMSO), methanol, and detergents, often found in extracts of natural products), reproducibility, and a low frequency of positives or hit rate (preferably under 1%, although the frequency can be slightly higher with some cell-based assays). Compound stocks are usually prepared and supplied in DMSO; appropriate dilutions of those compounds are then made for each specific screen, with the final compound concentration actually tested in a screen typically being within the range of 1–50 μg mL^{-1}. Assay technologies have different tolerances to solvents, and a compound test concentration is optimized to give a workable hit rate with good screen performance. Cell-based screens have a lower tolerance to DMSO (usually the concentration is kept at <1%) than other screening approaches. A common screening strategy first tests a "standard" or "universal" set of compounds (usually containing 10,000 to 20,000 distinct entities), which are selected to represent a diverse set of compounds and may thus give a higher proportion of hits than random screening. This is also an excellent method of determining hit rates before a full random screening campaign. After optimization of a micromolar lead, it is possible to generate a selective nanomolar lead.

Screen Sample Presentation

Screen sample presentation involves the integration of samples such as combinatorial libraries, natural products, and file compounds into the screening program. Every new sample will be added into the corporate sample database, and usually stocks or sample-containing microtiter plates will be bar-coded for data integrity and sample tracking. Traditionally, companies stored dry stocks of compounds (master stocks), small amounts of which were weighed out and dissolved in the appropriate solvent (working stock) just prior to screening. With the explosion in compound availability due to combinatorial chemistry, this approach is changing, and the master stocks are now being stored at low temperatures (to maintain compound stability) in liquids to eliminate the time-consuming, laborious weighing stage, and to allow faster transfer from the store to the screening plates using liquid-handling robotics. After a suitable lead is found, it is desirable to optimize this compound into a potent, selective lead. This can become a long, labor-intensive process, but advances in combinatorial chemistry have begun to decrease the time required for lead optimization.

Automated Microtiter Plate-Based Screening

The use of advanced robotics in high-throughput screening has become common-place in the past few years in most major pharmaceutical companies. The robotics systems usually use 96-well or 384-well plates for screening. Table II summarizes the assay throughputs per 24 hours for a variety of approaches.

A typical automated screening system currently in use is shown in Figure 2. A need has grown for automated systems that can fully utilize the explosion in the numbers of compounds made available by combinatorial chemistry technologies and the increasing collections of file compounds and natural extracts. Suppliers of fully integrated screening robotics platforms include CRS Robotics (Burlington, Ontario), Robocon (Vienna, Austria), Sagian (Indianapolis, IN), Scitec (Wilmington, DE) and Zymark (Hopkington, MA).

Data Management

Data management is a critical part of the screening process, especially with the increased throughputs and number of data points enabled by screen automation (*see* Chapter 13). The ideal data management process must involve compound tracking, usually by bar-coding, from the compound store to microtiter well and plate and the subsequent screen. The biological result in the screen for that specific compound must also be transferred into the corporate database for analysis and profiling. Several companies now offer commercially available screening and chemistry data management packages, including MDL Information Systems (San Leandro, CA), MLR Automation (Alameda, CA) and Tripos (St. Louis, MO). However, many established pharmaceutical firms continue to employ their own in-house system.

Impact of Combinatorial Chemistry

Primary Screening and Lead Optimization

The major impact of combinatorial chemistry on drug screening is the possibility of an unlimited supply of samples (*6–8*). At most pharmaceutical companies, the sam-

TABLE II. Assay Throughputs per 24 Hours for Various Screening Approaches

Procedure	Throughput
Manual	20 microtiter plates
Automated	60 microtiter plates (cell-based assays)
	120 microtiter plates (homogeneous biochemical assays)

NOTE: Number of actual compounds tested will depend on whether the screening regime tests a discrete compound or pools of compounds per well.

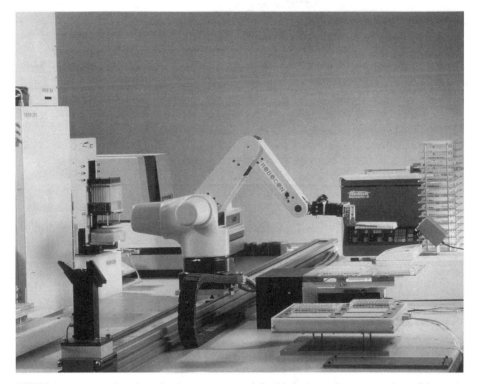

FIGURE 2. An example of a robotics system used for high-throughput screening of 96-well microtiter plates.

ple collection contains 80,000–1,000,000 discrete samples or natural product extracts, in addition to combinatorial libraries that may include millions of molecules (mainly peptides). With further advances in combinatorial chemistry, it is possible to increase the number of compounds in the libraries exponentially. The numbers of compounds in libraries have already overloaded the traditional 96-well microtiter plate screening process. The variety of formats used for combinatorial synthesis, including silicon or glass chips (*9, 10*), beads (*6, 7*), and pegs and arrays (*11*), have also broken with traditional approaches to chemical synthesis and screening (*see* Chapter 10).

These new formats have had an impact on the type of assays performed. Affymax (Palo Alto, CA) pioneered fluorescence-activated cell sorting (FACS)-based screening using peptide libraries on beads (*12*). This approach is capable of ultrahigh-throughputs (millions of beads per day) but is restricted to binding assays using fluorescently tagged proteins. Approaches using silicon or glass chips are again currently restricted to binding assays. Companies such as Affymetrix (Santa Clara, CA), Genometrix (The Woodlands, TX), and Nanogen (San Diego, CA) have produced peptide and oligonucleotide libraries that are useful for antibody screening, DNA analysis, or sequencing, respectively. Such libraries chiefly target diagnostic applications and, although they are possibly too expensive for primary screening applications, offer some potential for secondary screening. Orchid Biocomputer (Princeton, NJ) has stated its plans to make a 10,000-well plate for miniaturized automated combinatorial synthesis (13). This combinatorial format may also be applicable to screening with certain specialized biochemical screens.

Another recent compound screening approach is the free format screening system. This method can involve immobilizing the biological target (e.g., enzymes, cells, or receptors) to a format such as a gel or membrane, then dispensing the compound (either liquid or bead-based), and performing the assay in the gel or membrane matrix (14). Bead-based libraries can be treated to cause a partial release of the compound, allowing subsequent retesting of the beads for either compound characterization or in another screen. This approach has been used to identify receptor agonists for G-protein-coupled receptors expressed in frog melanophores immobilized in a gel matrix. A variant of this approach is to dispense or immobilize compounds onto a surface, then pour the biological target over it and look for, for example, zones of inhibition. Such approaches have the great advantage of high-throughputs (10,000 samples per 9×9 cm area), with the assay read by eye or image analysis (15). A disadvantage of these methods is that compound diffusion occurs so rapidly that assays with longer incubation times (>30 min) become problematical because of the formation of large zones of inhibition. Also, not every assay type works in this system, especially if the assay involves multiple reagent additions, incubation stages, or washing.

Decoding Strategy

Because of the explosion in sample numbers and subsequent pressure on screening compounds, some companies have adapted a pooling strategy for primary screening in 96-well microtiter plate formats in an attempt to screen more samples. Pools can include two to thousands of samples per well. The decoding strategy varies according to the sample set tested (see Chapters 6 and 9 for details). For example, if the hit well contains a pool of ten discrete compounds, each discrete compound would then be tested separately to identify active compound(s). The pooling strategy and sample locations used would be encoded in the computer database, then retrieved after screening. Differences of opinion exist about the utility of pooling, because it is common to find that some hits actually occur from summation effects (small amounts of activity from different compounds adding up to a larger effect that is thus actually a false positive result). In addition, a well may theoretically contain agonists and antagonists that compete with each other. If the well contained combinatorially derived compounds, then it may also be possible to predict the active compound structure by the pattern of hits and the synthetic route selected. Decoding of the hit can also be achieved using various tags such as DNA tags (Affymax, Palo Alto, CA), organic tags (Pharmacopoeia, Princeton, NJ) or radiofrequency tags (Irori, La Jolla, CA). By using a strategy such as the "positional scanning" approach of Trega Pharmaceuticals (La Jolla, CA), decoding of the structure can be accomplished simply by analyzing results obtained from a primary screen.

Future Trends in Screening of Combinatorial Libraries

Combinatorial Chemistry

An incredible increase has occurred in the numbers of compounds generated by combinatorial chemistry. Until recently, diversity among these compounds has been limited to peptides or peptidomimetics. Combinatorial synthetic protocols have now been developed for a widening class of chemicals, and difficult chemistries, such as carbohydrate synthesis, are now being tackled (*16*). The trend in the industry will be toward further research on structural diversity and developing combinatorial synthetic protocols for a wider range of chemistries. The actual content of a pharmaceutical company's library will probably shift even further away from file compounds toward combinatorial libraries.

Combinatorial Biology

The new approach of combinatorial biology uses microorganism hosts (primarily *Escherichia coli*) to produce combinatorial structures. This procedure recombines genes from a variety of diverse organisms and expresses them in a host strain. For example, the company ChromaXome (San Diego, CA) utilizes marine bacteria as a source of genetic material; other workers have used *Streptomyces* (*17, 18*). Theoretically, this approach has the potential to produce billions of different novel molecules. Before the recombinants can be made, however, the genes encoding every element of a synthetic pathway must be identified and the pathway elucidated. ChromaXome reportedly uses the reporter gene green fluorescent protein (GFP), engineered to report a specific biological activity in the host bacterium, and fluorescent-activated cell sorting (FACS) to identify potential anti-infective agents. Green fluorescent protein is a recently discovered protein that is inherently fluorescent, thus allowing simple assays (*19–21*).

Novel Bioassays and Detection Technologies

Most chip or FACS-based assay systems require fluorescent reporter molecules as readouts. Researchers are currently developing better fluorescent labels or probes for such assays: brighter, longer wavelength (further away from cellular autofluorescence, which occurs in the blue spectrum), and easily linkable to a variety of biological molecules. An example of this is the cyanine dyes from Amersham International (Little Chalfont, UK), which have been engineered to have emission in the far-red spectrum. Brighter mutants of GFP have now been isolated with improved properties and spectra over the wild-type (*22, 23*). The generation of different colored mutants of GFP (blue, cyan, and green) make multiparameter detection in FACS or in living cells possible (*24*). In yet further exploration of this novel molecule for possible screening applications, it has now been engineered as a real-time enzyme (kinase or protease) sensor for possible use in biochemical or cell-based screens (*23*). This was done by cloning in enzyme substrate sites that report enzyme activity by a change in the GFP fluorescence. Rapid functional assays for cells, such as fluorescent measurement of ion fluxes using calcium-sensitive dyes like fluo-3, are already in use for screening applications.

The requirement for faster screening has led to the recent development of homogeneous technologies (those with no washing, filtration, or centrifugation steps), which simplify automation by requiring the robot to perform fewer liquid addition and handling tasks and thus enable faster throughputs (Table III). The scintillation proximity assay was the first mainstream technique for homogenous assays. This assay uses beads containing scintillant and coated with a biological molecule such as an antibody, receptor, or substrate. Radiolabeled antigens or ligands are then added, and if they bind, the proximity of radioactivity to the scintillant results in

TABLE III. Homogeneous Assay Technologies Useful for High-Throughput Screening

Homogeneous Technology	Applications	Signal	Company or Supplier
Scintillation proximity assay (SPA)	Receptor–ligand Enzyme assays Radioimmunoassay	Radioactive (^{125}I and ^{3}H)	Amersham Intl. (Little Chalfont, UK)
Scintiplates	As per SPA Cellular assays	Radioactive (most radioisotopes)	Wallac (Finland)
FlashPlates	Receptor–ligand Enzyme assays Radioimmunoassay	Radioactive (^{125}I and ^{3}H)	NEN Dupont (Boston, MA)
Homogeneous time-resolved fluorescence	Receptor–ligand Protein–protein	Fluorescence	Packard Instruments (Meriden, CT)
Fluorescence polarization	Receptor–ligand Enzyme assays ELISA[a]	Fluorescence	Jolley Consultancy (Round Lake, IL)
Green fluorescent protein	Reporter gene assays	Fluorescence	Aurora Biosciences Corp. (La Jolla, CA)
	Enzyme assays (kinases and proteases)		Aurora Biosciences Corp. (La Jolla, CA)

[a]Enzyme-linked immunosorbent assay.

emission events. Drugs that interfere with binding can decrease this scintillation. Fluorescent methodologies have now appeared as the second wave of homogenous assay technologies. This development is driven by the wide variation of fluorescent probes available (over 2,000) for tagging molecules or cells, and by the potential of fluorescence for multiplexing, i.e., the measurement of two or more events or signals within one test sample (*25, 26*). Another trend in screening uses imaging systems that accelerate throughput by counting every sample in a 96- or 384-well plate simultaneously (Cambridge Imaging Limited, Cambridge, UK) thus reducing the plate reading time from minutes to seconds. Imaging is also a powerful approach that can capture rapid events (such as ion fluxes in living cells) and thus give important information on kinetics and pharmacology.

Another important force influencing the development of screening is the number of targets being generated by genomics and molecular biology initiatives. These developments have produced a need for more sophisticated and powerful approaches, especially for difficult or intractable targets. The most successful screening approaches for genomic-derived targets will undoubtedly be based on mammalian cells, as this will offer the ability to characterize the function of the target in a relevant physiological background.

Screen Automation

With current automated screening platforms for 96-well plates, little scope exists for further improvement in throughput. The technology is now very reliable, mature, and widely available to every laboratory. The suggestion to expand capacity by simply buying more robots is short-sighted because it will increase costs (of both reagents and compounds) and, when used with radioactivity, will cause waste storage and disposal problems. The 96-well format was a major advance over the test tube, but has now become restrictive. The way forward now appears to be further miniaturization.

Screen Miniaturization

Several of the leading companies are now looking toward screen miniaturization as the solution to increase throughputs while reducing costs. A few companies are exploring the use of the 384-well plate, which has four times as many wells in the same 12×8 cm footprint as a 96-well microtiter plate (MTP). Currently, there is still a lack of enabling technology for this screening format, such as 384-well MTP readers, appropriate liquid-handling systems, and even plates of sufficient quality and variety (e.g., tissue-culture treated).

Several new approaches for miniaturized assays have been proposed, such as FACS with bacterial assays (ChromaXome, San Diego, CA) and ultrahigh-throughput screening using nanoformats (Aurora Biosciences Corporation, La Jolla, CA). The latter involves the use of miniaturized wells (several thousand in an 8×12 cm format) that allow most sample types to be tested in a broad range of assays.

The physics of signal generation and detection, and the wide variety of fluorescent probes available (27), suggest that only fluorescent signals will ultimately have the flexibility, resolution, and strength of signal to allow miniaturization of a broad range of assay types. Improvements such as better hardware (optics, filters and detectors) and the use of time-resolved fluorescence techniques have allowed minimization of nonspecific fluorescence from samples. Improvements such as these have enabled fluorescence to now become one of the leading approaches for drug screening. Although it is possible to perform single molecule detection (28, 29) or single cell analysis with a variety of fluorescent probes (30), results with these systems are variable. In addition, single cells are inherently heterogeneous. For screening purposes, miniaturization will have to be done with a population (several hundred or thousands) of molecules or cells to enable a high quality, reproducible, average result. Aurora Biosciences Corporation is currently developing ultrahigh-throughput screening systems, which should work with fluorescent assays performed in nanoliter volumes. This approach typifies the technological changes that have occurred in screening, due mainly to the enormous supply of compounds made available by combinatorial chemistry.

Data Processing

A real need in the screening of combinatorial libraries is for fully integrated software solutions that perform data acquisition, distribution, and analysis, and display the results in real time in an integrated open format. The high-performance hardware should have a modular design, with scalable architecture to allow growth and integration with both current and future systems. Data reduction packages that transfer data directly into relational databases, use automated report generation, and include new approaches (neural nets, genetic algorithms, fuzzy logic) for database mining are also being developed. One company, 3-D Pharmaceuticals (Philadelphia, PA), has stated (31) that it may be possible to produce a computer-controlled iterative process to control the generation of chemical entities, screen these entities, obtain SAR, and then optimize any leads using another cycle of chemical synthesis and screening. One can envisage that drug discovery in real time may even become possible, with advances in informatics allowing results to be analyzed the instant they are produced, and a rapid, iterative synthesis and screening cycle then being used to optimize the lead.

Conclusions

The drug discovery process has undergone a technological revolution during the past few years, with the new technologies of combinatorial chemistry, automation, genomics, and informatics quickly coming into mainstream use. The most efficient and successful companies in the future will be those that are adept at managing state-of-the-art technologies and can combine them with high-quality chemical entities, biological targets, and screens. The ability to synergize combinatorial synthesis, miniaturized screens, automation, and state-of-the art informatics makes it possible to imagine real-time, ultrahigh-throughput lead generation and optimization. This approach may well become the leading drug discovery approach of the future.

Acknowledgments

Many thanks to Harry Stylli and Tim Rink for critical review of this document, and to Mary Jean Pramik for helping in its finalization.

References

1. Zenie, F. H. *Bio/Technology* **1994**, *12*, 736.
2. Yevich, J. P. In *A Textbook of Drug Design and Development*; Krogsgaard-Larsen, P.; Bundgaard, H., Eds; Harwood Academic: Philadelphia, 1990; Chapter 16, pp 607–630.
3. Mills, A.; Duggan, M. J. *Trends Pharmacol. Sci.* **1993**, *14*, 394–396.
4. Offermans, S.; Simon, M. I. *J. Biol. Chem.* **1995**, *268*, 10139–10144
5. Mayer, A.; Neuenhofer, S. *Angew. Chem. Int. Ed. Engl.* **1994**, *33*, 1044–1072.
6. Gallop, M. A.; Barrett, R. W.; Dower, W. J.; Fodor, S. P. A.; Gordon, E. M. *J. Med. Chem.* **1994**, *37*, 1233–1251.
7. Gordon, E. M.; Barrett, R. W.; Dower, W. J.; Fodor, S. P. A.; Gallop, M. A. *J. Med. Chem.* **1994**, *37*, 1385–1401.
8. Patel, D. V.; Gordon, E. M. *Drug Disc. Today* **1996**, *1*, 134–144.
9. Fodor, S. P. A.; Rava, R. P.; Huang, X. C.; Pease, A. C.; Holmes, C. P.; Adams, C. L. *Nature (London)* **1993**, *364*, 555–556.
10. Jacobs, J. W.; Fodor, S. P. A. *TIBTECH* **1994**, *12*, 19–26.
11. Hobbs DeWitt, S.; Kiely, J. S.; Stankovic, C. J.; Schroeder, M. C.; Reynolds Cody, D. M.; Pavia, M. R. *Proc. Natl. Acad. Sci. U.S.A.* **1993**, *90*, 6909–6913.
12. Needels, M. C.; Jones, D. G.; Tate, E. H.; Heinkel, G. L.; Kochersperger, L. M.; Dower, W. J.; Barrett, R. W.; Gallop, M. A. *Proc. Natl. Acad. Sci. U.S.A.* **1993**, *90*, 10700–10704.
13. Begg, G. S.; Simpson, R. J.; Burgess, A. W. U.S. Patent 5 516 698, 1996.
14. McClintock, T. S.; Graminski, G. F.; Potenza, M. N.; Jayawickreme, C. K.; Roby-Shemkovitz, A.; Lerner, M. R. *Anal. Biochem.* **1993**, *209*, 298–305.
15. Lerner, M. R. *Trends Neurosci.* **1994**, *17*, 142–146.
16. Kanie, O.; Barresi, F.; Ding, Y.; Labbe, J.; Otter, A.; Forsberg L. S.; Ernst, B.; Hindsgaul, O. *Angew. Chem. Int. Ed. Engl.* **1995**, *34*, 2720–2722.
17. Hutchinson, C. R. *Bio/Technology* **1994**, *12*, 375–380.
18. McDaniel, R.; Ebert-Khosla, S.; Hopwood, D. A.; Khosla, C. *Science (Washington, D.C.)* **1993**, *262*, 1546–1550.
19. Prasher, C. D.; Eckenrode, V. K.; Ward, W. W.; Prendergast, F. G.; Cormier, M. J. *Gene* **1992**, *111*, 229–233.
20. Inouye, S.; Tsuji, F. I. *FEBS Lett.* **1994**, *341*, 277–280.
21. Ward, W. W.; Prentice, H. J.; Roth, A. F.; Cody, C. W.; Reeves, S. C. *Photochem. Photobiol.* **1982**, *35*, 803–808.
22. Heim, R.; Prasher D. C.; Tsien, R. Y. *Proc. Natl. Acad. Sci. U.S.A.* **1994**, *91*, 12501–12504.
23. Cubitt, A. B.; Heim, R.; Adams, S. R.; Boyd, A. E.; Gross, L. A.; Tsien, R. Y. *Trends Biochem. Sci.* **1995**, *20*, 448–455.
24. Rizzuto, R.; Brini M.; De Giorgi F.; Rossi R.; Heim R.; Tsien R. Y.; Pozzan T. *Curr. Biol.* **1996**, *6*, 183–188.

25. Bright, G. R. In *Fluorescent and Luminescent Probes for Biological Activity. A Practical Guide to Technology for Quantitative Real-Time Analysis*; Mason, W. T., Ed.; Academic: London, 1993; Chapter 14, pp 204–215.

26. Tsien, R. Y. *Chem. Eng. News* **1994**, July 18, 34–44.

27. Kasten, F. H. In *Fluorescent and Luminescent Probes for Biological Activity. A Practical Guide to Technology for Quantitative Real-Time Analysis;* Mason, W. T., Ed.; Academic: London, 1993; Chapter 2, pp 12–33.

28. Kabata, H.; Kurosawa, O.; Arai, I.; Washizu, M.; Margarson, S. A.; Glass, R. E.; Shimamoto, N. *Science (Washington, D.C.)* **1993**, *262*, 1561–1563.

29. Nie, S.; Chiu D. T.; Zare R. N. *Science (Washington, D.C.)* **1994**, *266*, 1018–1021.

30. Tsien, R. Y. *Methods Cell Biol.* **1989**, *30*, 127–156.

31. Agrafiotis, D. K.; Bone, R. F.; Salemme, F. R.; Soll, R. M. U.S. patent 5 463 564, 1995.

15

Summary

Sheila H. DeWitt and Anthony W. Czarnik

A revolution in any discipline initially results in chaos and leads to an upheaval of previously accepted techniques or protocols. This situation exemplifies what the field of chemistry is experiencing today. With the integration of new tools, including reaction equipment, automation, and advanced information systems, comes the promise of the rapid generation and optimization of synthetic compounds. Although the chemical concepts remain the same, the methods of implementation vary dramatically. Therefore, practicing chemists are currently seeking to utilize new tools without knowing where to obtain information about the tools or how they are used.

The aim of this book and future books in this series is to provide chemists with the knowledge to make educated decisions to achieve their goals and objectives in emerging areas of technology. As a reference guide and tutorial, the reader can use this book to evaluate and select a course of action, recognizing that there may be an unsaid chasm between concept and practice. Fortunately, chemists have always been adept at reducing concepts to practice.

Limitations

The ultimate vision for a combinatorial or high-throughput chemistry laboratory is the synthesis of thousands or even millions of individual compounds of >99% purity each day. Although advances in molecular biology and automation have supplied the targets for screening and the means to rapidly synthesize compounds, bottlenecks have emerged. These emerging needs include a wider repertoire of solid-phase synthesis reactions, high-throughput purification and analysis methods, and better information management tools. As these new challenges are addressed, future needs will be identified. For example, technology is not available to purify or analyze hundreds or thousands of compounds in parallel. Advances in high-throughput serial methods such as solid-phase extraction (SPE), high-performance liquid chromatography (HPLC), and mass spectroscopy (MS) cannot meet the demand. Because of these limitations, currently only 5% of a library (e.g., 500 compounds out of a library of 10,000 compounds) is typically analyzed by only one or two semiquantitative techniques. As a result, the reliability of the corresponding biological data can be compromised (*1*). Alternative methods for the purification and analysis of products generated by combinatorial chemistry need to be developed to advance from library generation to high-quality optimization programs.

Enabling Technologies

Chemists have routinely advanced their practice through the use of continuous technologies, as in the introduction of flash chromatography to improve the throughput and resolution of traditional chromatographic methods employing gravity. The advent of combinatorial chemistry, however, represents the introduction of a discontinuous technology, a technology not closely related to previous methods, that significantly impacts the way that chemistry will be executed in the future (*2*). Discontinuous technologies require a change in philosophy and execution.

Some enabling technologies that have been applied to combinatorial chemistry initiatives include statistical experimental design (*3*), expert systems (*4*), ink-jet printing (*5*), microfabrication (*6*), and nanomaterials (*7*). Other technologies that have historically been applied to the computer, manufacturing, and medical industries may also find utility in the field of chemistry. For example, electronic notepads, parallel processing computers, digital imaging, neural networks, object-oriented programming, fuzzy logic, and endoscopic technology (*8*) may enable continued advances for chemistry.

Future Directions

It is noteworthy that the execution of combinatorial chemistry strategies can be and often is limited by the availability of equipment and instrumentation. Combinatorial chemistry efforts have begun to identify the limitations of traditional tools for synthetic chemists. The overriding goal is to increase productivity while maintaining high standards for both the quality and the quantity of final products. The laboratory of the future will draw upon a wider repertoire of tools, striving for ever-increasing efficiency and productivity while continuing to advance the forefront of all chemistry programs.

References

1.   MacDonald, A. A.; Nickell, D. G.; DeWitt, S. H. *Pharm. News* **1996**, *3*, 19–22.

2. Moore, G. A. *Crossing the Chasm: Marketing and Selling Technology Products to Mainstream Customers*; HarperBusiness: New York, 1993.

3. Martin, E. J.; Blaney, J. M.; Siani, M. A.; Spellmeyer, D. C.; Wong, A. K.; Moos, W. H. *J. Med. Chem.* **1995**, *38*, 1431–1436.

4. Lousse, F.; Iskra, J.-L.; Porte, C.; Delacroix, A. *Lab. Rob. Autom.* **1995**, *7*, 93–100.

5. Nishioka, G. M. U.S. Patent 5 449 754, 1995.

6. Begg, G. S.; Simpson, R. J.; Burgess, A. W. U.S. Patent 5 516 698, 1996.

7. Martin, C. R. *Science (Washington, D.C.)* **1994**, *266*, 1961–1966.

8. Burrus, D.; Gittines, R. *Technotrends: How to Use Technology to Go Beyond Your Competition*; HarperBusiness: New York, 1993.

Glossary

ASPECT — Chemically modified polyolefin particles; used as a support in solid-phase synthesis.

ATR — *See* Attenuated total reflectance.

Attenuated total reflectance (ATR) — Occurs when a sample is brought into contact with an internal-reflection element that has a higher refractive index than the sample. ATR is also called internal reflection or multiple internal reflectance. The sample must be brought near the optical element where the sample interacts with the evanescent wave. The effective path length for this interaction is typically a fraction of a wavelength, and it depends on many parameters, making this a powerful technique.

Bacteriophage libraries — Up to hundreds of millions of peptides displayed on the surface proteins of bacteriophage particles.

Bioactive compound — A compound that elicits a general response in a biological assay.

Biological assay (bioassay) — A biological test system, generally reduced to microtiter plate format for high-throughput screening, that permits the evaluation of a compound for level of activity.

Building blocks — The set of reagents used to generate the components of a combinatorial library.

Capacity factor — A chromatographic term defined as the number of moles of a solute in the stationary phase divided by the number of moles of the same solute in the moving liquid phase. The capacity factor is calculated on the basis of a comparison of the retention time (or migration time) of the solute compared with the void volume.

Capillary electrochromatography (CEC) — A high-resolution, nanoscale separation technique wherein a mixture of neutral solutes is separated into its individual components on the basis of differences in their partition between a moving liquid phase and a stationary solid support. The separation of charged solutes involves a combination of partitioning and electromigration mechanisms. The liquid phase is driven electroosmotically through the capillary under the application of an external electric field.

Capping – A procedure performed after the coupling step and prior to removal of the temporary protecting group in a multistep solid-phase organic synthesis protocol, in order to substitute deletion sequences with terminated sequences. It is usually done by reaction with acetic anhydride in SPPS and with acetic anhydride–DMAP or acetic anhydride–*N*-methylimidazole in oligonucleotide synthesis. Terminated sequences are usually easier to separate from the desired product than deletion sequences.

CEC – *See* Capillary electrochromatography.

Chemical database – A computer-based file of chemical compounds, represented by their two-dimensional (2D) chemical structure diagrams and, in an increasing number of cases, the three-dimensional (3D) atomic coordinates. The latter are either measured experimentally or, more commonly, obtained by calculation using a structure generation program.

Chemical library – A set of chemicals large enough to require organization to be used efficiently.

Closed-loop experimentation – An experiment or set of experiments that can be altered automatically in real-time based on the data that are being collected. Examples include increasing the temperature on a sluggish reaction or terminating a simplex search if the region of optimal response has been attained. Closing the experimental loop, where data serve as feedback for altering ongoing experiments, requires specific instructions in an experimental plan and an objective function. Compare with Open-loop experimentation.

Cluster analysis – A multivariate statistical technique for grouping the objects in a data set such that similar objects are in the same group, or cluster, whilst being separated from dissimilar objects in other clusters.

CMS algorithm – *See* Composite modified simplex algorithm.

Combinatorial chemistry – a new subfield of chemistry with the goal of synthesizing large numbers of chemical entities by condensing a small number of reagents together in all combinations allowed by a set of chemical reactions

Combinatorial development – The use of combinatorial chemistry principles to optimize lead series, especially in pharmaceutical research and development, in order to advance molecules more rapidly into and through preclinical development (e.g., studies of biodisposition, efficacy, and safety).

Combinatorial discovery – The use of combinatorial chemistry techniques to discover, for example, new drug candidates.

Combinatorial library – A subclassification of a chemical library in which every possible member that can be generated from the sets of reactants is present.

Composite modified simplex (CMS) algorithm – An eclectic combination of alterations made over the years to the original simplex algorithm, yielding a more robust and sophisticated algorithm. Features of the CMS algorithm include the ability of the simplex to change shapes, expand in regions of steadily improving response, contract in optimal regions, reflect from boundaries, avoid oscillating on ridges, and avoid getting stuck in valleys between peaks.

Compound clips – Mixtures of pure compounds deliberately prepared for the purpose of initial lead identification in a drug-discovery operation.

Connection table – A machine-readable representation of a 2D chemical structure diagram, in which the atoms comprising a molecule are listed together with their pendant bonds. This representation forms the basis for all existing 2D chemical databases (vide supra).

Convergent synthesis – Strategy involving stepwise synthesis of partially protected oligomeric segments, followed by condensation of the segments.

COSY – *See* Homonuclear correlated spectroscopy.

2D NMR – *See* Two-dimensional NMR.

DCR method – *See* Divide-couple recombine method.

Deconvolution – A method to identify one active molecule within a mixture of compounds by making and assaying one or several partitions (series of less complex mixtures) of this mixture.

Deletion peptide – A peptide lacking one or more amino acid residue(s) because of incomplete coupling reaction(s).

Dialkylamine tags – A tag that can be used for binary encoding. Upon removal from the bead and reaction with dansyl chloride, a fluorescent amine derivative is produced and can be analyzed by HPLC.

Divergent synthesis – A synthesis approach in which multiple final products are prepared from one starting material.

Diversity – As in molecular diversity, the multidimensional physicochemical property space (size, shape, electronic properties, etc.) that defines a set of molecules or a library.

Diversity index – A measure that tries to quantify the degree of structural diversity present within a set of compounds. Several such indices have already been suggested, but there is no widespread agreement as yet as to which are the most appropriate.

Diversomer – A trademarked name denoting multiple related compounds (a contraction of diverse oligomers), as in combinatorial libraries.

Divide-couple recombine method (DCR method) – A way of assembling a set of *n* building blocks on a substrate in a two-step procedure. The first step consists of dividing the substrate into *n* equal batches, each building block reacting individually with one of the batches. In the second step the different batches are recombined.

DNA encoding strand – A type of tag used in encoding a library. Usually two or three nucleotides (codons) encode for an amino acid building block. The DNA tag is read by polymerase chain reaction amplification and sequencing.

Electroosmotic flow (EOF) – The bulk flow of solution that occurs within a silica capillary under the application of an external electric field. Electroosmotic flow originates from the movement of hydrated buffer ions in the double layer region along the capillary wall in response to the applied field.

Electrophoric tags – Halogen-substituted phenoxyalkyl aryl ethers which may be attached to the solid support by carbene insertion, obviating the need for orthogonal protecting schemes in split synthesis. Tag detachment is accomplished by chemical oxidation, and subsequent analysis is conducted using electron-capture gas chromatography.

Encoded library – Typically, a solid-phase library of compounds in which each resin bead has attached to it not only the compound of interest, but also a chemical code (such as an oligonucleotide) that allows facile identification of the compound.

Encoding – The simplest strategy for determining the structure of a compound prepared on a single reaction carrier (e.g., bead). During encoding, chemical synthesis is performed on a solid support such as a bead, and simultaneously with the synthetic step a tag is coupled to the bead. Different tags are used for each different reaction. Encoding methods using chemical and radio-frequency tags are known.

End effector – A "hand" attached to a robotic arm to achieve specific mechanical or chemical operations. Robotic systems of this design typically can access a variety of end effectors to achieve specific functions.

Enumeration – The process of listing-out specific or subgeneric structures from generic structures. With complete enumeration, all of the fully specific structures are listed. With partial enumeration, some or all of the specific or subgeneric structures are listed.

EOF – *See* Electroosmotic flow.

Expansin – A cross-linked polyacrylamide support; used in solid-phase synthesis.

Experiment planner – The software for composing the set of all instructions (constituting the experimental plan) needed to implement an experiment on an automated chemistry workstation.

Experimental template – An experimental plan that incorporates several variable parameters (such as concentration of one or more species, temperature, type of solvent, etc.). By using different variable parameters, the experimental template can be used repetitively for a similar type of experimentation, as occurs in statistically designed experiments.

External standard quantitation – A quantitation technique wherein a calibration curve or single point standard is prepared separate from the sample of interest. Sample quantitation is based on the ratio of the sample response to the external standard response and an appropriate calibration factor. The external standard technique has the advantage that it does not complicate analysis of the sample. External standard quantitation, however, may be analyst-dependent and requires good analytical technique.

Factorial design experiment – A systematic set of experiments in a space defined by several factors (such as temperature or concentration), where each factor is examined at several different levels, and all combinations of different factors are examined. Thus, with three factors (l, m, n) defining the three-dimensional space, and with factor l, m, and n examined at 4, 3, and 5 levels, respectively, a total of $4 \times 3 \times 5 = 60$ experiments are performed. Factorial design experiments use the same experimental template with inscription of specific values of each factor to generate each experimental plan. Factorial design experiments are performed without concurrent data evaluation (open-loop) and are ideally suited for parallel implementation.

fid – *See* Free induction decay.

Fingerprint – A bit-string representation of a chemical molecule that encodes the presence of (predominantly 2D) substructural features in a molecule. Fingerprints are widely used for substructure searching (vide infra), where they provide a simple way of increasing the efficiency of searching, and similarity searching, where molecules that have large numbers of bits in their fingerprints in common are defined to be structurally similar.

Flower plot – A circular bar graph that represents diversity or other data, allowing a graphical view of multiple dimensions simultaneously.

Free induction decay (fid) – Output signal from an NMR experiment. The signal is Fourier-transformed to produce the NMR spectrum.

Gas chromatography–mass spectrometry (GC–MS) – A method in which a sample is fractionated using gas chromatography, and the column eluant is immediately characterized by mass spectrometry

GC–MS – *See* Gas chromatography–mass spectrometry.

Gel-phase NMR – NMR of organic compounds covalently bound to a polymeric resin, typically a modified 1% polystyrene support, suspended in an organic solvent. NMR spectra are taken in the normal manner.

General screening library – A collection of compounds, typically greater than 10,000 in number, utilized for the purpose of initial lead identification in a drug-discovery operation.

Generic structure – A "Markush" structure describing one or more specific structures. This well-characterized structure consists of a root, or scaffold, a series of R-group labels, and explicitly defined R-group components.

Handle – A bifunctional spacer designed to incorporate features of a smoothly cleavable protecting group on one end and a functional group that is used for attachment to the support on the other end. Synonymous with *linker*.

Head-to-tail cyclization – Cyclization of a peptide that is side-chain-anchored to the solid support.

Heteronuclear multiple bond correlation (HMBC) – This 2D NMR experiment correlates protons that are *J*-coupled over two to three bonds with carbons generally, using proton NMR detection.

Heteronuclear multiple quantum correlation (HMQC) – This 2D NMR experiment correlates proton signals with carbon signals, using proton NMR detection.

High-performance liquid chromatography (HPLC) – A high-resolution technique used to separate the components of a chemical mixture on the basis of differences in each component's partition coefficient between a moving liquid phase and a stationary solid support. The liquid phase is pumped under high pressure through a cylindrical column packed with 3–10 μm particles.

High-throughput screening – The rapid screening of compounds against a representative set of assays, usually without bias toward a particular project. The effort is often centralized, requiring compound submittal. The assays are typically less expensive, rapid, and easily evaluated. The procedure is often highly automated and under robotic control.

Hit pool – A pool or mixture of compounds that is active in an assay.

HMBC – *See* Heteronuclear multiple bond correlation.

HMQC – *See* Heteronuclear multiple quantum correlation.

Holmes resin – Resin with 4-(2′-aminoethyl)-2-methoxy-5-nitrophenoxypropionic acid handle.

Homonuclear correlated spectroscopy (COSY) – A homonuclear 2D NMR method that correlates proton spin coupling partners in a molecule in a two dimensional array.

HPLC – *See* High-performance liquid chromatography.

HTML – *See* Hypertext markup language.

Hypertext markup language (HTML) – An implementation of the use of markup (information added to content) for distributing information on the World Wide Web. HTML is an implementation of SGML.

Informatics – Field dealing with information capture, storage, retrieval, and analysis, particularly associated with database technologies, as in bioinformatics and the current genomics (applied genetics) revolution.

Internal standard quantitation – A quantitation technique wherein a reference standard of known concentration is added to the sample. The sample concentration is then determined from the ratio of the sample response to the internal standard response and an appropriate calibration factor. Because the standard and sample responses are measured within the same analysis, the technique is less prone to precision problems than external standard quantitation.

Ionization efficiency – A term encountered in mass spectrometry to describe the ratio of the number of sample ions formed to the number of electrons, photons, or particles that are used to produce ionization.

Iterative deconvolution – A methodology based on the progressive identification of the most active builiding block for each position of a combinatorial library. Considering a library involving three coupling steps, the general formula is $A_{nA}B_{nB}C_{nC}$ (where A, B, and C are mixtures of nA, nB and nC building blocks). The first step is to determine which building block is required at one of the three positions. To define position A, nA sublibraries represented as $OB_{nB}C_{nC}$ are prepared (where O is one of the nA building blocks) and assayed. The most active sublibrary is noted, for example $A_3B_{nB}C_{nC}$, and nB sublibraries, A_3OC_{nC}, are then prepared (where O is one of the nB building blocks) and assayed to define position B. The process is repeated until all positions are defined.

J-resolved spectroscopy – This 2D experiment separates the chemical shift axis from the homonuclear coupling constant axis. Therefore, multiplets that may overlap in the 1D NMR spectrum will be resolved in the 2D experiment, allowing coupling patterns and coupling constants to be determined.

Jarvis-Patrick clustering – A method of cluster analysis (vide supra) in which objects are clustered together if they have large numbers of nearest neighbours in common. This method has been extensively used for clustering databases of chemical structures represented by 2D fingerprints.

LC-MS – *See* Liquid chromatography–mass spectrometry.

Lead explosion library – A collection of compounds lying within the diversity neighborhood of a biological lead compound. The purpose of such a library is the identification of additional active compounds.

Lead optimization library – A specific set of compounds designed to optimize the biological activity and selectivity of a lead compound.

Lead series – A specific lead molecule or series of leads that can serve as a starting point for follow-up assays or optimization studies.

Libraries from libraries – Making a new library by modifying a given building block simultaneously across an entire existing library.

Library – Any collection of items large enough to require organization to be used efficiently.

Limit of detection – The detector response below which a solute signal cannot be distinguished from background noise. The limit of detection is generally defined as a signal-to-noise ratio (S/N) of 2 or 3.

Linker – *See* Handle.

Liquid chromatography-mass spectrometry (LC-MS) – A method in which a sample is fractionated using liquid chromatography, and the column eluant is immediately characterized by mass spectrometry

Low-medium-throughput screening – The less rapid screening of compounds through one or more assays typically biased toward a particular project. The assays are typically more expensive, time-consuming, and difficult to evaluate. The procedure, while often automated, usually requires frequent human intervention.

Magic-angle spinning NMR – This NMR technique provides high-resolution NMR spectra from solid samples. It involves spinning the sample at a relatively high speed (kHz), and at an angle of 54.7° relative to the magnetic field (magic angle).

MEKC – *See* Micellar electrokinetic capillary chromatography.

Merrifield synthesis – Used broadly to refer to the solid-phase synthesis of peptides.

Micellar electrokinetic capillary chromatography (MEKC) – A high-resolution, nanoscale, chromatographic separation technique wherein solutes are separated on the basis of differences in their partition between a moving liquid phase and a micellar pseudophase. Liquid-phase flow is driven electroosmotically.

Mixed resin – A method in which resin particles are divided into multiple wells in order to carry out a given reaction, after which the contents of the different wells are recombined, mixed, separated again, and so on, in an iterative fashion, which can result in large libraries of roughly equimolar mixtures of molecules.

Mixture library – A library of compounds in which each sample (well or pool) contains multiple compounds, in contrast to a library in which each sample contains predominantly one compound.

Mixture synthesis – A method for preparing a combinatorial library in which defined groups of synthons are coupled to a solid support and then the process is repeated until the library is complete. For example, a hexapeptide library may be prepared as 400 sublibraries each containing 160,000 discrete members. Each sublibrary in turn is composed of 400 defined dipeptides coupled with 4 sets of 20 amino acids.

Molecular diversity – A property of a collection of compounds, signified by widely ranging molecular properties of size, shape, polarity, polarizability, acidity, and basicity.

Molecular tags – Chemically robust tags used in a binary encoding scheme for split synthesis. These include haloaromatic and secondary amine tags. Molecular tags remained associated with the solid support during ligand removal and are detached in a separate chemical step for analysis by sensitive chromatographic techniques.

Multipin – Typically, a thin plastic holder containing 96 pins arranged to fit into the now familiar 96-well plates used in many biological laboratories, whereupon solid-phase synthesis may be conducted on the surface of the pins through the use of polymer grafts, spacers, and linkers.

Objective function – The goal of a closed-loop experiment. In an automated chemistry workstation, the objective function is described in software such that analytical data collected during experimentation are processed appropriately in the decision-making process concerning ongoing or future experiments.

Open-loop experimentation – An experiment or set of experiments where the data are not evaluated until after experimentation is complete. In open-loop experimentation with an automated chemistry workstation, no decisions can be taken during the course of experimentation based on the nature of the data. Contrast with Closed-loop experimentation.

Orthogonal libraries – The principle of orthogonal libraries is to screen a couple of libraries, A and B, containing the same molecules but distributed in two different sets of sublibraries. In the case of orthogonal libraries, any given sublibrary of A shares only one defined compound with any sublibrary of B. An active compound will show activity in both a sublibrary of A and a sublibrary of B and will unequivocally be identified as the unique molecule shared between these two sublibraries.

Orthogonal protection – Protection strategy employing completely independent classes of protecting groups which can be removed in any order and in the presence of all other classes.

Osborn–Robinson resin – Resin with 2-azidomethyl-*N*-hydroxymethyl-6,*N*-dimethyl-benzamide-4-oxo handle.

Parallel experimentation — The performance of more than one experiment at the same time by interleaving operations from different experiments, yielding enhanced throughput and more rapid acquisition of data. In general, this approach requires a scheduler and resource manager. Contrast with Serial experimentation.

Parallel synthesis — A method for preparing a library whereby the compounds are made individually by automated or semiautomated methods.

Partial factorial design experiment — A factorial design experiment where the total number of experiments performed is less than expected for a regular factorial design experiment (where all factors of each dimension are examined). Examples include star designs, face-centered designs, core and face-centered designs, and so forth.

Partition — A partition of A may be any set P(A) of subsets $a_i \subset$ A that verifies the three following conditions. 1). None of these subsets a_i is empty ($a_i \neq \varnothing$). 2) The union of all of these subsets equals A ($U_i\ a_i$ = A). 3) The intersection of any two different subsets is empty ($a_i \cap a_j = \varnothing$ if $i \neq j$).

Pepsyn — Cross-linked polyacrylamide support; used in solid-phase synthesis.

Pepsyn K — Cross-linked polyacrylamide within macropores of kieselguhr matrix; used as a support in solid-phase synthesis.

Peptide encoding strand — A type of tag used in encoding a library. It is usually composed of naturally occurring amino acids strung together simultaneously with ligand synthesis. The peptide strand may be microsequenced by Edman degradation thus establishing the identity of the complete ligand.

Peptoid — Referring to modified peptides, particularly oligopeptides that are multiply *N*-substituted (a backbone containing N–R instead of N–H at multiple positions).

"Permanent" protecting group — Used to protect functional group(s) that are not involved in coupling reactions, in order to avoid unwanted branching or other side reactions. Often removed at the same time as the final product is cleaved form the solid phase.

Pharmacophore — A recognized portion of a molecule responsible for imparting bioactivity.

Phosphoramidite — The most common monomer used in solid-phase synthesis of oligonucleotides.

Plate height — The plate height for a column is a measure of the column's separation efficiency. Columns with small plate heights have higher efficiencies than columns with greater plate heights.

Polyethylene pins — Rods with polyethylene crowns, grafted with acrylic acid or 2-hydroxyethyl methacrylate.

Polyhipe – Polyacrylamide within macropores of cross-linked polystyrene; used in solid-phase synthesis.

Polystyrene-Kel-F – Pellicular impermeable core surrounded by a mobile layer of linear polystyrene chains.

Positional scanning – A methodology based on the direct identification of the most active building block for each position of a combinatorial library. It involves the initial synthesis of series of sublibraries in which a position is defined with a single building block while the other positions consist of mixtures. Considering a library involving three coupling steps, the general formula is $A_{nA}B_{nB}C_{nC}$ (where A, B, and C are mixtures of nA, nB, and nC building blocks). To define the preferred residue at each position, three sets of respectively nA, nB, and nC sublibraries are prepared: $OB_{nB}C_{nC}$ (where O is one of the nA building blocks), $A_{nA}OC_{nC}$ (where O is one of the nB building blocks), and $A_{nA}B_{nB}O$ (where O is one of the nC building blocks) and assayed. The first set of sublibraries defines position A; the second set, position B; and the third set, position C.

Postmodification – *See* Libraries from libraries.

Preformed handle – Prepared by initial attachment of the terminal monomer to the handle. The preformed handle is coupled to the support.

"Preview" sequencing – A method of checking for deletions in long peptides by analyzing the peptide-resin by automated Edman degradation during the synthesis.

Prior art – Existing public information, usually in reference to intellectual property.

Procedural event – The instruction for a specific operation in an automated chemistry workstation. A list of procedural events constitutes the instructions for an experiment (the experimental plan).

Process optimization – The activities involved in defining and refining a chemical reaction (or set of reactions) in preparation for industrial scale-up. These activities include establishing standard conditions for performing the reaction to achieve a specified quality level; identifying appropriate solvents and reagents for environmental, economic, and safety considerations; examining calorimetric parameters of the reaction; developing methods so that the reaction can be monitored and implemented in an automated manner; and investigating the reaction at a scale intermediate between that typical of bench chemists and that of the industrial plant.

Pseudodilution – A kinetic phenomenon based on the fact that substrates that are bound covalently to polymeric supports are less likely to undergo intermolecular reactions because of relative isolation of sites.

Quantitative structure–activity relationship – A quantitative determination of the effects of variation in molecular structure with biological activity.

Racemization — A process that results in loss of optical activity of a compound that was initially optically active.

Radio-frequency tags — A nonchemical method of encoding a library. A glass-coated radio-frequency transponder is enclosed in a porous bag of solid support. The information, either emitted or recorded on a transponder, is uploaded into a computer each time a building block is added or a synthetic step is carried out.

Ramage resin — Resin with 3-(4-hydroxymethylphenyl)-3-trimethylsilylpropionic acid handle.

Reaction matrix — The ratio of starting materials used to prepare a set of reaction products. For example, in a bimolecular reaction, a 10×10 matrix would afford a matrix of 100 reaction products.

Reaction space — The multi-dimensional space defined by the set of all parameters that describe a chemical reaction, such as solvent, temperature, reactants, reagents, catalysts, concentrations of reactants, reagents, or catalysts, etc.

Reaction station — The equipment for performing reactions, usually involving sets of reaction vessels and associated devices for stirring, temperature control, filtration, precipitate-sensing, etc.

Resolution — The degree of separation of a mixture of solutes into its individual components. Chromatographic resolution is influenced by the efficiency and selectivity of the separation as well as by the capacity factors of the solutes in the mixture.

Resource manager — A software module that handles the bookkeeping concerning all resources for experimentation, including chemical materials (reactants, solvents, and reagents) and containers (reaction vessels, sample vials, workup vials).

Rink linker — 4-(2',4'-Dimethoxyphenylaminomethyl)phenoxymethyl handle, or 4-(2',4'-dimethoxyphenylhydroxymethyl)phenoxymethyl handle.

"Safety-catch" linkers — A linker containing an otherwise stable anchor which is activated in a separate orthogonal step prior to cleavage.

Scaffold — Template or core structure upon which various pharmacophores are appended.

Scheduling heuristics — Rules-of-thumb for generating a schedule of experiments. Finding the optimal solution in scheduling can be computationally prohibitive, thus successful solutions and principles borne from experience are relied upon to generate what generally are good, albeit not necessarily the best, schedules.

Search space – The multidimensional space defined by the set of all parameters that describe the region in which an experimental search is performed. In general, the search space is a subset of the reaction space.

Segment condensation – Condensation of preformed oligomers, as opposed to stepwise condensations that involve condensations of monomers.

Selectivity – A measure of the ability of two components to separate from each other. Chromatographic selectivity is influenced by differences in the degree to which solutes interact with the liquid phase versus their interaction with the solid support.

Separation efficiency – A chromatographic term which provides a measure of the relative ability of a set of separation conditions to provide narrow, well-defined solute bands. Separation efficiency (N) is calculated from the ratio of solute retention time to width of the solute band and is measured in theoretical plates per meter.

Serial experimentation – The performance of one experiment after another, with no interleaving of operations from different experiments. Contrast with Parallel experimentation.

SGML – *See* Standard generalized markup language.

Shaving – A selective proteolysis procedure that differentiates between interior and surface sites on microporous beaded supports.

Signal-to-noise (S/N) ratio – The ratio of a detector's signal originating from a solute to the background signal present in the absence of any solute of interest.

Similar property principle – The principle that molecules that are structurally similar will have similar properties. It has been extensively used for validating the effectiveness of computational procedures for searching and clustering chemical databases.

Similarity coefficient – A mathematical function that quantifies the similarity between two multivariate objects. There are many different types of coefficients: that used most frequently in chemical applications is the Tanimoto coefficient, which has values between zero and unity for a pair of molecules having no features in common and having identical representations, respectively.

Similarity searching – A way of searching a chemical database (vide supra) that enables the retrieval of those molecules that are most similar to a user-defined query, or target, structure, such as a weak lead in a drug- or pesticide-discovery program.

Simplex – A geometrical object, whose vertices describe the values of the variable parameters for specific experiments, that moves through the search space in a simplex search. The simplex is an $(n + 1)$-dimensional object in an n-dimensional search space.

Simplex experiment — A series of experiments aimed at finding a region in a search space having improved response. The simplex algorithm involves an evolutionary hill-climbing approach by comparison of experimental responses from successive points in the search space. The simplex algorithm examines new points by moving in the direction opposite of the worst response among the points constituting the current simplex. By necessity, the simplex algorithm entails sequential, rather than parallel, experimentation. Simplex algorithms of varying degrees of sophistication have been developed, ranging from the original simplex algorithm to the composite modified simplex algorithm. *See also* Composite modified simplex algorithm.

S/N ratio — *See* Signal-to-noise ratio.

Sparrow amide resin — Cross-linked polyacrylamide support; used in solid-phase synthesis.

Spin echo spectroscopy (CPMG) — An NMR technique that helps to minimize the peaks due to the polymer support.

Split resin — *See* Mixed resin.

SQL — *See* Structured query language.

Standard generalized markup language (SGML) — An international standard designed to facilitate the exchange of information across systems, devices, languages, and applications. HTML is an implementation of SGML.

Statistically designed experiments — Experiments designed in accord with statistical principles or performed in a systematic manner, in order to obtain a dataset of desired quality. Factorial design, partial factorial design, and simplex experiments are examples of statistically designed experiments.

Stepwise solid-phase synthesis — Coupling of the monomers in discrete steps, as opposed to segment condensations that involve coupling of oligomers.

Structured query language (SQL) — A high-level language for structuring, querying, and changing information in a relational database.

Subgeneric structure — A "less generic" structure derived from (i.e., enumerated from) a more generic structure. A subgeneric structure represents a subset of the structures represented by the parent generic structure.

Sublibrary — One set (of two or more sets) of library compounds generally containing a common synthon or building block, as produced during mixture or split synthesis. For example, pooled resin prepared by split synthesis containing a free amine is apportioned equally into five reaction vessels to which five different acid chlorides are added. The acylated resin so obtained in each of the reaction vessels is considered a sublibrary.

Substructure searching – A way of searching a chemical database (vide supra) that enables the retrieval of all molecules that contain some user-defined query pattern. This can either be a pattern of atoms and bonds, such as a ring system for which all of the analogues are required, in a search of a 2D database, or a pharmacophore consisting of a pattern of atoms and associated geometric constraints, such as interatomic distances, in a search of a 3D database.

T-bag – A Teflon bag, not unlike a tea bag, in which resin particles are placed to facilitate the handling of multiple, simultaneous, solid-phase syntheses.

Temporary protecting groups – Protection groups that can be removed during stepwise solid-phase synthesis without cleaving the product from the support.

TentaGel – Polyethylene glycol polymerized onto cross-linked polystyrene (POE–PS); used in solid-phase synthesis.

Terminated peptides – Formed when a peptide chain stops growing at some point during the assembly of the peptide, often because of deliberate capping.

Thin-layer chromatography (TLC) – An analytical method for examining the number of components in a mixture. A sample is applied to a plate (or tube) coated with a solid chromatographic medium, and solvent is allowed to pass over the surface of the plate. Separation is achieved on the basis of the partitioning of the components between the mobile solvent phase and the stationary solid phase. The plate can then be viewed visually or scanned automatically to determine the retention times of the various components on the plate. This is a type of development chromatography, where all material in the sample placed on the plate is available for analysis. In contrast, high-pressure liquid chromatography is a form of elution chromatography, where only the components that traverse the entire chromatography column are detected. TLC is typically used as the first means of analyzing crude reaction mixtures.

TLC – *See* Thin-layer chromatography.

TOCSY – *See* Total correlation spectroscopy.

Total correlation spectroscopy (TOCSY) – A homonuclear 2D NMR method by which the complete proton-proton spin topology can be observed.

Two-dimensional NMR (2D NMR) – A pulse sequence (special NMR program) is used to produce and detect an NMR signal as a function of two variables. These variables allow for easier interpretation of complicated spectra. Typically, these data are viewed in a matrix presentation, where the peaks are represented as contour plots.

Utilization – The fraction of time (0–100%) during a schedule in an automated chemistry workstation when the robot is active. A good schedule, where parallel experimentation is achieved efficiently, has a high degree of utilization.

Virtual library — A computer-generated library of compounds, useful in the planning of combinatorial libraries. A series of compounds generated from the enumeration of a generic structure.

Void volume — The retention time (or volume) of a solute that does not partition into the stationary phase and remains in the liquid phase.

Wang resin — Resin with 4-alkoxybenzyl alcohol handle.

Zero dead volume — The term zero dead volume describes the hydrodynamic properties of a device that has been designed to mimimize the dispersion of a solute band passing through the device in a moving solvent stream.

Abbreviations

AA – amino acid

ACD – Available Chemicals Database

ACE – angiotensin converting enzyme

AMS – accelerator mass spectrometry

APcI – atmospheric pressure chemical ionization interface

AS – affinity selection

Asp – aspartic acid

ATR – attenuated total reflection

BAL – backbone amide linker

BHA – benzhydrylamide resin

Boc – *tert*-butyloxycarbonyl

BOP – benzotriazolyl *N*-oxytris(dimethylamino)phosphonium hexafluorophosphate

Bz – benzoyl

Bzl – benzyl

cAMP – cyclic adenylic acid

CAS – Chemical Abstracts Service

CEC – capillary electrochromatography

CHA – 5-{[(*R,S*)-5-amino-10,11-dihydrodibenzo[*a,d*]cyclohepten-2-yl]oxy}valeric acid handle

cHex – cyclohexyl

CLEAR – cross-linked ethoxylate acrylate resin; used as a support in solid-phase synthesis

CMS – composite modified simplex

COS – combinatorial organic synthesis

CPG – controlled pore glass

CPU – central processing unit

CSD – Cambridge Structural Database

CZE – capillary zone electrophoresis

DAL – dimethoxy acido-labile handle; 3-(2′-methoxyphenylaminomethyl)-4-methoxyphenoxyacetic acid

DBU – 1,8-diazabicyclo[5.4.0]undec-7-ene

DCC – *N,N′*-dicyclohexylcarbodiimide

DCM – dichloromethane

DCR – divide-couple recombine method

DHFR – dihydrofolate reductase

DHPP – 4-(1′,1′-dimethyl-1′-hydroxypropyl)phenoxyacetic acid handle

DIC – *N,N′*-diisopropylcarbodiimide

DIEA – *N,N*-diisopropylethylamine

DIPCDI – see DIC

DMAP – 4-dimethylaminopyridine

dmf – dimethylformamidine

DMF – *N,N*-dimethylformamide

DMSO – dimethyl sulfoxide

DMTr – dimethoxytrityl; di-*p*-anisylphenylmethyl

DNA – oligodeoxyribonucleotides; deoxyribonucleic acid

Dod – 4-(4′-methoxybenzhydrylamine)phenoxyacetic acid handle

DPPA – diphenylphosphoryl azide

Dpr(Phoc) – N^3-phenyloxycarbonyl-L-2,3-diaminopropionic acid handle

ECD – electron-capture detector

EDT – 1,2-ethanedithiol

ELSD – evaporative light scattering detection

EOF – electroosmotic Flow

ESI – electrospray ionization

ESMS – electrospray mass spectrometry

EVAL membrane – poly(ethylene-*co*-vinyl alcohol); used as a support in solid-phase synthesis

FABMS – fast atom bombardment mass spectrometry

fid – free induction decay. Output signal from an NMR experiment. This signal is Fourier-transformed to produce the NMR spectrum.

Fmoc – 9-fluorenylmethyloxycarbonyl

FTIR – Fourier transform infrared spectroscopy

GA – genetic algorithm

GC – gas chromatography

GC–MS – gas chromatography–mass spectrometry; a method in which a sample is fractionated using gas chromatography, and the column eluant is immediately characterized by mass spectrometry

HATU – *N*-[(dimethylamino)-1*H*-1,2,3-triazolo[4,5-*b*]pyridino-1-ylmethylene]-*N*-methylmethanaminium hexafluorophosphate *N*-oxide

HBTU – *N*-[1*H*-benzotriazol-1-yl(dimethylamino)methylene]-*N*-methylamethanaminium hexafluorophosphate *N*-oxide

HFMS – *N*-[(9-hydroxymethyl)-2-fluorenyl]succinamic acid handle

HMBC – heteronuclear multiple bond correlation

HMFA – 9-(hydroxymethyl)-2-fluorenecarboxylic acid handle, or 9-(hydroxymethyl)-2-fluoreneacetic acid handle

HMPA – 4-hydroxymethylphenoxyacetic acid linker

HMPB – 4-(4-hydroxymethyl-3-methoxyphenoxy)butyric acid linker

HMQC – heteronuclear multiple quantum correlation

HOAt – 1-hydroxy-7-azabenzotriazole

HOBt – 1-hydroxybenzotriazole

HOMO – highest occupied molecular orbital

HP – Hewlett Packard

HPA-PP membrane – polypropylene coated with poly(hydroxypropyl acrylate); used as a support in solid-phase synthesis

HPDI – 2-hydroxypropyl-dithio-2′-isobutyric acid handle

HPLC – high-performance liquid chromatography, or high-pressure liquid chromatography

HR-MAS NMR – high-resolution magic-angle spinning nuclear magnetic resonance

HTML – Hypertext Markup Language

HTS – high-throughput screening

HYCRAM – 4-hydroxycrotonic acid handle

HYCRON – 3-[(*trans*-4′-bromobut-2′-enyloxy)triethylene glycol]propionic acid handle

ICl – iodine monochloride

IRAA – internal reference amino acid

LC – liquid chromatography

LC-MS – liquid chromatography–mass spectrometry; a method in which a sample is fractionated using liquid chromatography, and the column eluant is immediately characterized by mass spectrometry

LCAA – long-chain alkylamine support

LLE – liquid–liquid extraction

LUMO – lowest unoccupied molecular orbital

Lys – lysine

MALDI-TOF MS – matrix-assisted laser desorption/ionization time-of-flight mass spectrometry

MAS – magic-angle spinning

MBHA resin – 4-methylbenzhydrylamide resin

MBS – Multiple Biomolecular Synthesizer

MDS – multidimensional scaling

MEKC – micellar electrokinetic capillary chromatography

MS – mass spectrometry

NMM – *N*-methylmorpholine

NMR – nuclear magnetic resonance

Nonb – 3-nitro-4-aminomethylbenzoic acid handle

NPE – 3-nitro-4-(2-hydroxyethyl)benzoic acid handle

NPY – neuropeptide Y

ONb – 3-nitro-4-hydroxymethylbenzoic acid handle

PAB – *p*-alkoxybenzyl

PAL – tris(alkoxy)benzylamide linker, 5-(4-aminomethyl-3,5-dimethoxyphenoxy)-valeric acid

PAM – 4-(hydroxymethyl)phenylacetic acid handle

Pbf – 2,2,4,6,7-pentamethyldihydrobenzofuran-5-sulfonyl

Pbs – 2,4,5-trichlorophenyl-*N*-[[4-(hydroxymethyl)phenoxy]-*tert*-butylphenylsilyl]-phenyl pentanedioate monoamide handle

PCR – polymerase chain reaction

PDQ – pharmacophore-derived query

PEG – poly(ethylene glycol)

PEG-PS – poly(ethylene glycol) grafted to a cross-linked polystyrene support; used in solid-phase synthesis

PEGA – cross-linked poly(ethylene glycol) acrylamide support; used in solid-phase synthesis

PEO-PEPS – 3,6,9-trioxadecanoic acid coupled to PEPS films; used in solid-phase synthesis

PEPS films – polystyrene grafted onto polyethylene films; used as a support in solid-phase synthesis

Pmc – 2,2,5,7,8-pentamethylchroman-6-sulfonyl

PNA – peptide nucleic acid

POE-PS – polyethylene glycol polymerized onto cross-linked polystyrene (TentaGel); used in solid-phase synthesis

PS – cross-linked polystyrene support; used in solid-phase synthesis

QSAR – quantitative structure–activity relationship

RDBMS – relational database management system

RNA – oligoribonucleotides; ribonucleic acid

RNN – reciprocal nerest neighbors

RPR – Rhone Poulenc Rorer

SAC – 4-[1-hydroxy-2-(trimethylsilyl)ethyl]benzoic acid ("silyl acid") handle

SAL – 4-[1-amino-2-(trimethylsilyl)ethyl]phenoxyacetic acid handle

SASRIN – resin with 2-methoxy-4-alkoxybenzyl alcohol linker

SBFTIRM – single-bead Fourier-transform infrared microspectroscopy

SCAL – safety-catch anchoring linker; 4-[4,4′-bis(methylsulfinyl)-2-oxybenzhydryl-amine]butanoic acid handle

SDS – sodium dodecyl sulfate

SGML - standard Generalized Markup Language; an international standard designed to facilitate the exchange of information across systems, devices, languages, and applications. HTML is an implementation of SGML.

SMILES – Simplified Molecular Input Line Entry Specification

SPF – suboptimal binding factor

SPOC – solid-phase organic chemistry

SPOS – solid-phase organic synthesis

SPPS – solid-phase peptide synthesis

SPS – solid-phase synthesis

SQL – structured query language; a high-level language for structuring, querying, and changing information in a relational database

*t*Bu – *tert*-butyl

TEA – triethylamine

TEGDA-PS – polystyrene cross-linked with tetraethyleneglycol diacrylate; used in solid-phase synthesis

TFA – trifluoroacetic acid

TFFH – tetramethylfluoroformamidinium hexafluorophosphate

TFMSA – trifluoromethanesulphonic acid

THP – tetrahydropyran

TLC – thin-layer chromatography

Tmob – 2,4,6-trimethoxybenzyl

TOCSY – total correlation spectroscopy

Tos – *p*-toluenesulfonyl

Trt – trityl

WWW – World Wide Web

XAL – 5-(9-aminoxanthen-3-oxy)valeric acid handle

Xan – 9-xanthenyl

Index

Acquisitions editor: Cheryl Shanks
Development editor: William Wells, Biotext, Inc.
Copy editor and indexer: Jay C. Cherniak
Production editor: Amie Jackowski

Text designed and typeset by Betsy Kulamer, Washington, DC
Printed and bound by Maple Press Inc., York, PA